Glossary of Symbols

$\bar{A}$	Complement of set A
AD	Absolute deviation
ANOVA	Analysis of variance
α (alpha)	Probability of a type I error
β (beta)	Probability of a type II error
$1 - \beta$	Power of a statistical test
β_0	y-intercept of the true linear relationship
β_1	Slope of the true linear relationship
b_0	y-intercept for the line of best fit for the sample data
b_1	Slope for the line of best fit for the sample data
c	Column number or class width
d	Difference in value between two paired pieces of data
df or df ( )	Number of degrees of freedom
d_i	Difference in the rankings of the ith element
E	Expected value or maximum error of estimate
$e = y - \hat{y}$	Error (observed)
ε (epsilon)	Experimental error
ε_{ij}	Amount of experimental error in the value of the jth piece of data in the ith row
F	F distribution statistic
$F(\mathrm{df}_n, \mathrm{df}_d, \alpha)$	Critical value for the F distribution
f	Frequency
H	Value of the largest-valued piece of data in a sample
H_a	Alternative hypothesis
H_0	Null hypothesis
i	Index number when used with Σ notation
i	Position number for a particular data
k	Identifier for the kth percentile
k	Number of cells or variables
L	Value of the smallest-valued piece of data in a sample
m	Number of classes
MAD	Mean absolute deviation
MS ( )	Mean square
MSE	Mean square error
MPS	Marginal propensity to save
μ (mu)	Population mean
μ_d	Mean value of the paired differences

(continued on back cover)

Elementary Statistics
4th Edition

Robert Johnson
Monroe Community College

Duxbury Press · Boston

PWS PUBLISHERS

Prindle. Weber & Schmidt • 🐿 • Willard Grant Press • **WG** • Duxbury Press • ◆
Statler Office Building • 20 Park Plaza • Boston. Massachusetts 02116

PWS Publishers is a division of Wadsworth, Inc.

Printed in the United States of America

85 86 87 88—10 9 8 7 6 5 4 3

Library of Congress Cataloging in Publication Data
Johnson, Robert Russell, 1939–
    Elementary statistics.

    Includes index.
    1. Statistics. I. Title.
QA276.12.J64   1984   519.5   83-11557
ISBN 0-87150-406-5

ISBN 0-87150-406-5

The cover art, etching "IV," is an artistic collaboration by Caroline Stone and George Chaikin. Mr. Chaikin created the design on a computer in the language, "Artspeak." Ms. Stone converted the 12-inch images to Kodaliths, photo-transferred them to sensitized plates, etched, and then printed them in the traditional manner. Used by permission of the artists.

Cover and text design by Sara Waller. Composition by Composition House Limited. Artwork by Anco. Text Printed and bound by Maple-Vail Book Manufacturing Group. Covers printed by John P. Pow Company, Inc.

To my mother
and to the
memory of my father

Preface

The primary objective of this book is to present a truly readable introduction to statistics that will motivate students by presenting statistics in a context that relates to their personal experiences and that is organized to promote learning and understanding.

Statistics is a practical discipline that responds to the changing needs of our society. Today's student is a product of a particular cultural environment and as such is motivated differently from the student of a few years ago. Statistics is presented in this text as a very useful tool in learning about the world around us. While studying descriptive and inferential concepts, students will become aware of the practical application of these concepts to such fields as business, biology, engineering, industry, and the social sciences.

This book was written for use in an introductory course for non-mathematics students who need a working knowledge of statistics but do not have a strong mathematics background. Statistics requires the use of many formulas, so those students who have not had intermediate algebra should complete at least one semester of college mathematics before beginning this course. With students from such a mathematical background in mind, I have emphasized the readability, motivation, and promotion of learning as the primary goals for this book.

The Changes in the Fourth Edition

The teaching objectives of this edition are the same as those of the previous editions. The following significant changes made in this revision should be helpful in attaining these teaching objectives:

1. A discussion of data collection and the role of the computer in statistics has been added to chapter 2.

2. The descriptive statistics for single variable data have been reorganized and combined into one chapter with a "two-unit" approach. Unit 1 deals with graphical techniques; unit 2 deals with calculated techniques.

3. Chapter 4 on Probability has been reworked by merging the General and Specific Addition (and Multiplication) Rules under one heading. The exposition has also been divided into additional sections to clarify the discussion and increase the number of drill exercises.

4. *P*-values have been added to hypothesis testing.

5. Coverage of unequal replicates has been added to ANOVA.

6. Minitab Computer Output has been included for Descriptive statistics, Hypothesis-testing, Chi Square, ANOVA, and Regression Analysis.

7. Four "Working with Your Own Data" sections have been included for student exploration. These provide a more personalized learning experience by directing students to collect their own data and apply the techniques they have been studying by use of those data.

8. The exercise sets have been altered substantially. Many of the exercises are new, and others have been modified so that solutions are different from solutions in the third edition.

To the Instructor: The Text As a Teaching Tool

As stated earlier, one of my primary objectives in writing this book was to produce a truly readable presentation of elementary statistics. It is this specific intent that brought about the chapter format. Each chapter is designed to interest and involve students, and guide them step-by-step through the material in a logical manner. Each chapter includes:

A **Chapter Outline** that shows students what to look for in the following material.

A **News Article** that shows students how statistics are actually applied in the real world and demonstrates the type of statistic to be studied.

Chapter Objectives that tell students the specific information to be learned upon completion of the chapter.

Worked-out Examples with solutions to illustrate concepts as well as demonstrate the applications of statistics in real-world situations.

End-of-Section Exercises to facilitate practice in the use of concepts as they are presented.

In Retrospect, a section that provides a summary of the material just provided and relates the material to the chapter objectives.

Chapter Exercises that give students further opportunity to master conceptual and computational skills.

A **Vocabulary List** to help students review key terms.

A **Chapter Quiz** that helps students evaluate their mastery of the material.

The first three chapters are introductory in nature. Chapter 3 is a descriptive (first-look) presentation of bivariate data. I present this material at this point in the book for two reasons. First, students often ask about the relationship between two sets of data (such as heights and weights). Second, it affords me an opportunity to present a decision-making process (a hypothesis test with critical value) without confusion. This seems to reduce the resistance that always persists later when the formal hypothesis test procedure is introduced. (Of course, the instructor must make reference to this previous decision-making process.)

In the chapters on probability (4 and 5), I deliberately avoid the concepts of permutations and combinations. They are of no help in understanding statistics. Thus only the binomial coefficient is introduced in connection with the binomial probability distribution.

The instructor has several options in the selection of topics to be studied in a given course. I consider chapters 1 through 9 to be the basic core of a course (some sections of chapters 2, 3, and 6 may be omitted without affecting continuity). Following the completion of chapter 9, any combination of chapters 10 through 14 may be studied. There are two restrictions, however: chapter 3 must be studied prior to chapter 13, and chapter 10 must precede chapter 12.

The suggestions of instructors using the previous editions have been invaluable in helping me improve the text for the present revision. Should you, in teaching from this edition, have comments or suggestions, I would be most grateful to receive them. Please address such communications to me at Monroe Community College, Rochester, New York 14623.

To the Student: The Text As a Learning Tool

I believe that plain talk and a stress on common sense are this book's main merits as a learning tool. Such a treatment should allow you, provided that you have the necessary basic mathematics skills, to work your way through the course with relative ease. Examples of this procedure are (1) illustration 1–4 (p. 8), which is used to reemphasize the meanings of the eight basic definitions presented in chapter 1, and (2) the use of previously completed homework exercises to introduce a new concept (see the introduction of hypothesis testing on page 262).

It is my aim to motivate and involve you in the statistics that you are learning. The chapter format reflects these aims and can best promote learning if each part of each chapter is used as indicated.

1. Read the annotated outline to gain an initial familiarity with several of the basic terms and concepts to be presented.

2. Read the news article, which puts to practical use some of the concepts to be learned in the chapter.

3. Use the chapter objectives as a guide to map out the direction and scope of the chapter.

4. Learn and practice using the concepts of each section by doing the exercise set at the end of the respective section. Answers and partial solutions are provided at the back of the book to complement the study illustrations and to enable you to work independently. While working within a chapter, it will be helpful to save the results of the exercises, since some results will be used again in later exercise sets of the chapter. When this occurs, the later use has been cross-referenced.

5. Use the ln Retrospect section to reflect on the concepts you have just learned and the relationship of the material in this chapter to the material of previous chapters. At this point it would be meaningful to reread the news article.

6. The Chapter Exercises at the end of the chapter offer additional learning experiences, since in these exercises you must now identify the technique to be used and must be able to apply it. The exercises are graded; everyone will be able to complete the first exercises with reasonable ease, but succeeding exercises become more challenging. Occasionally the results of an exercise should be saved for use in later exercises.

7. The Vocabulary List and Quiz are provided as self-testing devices. You are encouraged to use them. The correct responses for the Quiz may be found in the back of the book.

8. The "Working with Your Own Data" sections direct you to collect a set of data, often of your own interest, and to apply the techniques you have studied. This opportunity should (i) reinforce the concepts studied and (ii) result in an interesting and informative statistical experience with real data.

Supplements

The complete instructional package that accompanies this book includes the following:

1. The **Study Guide with Self-Correcting Exercises** offers students an alternative approach to mastering difficult concepts. For each chapter in the text, the study guide provides alternative explanations of difficult, numerous worked-out examples, and self-correcting exercises. There is a review of elementary algebra.

2. A **Test Item Card File** for testing each chapter. These are printed on cards to make them easy to reproduce.

3. The **Solutions Manual** shows at least one complete solution to each exercise in the textbook. Occasionally I have added some parenthetical comments to aid the teacher in such areas as when to assign specific problems and how some problems can be of greatest use.

4. The **Minitab Student Supplement** by James Scott and Kenneth Bond of Creighton University. For those interested in teaching or learning the course interactively with the computer, this supplement is a text-specific introduction to the MINITAB Statistical Software Package and is keyed to text discussion and examples.

Acknowledgments

I owe a debt to many other books. Many of the ideas, principles, examples, and developments that appear in this text stem from thoughts provoked by these sources.

It is a pleasure to acknowledge the aid and encouragement I have received throughout the development of this text from my students and colleagues at Monroe Community College. Special thanks go also to those who read and offered suggestions about the previous editions. I also want to acknowledge and thank the reviewers for this edition: Paul Alper, College of St. Thomas; Nancy Adcox, Mt. San Antonio College; Joan Garfield, University of Minnesota General College; Carroll Hall, New Mexico State University; Hank Harmeling, North Shore Community College; Michael Karelius, American River College; Raymond Knodel, Bemidji State University; Larry Ringer, Texas A & M University; Thomas Sturm, College of St. Thomas; William Tomhave, University of Minnesota; and Jean Vincenzi, Saddleback College. In addition, Dr. Susan Reiland did a very thorough check of the entire manuscript.

The author and publisher also wish to acknowledge the following individuals who completed a research questionnaire for the fourth edition: Kenneth Wantling, E. William Chapin, John J. Avioli, Ronald Lee Persky, E. H. Bryan, Margaret Biggerstaff, Carlton Lee Evans, Glen Weber, Anthony Pruchnicki, H. Ajakie, Charles Phillips, Ann Watkins, Mohammed Rajah, Kenneth Eberhard, Jeff Mock, Rudy Freund, John Scarlett, and Ross A. Gollan.

I would also like to express my appreciation for the fine work that Sara Waller and associates have put into the production of the book. To the Duxbury Staff, and to Kim Kist, Editor, I would like to say thanks for all your assistance and encouragement.

Thanks also to the many authors and publishers who so generously extended reproduction permissions for the news articles and tables used in the text. These acknowledgments are specified individually throughout the text.

Robert R. Johnson

Contents

Appendixes 505

PART I

Descriptive Statistics

When one embarks on a statistical solution to a problem, a sequence of events must develop. The order in which these events occur should be: (1) the situation investigated is carefully and fully defined, (2) a sample of data is collected from the appropriate population following an established and appropriate procedure, (3) the sample data are converted into usable information (this usable information, either numerical or pictorial, is called the descriptive statistics*), and (4) the theories of statistical inference are applied to the sample information in order to draw conclusions about the sampled population (these conclusions or answers are called* inferences*).*

The first part of this textbook, chapters one through four, will concentrate on the first three of the four events identified above. The second part of the textbook will deal with probability theory, the theory on which statistical inferences rely. The third and fourth parts of the text will survey the various types of inferences that can be made from sample information.

1 Statistics

The Lasting Changes Brought by Women Workers

More women will be
entering the labor force
than ever before. . .

1970 1975 1980 1985 1990
▲ Percent   └Est.┘
*Number of women employed or looking
for work as a percentage of all women
of working age

. . .which will translate
into more two-earner
families

1970 1975 1980 1985 1990
▲ Percent   └Est.┘
Data: Bureau of Labor Statistics,
Data Resources Inc.

. . .widening the income
gap between dual- and
single-earner families

1970 1975 1980 1985
▲ Median income,   └Est.┘
 thousands of 1972 dollars

McDonald's Restaurant Chain Has Sold Over 43,000,000,000 Hamburgers!

4 out of 5 doctors recommend

Average Gain on Today's Stockmarket Was 4.14

"Oh, I was just an average guy . . . . . ."

It's Your Life

Statistics about the American teen

What's your biggest problem? If it's school, you're not alone. Dr. Myron Harris and Jane Norman have discovered that school weighs heaviest on the minds of young people. They surveyed over 160,000 teen-agers and published the results in *The Private Life of the American Teenager* (Rawson, Wade). Here's a list of some of their findings:

• More than eight out of ten teens with a working mother feel happy and proud of her.

• Eighty percent of the girls want a career.

• Ninety percent of the teens surveyed believe in marriage.

• Seventy-four percent say they would live with someone before marrying him or her.

• Seven out of ten high school students have tried marijuana.

• Sixty percent study only to pass tests, not to learn. Fifty-five percent admit that they cheat.

• Eighty-three percent say they can usually tell one or both parents how they feel about an issue.

• Almost sixty percent fear their parents' death—even more than they fear their own.

• Seventy-five percent of the teens believe that divorce is justified if parents argue often, if physical violence is involved, or if one or both parents are unfaithful.

Erdice Court

Chapter Objectives

The purpose of this introductory chapter is to present (1) an initial image of the field of statistics, (2) several of the basic vocabulary words used in studying statistics, and (3) the basic ideas and concerns about the processes used to obtain sample data.

Section 1–1 What is Statistics?

Statistics is the universal language of the sciences. Statistics is more than just a "kit of tools." As potential users of statistics we need to master the "art" of using these tools correctly. Careful use of statistical methods enables us to (1) accurately describe the findings of scientific research, (2) make decisions, and (3) make estimations.

Statistics involves numbers, subjects, and the use of these numbers and subjects. The word "statistics" has different meanings to people of varied backgrounds and interests. To some people it is a field of "hocus-pocus," whereby a person in the know overwhelms the layperson. To other people it is a way of collecting and displaying large amounts of numerical information. And to still another group it is a way of "making decisions in the face of uncertainty." In the proper perspective each of these points of view is correct.

descriptive and inferential statistics

The field of statistics can be roughly subdivided into two areas: descriptive statistics and inferential statistics. **Descriptive statistics** is what most people think of when they hear the word "statistics." It includes the collection, presentation, and description of numerical data. The term **inferential statistics** refers to the technique of interpreting the values resulting from the descriptive techniques and then using them to make decisions.

Statistics is more than just numbers—it is what is done to or with numbers. Let's use the following definition:

statistics

Statistics

The science of collecting, classifying, presenting, and interpreting numerical data.

Before we begin our detailed study, let's look at a few examples of how and when statistics can be and is applied.

Illustration 1–1

State University is planning an expansion program of its physical facilities. To draw up an effective course of action, the board of trustees decides that it needs the answer to this question: How many college students will we need to accommodate over the next ten years? This question can immediately be broken down into a number of smaller questions: How many college students will there be? How many will want to attend State U? And so on. To answer these questions the researcher will need to obtain data that will indicate the proportion of future high school graduates that will desire to attend State U, and then he will somehow need to project or predict the number of high school graduates that will be available over the next ten years.

Let's consider the question of the proportion of high school graduates that will wish to attend State U. The "best" way to answer this question is to find out what proportion have attended in the past. Note that this "best" answer assumes that there is a relationship between past and future. This does not always hold true, however. Wars, economic depressions, and other events will alter the "natural" progression of events. ☐

As you can see, when accurate answers are desired, many problems must be overcome. One rather obvious problem is the obtaining of historical data. Do we need to consider every student who has graduated from high school for the last several years? Do we need to account for all schools within a 500-mile radius? No, we do not; it would be impossible to research these questions completely. So we will test only some of this population—in other words, we will **sample** the population.

sample

There are other considerations, too. How accurate are our results? What is the probability that there will be a larger proportion of students wanting to attend State U? And so on.

We have not begun to exhaust the questions that may be relevant. At this time I only wish to start you on your way to thinking about and considering some of the problems involved in answering this type of question.

Illustration 1–2

Pickadilly Pete

"Everybody loves Pickadilly Pete!" Well, that is what Pickadilly Pete claims, and so he has decided to run for mayor of Skunk Hollow (population 279). But his campaign manager is not sure that he understands exactly what Pickadilly Pete means. Does Pete mean that all 279 inhabitants of Skunk Hollow love him, or does he mean that the majority of the 279 (at least 140) love him, or does he mean that at least half of the inhabitants of voting age love him? Pete could have meant any or all of these, and what he means could make a big difference in the campaign. (Suppose the 140 who love him are all children and can't vote.) Suppose you were hired as an independent research agent. What would you do to test Pickadilly Pete's claim? ☐

Illustration 1–3

How tall are sports car drivers? This is the question posed by the owners of Custom Sport Coupe, a local manufacturer of the world's finest sports car. They want to design and build a new model that is truly comfortable for the driver. Their present model is designed for people between 5 feet 2 inches and 5 feet 8 inches tall. The manufacturers are concerned because they have heard rumors that their car is uncomfortable for a large proportion of sports car enthusiasts. (If you have ever ridden in a CSC, you will understand the rumor, "built for short people with very short necks.") How might you go about obtaining an answer to the original question? What special considerations would you give to the process of obtaining samples? □

Each of these illustrations poses questions that should make you think about the situation and at the same time give you a feeling for statistics.

Section 1-2 Uses and Misuses of Statistics

The uses of statistics are unlimited. It is much harder to name a field in which statistics is not used than it is to name one in which statistics plays an integral part. The following are a few examples of how and where statistics is used:

1. In education descriptive statistics are frequently used to describe test results.
2. In science the data resulting from experiments must be collected and analyzed.
3. In government many kinds of statistical data are collected all the time. In fact, the U.S. government is probably the world's greatest collector of statistical data.

Misuses of statistics are often colorful and sometimes troublesome. Many people are concerned about the indifference of statistical descriptions; other people believe all statistics are lies. Most statistical lies are innocent, however, and result from using an inappropriate statistic, or an open, non-specific statement (such as Pickadilly Pete's claim), or data derived from a faulty sample. All these lead to a common result—misunderstanding of the information on the part of the consumer. Specific illustrations of the misuse of statistics will be given in chapter 2.

Section 1-3 Introduction to Basic Terms

To study statistics we need to be able to speak its language. Let's first define a few basic terms that will be used throughout the text. (These definitions are descriptive in nature and are not necessarily mathematically complete.)

population

Population

A collection, or set, of individuals, objects, or measurements whose properties are to be analyzed.

The population is the complete collection of individuals, objects, or measurements that are of interest to the sample collector. The concept of a population is the most fundamental idea in statistics. The population of concern must be carefully defined and is considered fully defined only when its membership list of elements is specified. The set of "all students who have ever attended a U.S. college" is an example of a well-defined population.

Typically we think of a population as a collection of people. However, in statistics the population could be a collection of animals, of manufactured objects, or of measurements. For example, the set of all values for the "heights" of all redwood trees in California could be the population.

---

sample

Sample

A subset of a population.

A sample consists of the individuals, objects, or measurements selected by the sample collector from the population.

---

response variable

Response Variable (or, simply, Variable)

A characteristic of interest about each individual element of a population or sample.

A student's age at entrance into college, the color of her hair, her height, weight, and so on, are all response variables.

---

data (singular)

Data (singular)

The value of the response variable associated with one element of a population or sample.

For example, Bill Jones entered college at the age of "23," his hair is "brown," he is "6 feet 1 inch" tall, and he weighs "183 pounds." These four pieces of data are the values of the response variables for Bill Jones.

---

data (plural)

Data (plural)

The set of values collected for the response variable from each of the elements belonging to the sample.

The set of 25 heights collected from each of 25 students is an example of a set of data.

experiment

Experiment

A planned activity whose results yield a set of data.

parameter

Parameter

A numerical characteristic of an entire population.

The "average" age at time of admission of all students who have ever attended our college or the proportion of students who were over 21 years of age when they entered college are examples of population parameters. A parameter is a value that describes the entire population. It is a common practice in statistics to use a Greek letter to symbolize the names of the various parameters. These symbols will be assigned as we study individual parameters.

statistic

Statistic

A numerical characteristic of a sample.

The "average" height found by using the set of 25 heights is an example of a sample statistic. A statistic is a value that describes a sample. Most sample statistics are found with the aid of formulas and are assigned symbolic names using letters of the English alphabet (for example, $\bar{x}$, s, and r).

Illustration 1-4

A statistics student is interested in finding out something about the dollar value of the typical car owned by the faculty members of our college. Each of the eight terms just described can be identified in this situation.

1. The *population* is the collection of all cars owned by the faculty members of our college. (Or it could be the collection of all dollar values of those cars.)
2. A *sample* is any part of that population. For example, the values of the cars owned by members of the mathematics department would be a sample.
3. The *response variable* is the actual value of each individual car.
4. One *data* would be the value of a particular car.

5. The *data* would be the set of values that correspond to the sample obtained.

6. An *experiment* would be the method of determining the value of each car in the sample. It could be accomplished by questioning each owner, or in other ways.

7. The *parameter* about which we are seeking information is the "average" value in the population.

8. The *statistic* that will be found is the "average" value in the sample.

□

There are basically two kinds of data: (1) data obtained from qualitative information and (2) data obtained from quantitative information.

attribute or qualitative data

Attribute or Qualitative Data

Results from a variable that asks for a quality type of description of the subject.

For example, color is an attribute of a car. A sample of the colors of the cars in a nearby parking lot would result in such data as blue, red, yellow, and so on. In general, the resulting data are a collection of word responses (as opposed to number responses).

variable or quantitative data

Variable or Quantitative Data

Result from obtaining quantities—counts (of how many) or measurements (length, weight, and so on).

discrete
continuous

Variable data can be subdivided into two classifications: (1) **discrete** variable data and (2) **continuous** variable data. In most cases the two can be distinguished by deciding whether the data result from a count or from a measurement. A count will always yield discrete variable data. The number of bugs on the right headlight(s) of an automobile will be 0, 1, 2, ...,* but it cannot be 1.9 or 3.25. (The partial remains of any bug will be counted as a whole bug.) The idea of discontinuous numerical values is somewhat synonymous with that of discrete numerical values.

A measure of a quantity will usually be continuous. John says that he weighs 162 pounds, to the nearest pound. But all we can be sure of is that his weight is some value between 161.5 and 162.5 pounds. He could actually weigh 162.000, 161.789, or any value in the interval 161.5–162.5. This is the basic concept of a continuous variable. There are many continuous variables

* Note: The notation "..." means that the listing of numbers continues on indefinitely.

that receive responses that appear to be discrete: for example, a person's age. When asked how old he was, Steve responded "19." He was 19 years old on his last birthday, but he is 19 plus some part of a year.

The use of fractions or decimals does not necessarily imply that data are continuous. The scoring of competitive diving is an example of a case in which decimals appear and the variable is actually discrete. A contestant can receive scores by halves only (5.5, 6.0, 6.5, and so on). These scores are actually discrete, because values *between* 6.0 and 6.5 cannot occur. There are other illustrations of this situation, but in this text a *count* or a *measurement* is about the only distinction that we will need to make.

There are situations in which data are measured in variable form and reported and discussed in attribute form. Two such situations are (1) the measurement of air pollution, as reported in the newspaper (measured in amount and reported as low, medium, or high), and (2) the Richter scale and the measurement of an earthquake.

Let's explore the difference between these terms and the types of data. If I were to go to a nearby parking lot right now and obtain some data, I would most likely think of it as a sample. However, it could be considered to be the population if I defined my population of concern to be the cars in that parking lot right now. It is more likely that it would be considered to be only part of a larger collection of cars, say of all cars that ever park in this lot or a sample of all cars parked in all parking lots at this time. (You must define your population and your sample carefully.) If I were to observe the make of each car, I would be collecting attribute data. If I were to record the number of people riding in each car as it left the parking lot, I would be collecting discrete data. (This clearly is a count of people.) If each car were to be weighed, continuous data would result, because all the various fractional parts of weight units could result from the weighing (see section 1–4).

Don't let the appearance of the data fool you in regard to their type. For example, suppose that after surveying this parking lot, I have a sample of 5 red, 8 blue, 6 green, and 2 yellow cars. These data appear to be the count of cars; therefore, one would think them to be discrete. This is not the case, however; they are attribute data. **You must look at each individual source to determine the kind of variable being used.** One specific car was red; "red," then, constituted the data. And red is an attribute. Thus the collection above (5 red, 8 blue, etc.) is a *summary* of the attribute data.

Another example of a deceiving variable is the license plate number (number part only) of each car. The numbers appear to be a discrete variable since only whole number values occur; however, these numbers are merely identification numbers. Identification numbers are not variables because they do not measure anything; they serve only to identify. Consider the weights of the cars as recorded on the registration: 3485, 3860, 2091, 4175, and so on. They are all whole numbers, but that does not make the variable discrete. The variable is "weight," and these data are measured to the nearest pound. Thus the variable "weight" is continuous. As you can see, the appearance of the data *after* they are recorded can be misleading in respect to their type. Remember to inspect an individual piece of data and you should have little trouble in distinguishing among attribute data and discrete and continuous data.

Exercises

1-1   Perform the "first-ace" experiment five times (instructions below) and observe the value of three different variables each time:

Variable 1: The color of the first ace to appear.

Variable 2: The longest distance across the dealt pile of cards.

Variable 3: The count of the card on which the first ace appears.

To perform the first-ace experiment, shuffle an ordinary deck of 52 playing cards containing 4 aces. Deal the cards one at a time onto a pile, stopping when the first ace appears. After the first ace has been placed on the pile, record the color of the ace. Now measure the longest distance across the scattered pile, before it is disturbed, and record it. Count and record the number of cards in the pile; repeat the experiment four more times.

Trial	x = Color of First Ace	d = Distance Across Pile	y = Count of Cards, Including First Ace
1	$x_1 =$	$d_1 =$	$y_1 =$
2	$x_2 =$	$d_2 =$	$y_2 =$
3	$x_3 =$	$d_3 =$	$y_3 =$
4	$x_4 =$	$d_4 =$	$y_4 =$
5	$x_5 =$	$d_5 =$	$y_5 =$
		$\sum_{i=1}^{5} d_i =$	$\sum_{i=1}^{5} y_i =$

NOTE: See appendix A for information about the $\sum$ notation.

1-2   Each of the three variables in the first-ace experiment produce a different type of data. What type of data results from each variable?

1-3   We want to investigate the cost of education. One of the expenses that a student must contend with is the cost of textbooks. Let x be the cost of all textbooks purchased by each student this semester at your college.

(a) Carefully describe the population.

(b) Carefully describe the variable.

1-4 The office of admissions wishes to estimate the average cost of textbooks per student per semester. (See exercise 1–3.)

(a) Describe the population parameter.

(b) The admissions office identified 50 students who registered and asked each of them to keep track of his or her textbook expenses and to report the total amount. The resulting 50 amounts form a sample. Describe the sample statistic that is of interest to the admissions office.

(c) Describe how you would use the 50 data in the sample to calculate the value of the sample statistic discussed in (b).

1-5 Which of the following may be considered examples of discrete variables?

(a) the breaking strength of a given type of string

(b) the hair color of children auditioning for *Annie*

(c) the number of stop signs in towns with less than 500 people

(d) the number of trains using a given track during a 24-hour time period

(e) whether or not a faucet is defective

1-6 Which of the following may be considered examples of continuous variables?

(a) the number of questions answered correctly on a Police Academy exam

(b) the number of telephone calls arriving at a switchboard per 10-minute period

(c) the length of time per telephone call to a certain real estate office

(d) the cost of Christmas trees during December

(e) the heights of heels on women's shoes

1-7 Which of the following situations demonstrate attribute (qualitative) data?

(a) you are polling people to see which candidate they are supporting

(b) you are making a study to see if the use of a new medicine reduces the length of time it takes a wound to heal

(c) you are checking the number of pages per job coming out of a computer

(d) you are studying the importance of certain job enrichments for hospital employees

(e) you are concerned with the distance that female college freshmen can kick a football

1-8 Suppose a twelve-year old asked you to explain the difference between a sample and a population;

(a) what information should your answer include?

(b) what reasons would you give him or her for why one would take a sample instead of surveying every member of the population?

Section 1–4 Measurability and Variability

Within a set of experimental data, we *always* expect variation. If little or no variation is found, we would guess that the measuring device is not sufficiently calibrated. For example, we take a carton of a favorite candy bar (24) and weigh each bar individually. We observe that all 24 candy bars weigh $\frac{7}{8}$ of an ounce to the nearest $\frac{1}{8}$ ounce. Does this mean that they are all identical in weight? Not really. Suppose that we were to weigh them on an analytical balance that weighs to the nearest milligram. Now their weights will be variable.

variability It does not matter what the response variable is; there will be **variability** in the numerical response if the tool of measurement is precise enough. A primary objective in statistical analysis is that of measuring variability.

Section 1–5 Comparison of Probability and Statistics

Probability and statistics are two separate but related fields of mathematics. It has been said that "probability is the vehicle of statistics." That is, if it were not for the laws of probability, the theory of statistics would not be possible.

Let's illustrate the relationship and the difference between these two branches of mathematics by looking at two boxes. We know the probability box contains 5 blue, 5 red, and 5 white poker chips. The subject of probability tries to answer such questions as this: If one chip is drawn from this box, what is the chance that it is blue? On the other hand, in the statistics box we don't know what the combination of chips is. We draw a sample and from this make conjectures about what we believe to be in the box. Note the difference: Probability asks about the chance that something (a sample) will happen when you know the possibilities (i.e., you know the population). Statistics asks you to draw a sample, describe the sample (descriptive statistics), and then make inferences about the population based on the information found in the sample (inferential statistics).

Probability
5B, 5R, 5W

Statistics
? ? ?

Exercises

1–9 Which of the following are best classified as "probability" problems and which are best classified as "statistics" problems?

(a) you are trying to determine whether there is an advantage when a horse starts from the #1 post position

(b) you are trying to determine the chance that a head will occur on a single flip of a coin

(c) you are trying to determine the chance of rain tonight

(d) you are trying to determine the likelihood that tonight's church barbeque attendance will exceed 400

(e) you are trying to determine the chance that a blue ball will be drawn from a bowl that contains 15 balls of which 5 are blue

Section 1-6 Data Collection

One of the first problems a statistician faces is collecting data. Data doesn't just happen. It must be collected. We must realize the importance of good sampling techniques, since the inferences that we ultimately make will be based on the statistics calculated from our sample data.

The collection of data for statistical analysis is an involved process and includes the following important steps:

1. Defining the objectives of the survey or experiment.
   Examples: (a) comparing the effectiveness of a new drug to the effectiveness of the standard drug, (b) estimating the average household income in our county.
2. Defining the variable and the population of interest.
   Examples: (a) length of recovery time for patients suffering from a particular disease, (b) total income for households in our county.
3. Defining the data collection and measuring schemes.
   This includes sampling procedures, sample size, and the data-measuring device (questionnaire, telephone, and so on).
4. Determining the appropriate descriptive or inferential data-analysis techniques.

Very often an analyst is stuck with data already collected, possibly even for other purposes, which makes it impossible to determine whether the data is "good" (representative) or not. Collecting the data yourself using approved techniques (as outlined above) is much preferred. While this text will be chiefly concerned with various data-analysis techniques, you should be aware of the problems of data collection.

The following illustrations present populations defined for specific investigations.

Illustration 1-5

The admissions office wishes to estimate the cost of college education at a local college. One of the components of total cost per semester is the cost of textbooks. Specifically, the admissions office would like to estimate the current "average" cost of textbooks per semester per student. The population of interest is the currently enrolled student body. □

Illustration 1-6

The cost of a one-minute television advertisement varies drastically according to the station, the day, and the time of day. This rate is determined by the proportion of potential viewers who are tuned to a particular station at a particular time. A national department store chain is opening a new store in a nearby city and is in the process of selecting time slots for advertising on WRQC-TV. They want to choose the time slots that will be most cost effective in attracting new customers. The population of interest is the daily television viewing habits of all persons who live within the broadcasting area. □

census

If every element in the population is listed, or enumerated, a **census** is compiled. A census for the population in illustration 1–5 could probably be obtained from the college's registrar in the form of a computer printout listing all registered students. However, censuses are seldom used because they are difficult to compile, very time-consuming, and therefore expensive. Imagine the task of compiling a census of every person who lives within the broadcasting area of WRQC-TV (illustration 1–6). When it is unrealistic to compile a census, only part of the population, a sample, will be analyzed.

When selecting a sample, it is necessary to construct a *sampling frame*.

---

sampling frame

Sampling Frame

A list of the elements belonging to the population from which the sample will be drawn.

The sampling frame should be identical to the population. Ideally the frame will have every element of the population listed once and only once. However, this is not always possible because it may be impractical or impossible to select directly from the total population. Since only the elements in the frame

representative sampling frame

have a chance to be selected as part of the sample, **it is important that the sampling frame be representative of the population**.

In illustration 1–5, the registrar's computer list will serve as a sampling frame for the admissions office. Lists of registered voters or the telephone directory are sometimes used as sampling frames of the general public. One of these might be used in illustration 1–6. Depending on the nature of the information sought, the list of registered voters or the telephone directory may or may not serve as a good sampling frame.

Having carefully defined the sampling frame, we are now prepared to select the elements from the sampling frame. This selection process is defined by the sample design.

---

sample design, or sampling plan

Sample Design, or Sampling Plan

The procedures used to collect sample data.

There are many different types of sample designs. However, all sample designs fit into two categories: judgment samples and probability samples.

---

judgment samples

Judgment Samples

Samples that are selected on the basis of being "typical."

When a judgment sample is drawn, the person selecting the sample chooses items that he or she thinks are representative of the population. The validity of the results from a judgment sample reflects the soundness of the collector's judgment.

probability samples	**Probability Samples**
	Samples in which the elements to be selected are drawn on the basis of probability. Each element in a population has a certain probability of being selected as part of the sample.

NOTE: **Statistical inference** (tests of hypothesis and confidence interval estimates) **requires that the sample design be a probability sample.**

The random sample is the most familiar probability sample.

random samples	**Random Samples**
	Each possible sample of fixed size *n* has an equal probability of being selected.

Let's look at a few of the simpler and easier to use sample designs.

One of the most common methods used to collect data for a random sample is the simple random sample.

simple random sample	**Simple Random Sample**
	A random sample selected in such a way that every element in the population has an equal probability of being chosen.

When a simple random sample is drawn, an effort must be made to ensure that each element has an equal probability of being selected. Mistakes are frequently made because the term *random* (equal chance) is confused with *haphazard* (without pattern). The proper procedure for selecting a simple random sample is to use a random number generator or a table of random numbers.

To select a simple random sample, first assign a number to each element in the sampling frame. This is usually done sequentially. Then go to a table of random numbers and select as many numbers as are needed for the sample size desired. (See appendix B for specific information regarding use of the random number table.) Each numbered element in the sampling frame that corresponds to a selected random number is chosen for the sample.

Illustration 1–7

Let's return to illustration 1–5. Mr. Clar, who works in the admissions office, has obtained a computer list of this semester's full-time enrollment.

There are 4265 student names on the list. He numbered the students 0001, 0002, 0003, and so on, up to 4265. Then, using four-digit random numbers, he identified a sample. (See appendix B for a discussion of the use of the random number table.) □

One of the easiest-to-use methods is the systematic sampling method.

---

systematic sample

Systematic Sample

A sample in which every kth item in the sampling frame is selected.

This method of selection uses the random number table only once, to find the starting point (first data). Moreover, it is not necessary to number the elements in the sampling frame or to know the total count of items in the sampling frame. Thus this is a good procedure for sampling a percentage of a large population. However, there are dangers inherent in using the systematic sampling technique. When the population is repetitive or cyclical in nature, systematic sampling should not be used.

To select an $x\%$ systematic sample, 1 out of every $100/x$ items will be selected by selecting every $100/x$th item. (If $100/x$ is not an integer, simply drop the fractional part and use only the integer part of $100/x$.) Suppose we desire a 3% systematic sample. The first item (starting point) is randomly selected, using the random number table, from the first 33 elements ($100/3 = 33\frac{1}{3}$, which becomes 33) in the sampling frame. Every 33rd element in the sampling frame, starting from the location of the first element, also belongs to the sample.

When sampling very large populations, sometimes it is possible to divide the population into subpopulations on the basis of some characteristic. These subpopulations are called **strata**. These smaller, easier-to-work-with strata are sampled separately. One of the sample designs that starts by stratifying the sampling frame is the stratified sampling method.

strata

---

stratified sample

Stratified Sample

A sample obtained by stratifying the sampling frame and then selecting a fixed number of items from each of the strata.

When a stratified sample is to be drawn, the population is subdivided into the various strata and then a subsample is drawn from each stratum. These subsamples may be drawn randomly, systematically, or by whatever method desired. Then the subsamples are combined into one sample of data for further use.

A slight modification to the stratified method results in a quota or a proportional sample.

---

quota sample or proportional sample

Quota Sample or Proportional Sample

A sample obtained by stratifying the sampling frame and then selecting a number of items from each stratum according to a quota or in proportion to the size of the stratum.

If the sample collector were to choose "one item for every 500 members," he would be collecting a "0.002 sample." The "one for every 500" is a quota. To select a "0.002 ($\frac{1}{500}$) sample" is to express the same information as a proportion. Each stratum is sampled and the subsamples are combined to form the sample.

Another sampling method that starts by stratifying results is a cluster sample.

---

cluster sample

Cluster Sample

A sample obtained by stratifying the sampling frame and then selecting sample items from some, but not all, of the strata.

The cluster sample is obtained by using either random numbers or a systematic method to first identify the strata (clusters) to be sampled and then to select the items from within these strata. Then the subsamples are combined to form the sample.

The sampling plan used in a particular situation depends on several factors, such as: (1) the nature of the population and the variable, (2) the ease (or difficulty) of sampling, and (3) the cost of sampling. These and other factors must be weighed, with tradeoffs usually occurring, before the exact sampling technique is determined.

Exercises

1–10 (a) What is a sampling frame?

(b) What did Mr. Clar use for a sampling frame in illustration 1–7?

1–11 The article at the beginning of chapter 4 reports the results of a poll taken in New York State in regard to restrictions on public smoking.

(a) What population was sampled?

(b) What was used as a sampling frame?

(c) Do you think this sampling frame is representative of the population? Explain.

(d) Was a simple random sample used? Explain.

1–12 A simple random sample could be very difficult to obtain. Why?

18   Chapter 1   Statistics

Section 1–7 Statistics and the Computer

In recent years, the computer has had a tremendous effect on almost every aspect of life. The field of statistics is no exception. As you will see, the field of statistics uses many techniques that are repetitive in nature; formulas used to calculate descriptive statistics, procedures for constructing graphic displays of data, and procedures that are followed to formulate statistical inferences. The computer is very good at performing these repetitive operations. It is fairly common for someone with a set of sample data that needs to be analyzed to seek out someone who knows how to use and has access to a computer. If that computer has one of the standard statistical packages on line, it will make the analysis easy to complete. Your local computer center can provide you with a list of what is available. Some of the more readily available packaged programs are: MINITAB, Biomed (Biomedical Programs), SAS (Statistical Analysis System), IBM Scientific Subroutine Packages, and SPSS (Statistical Package for the Social Sciences). You will see examples of computer output throughout this textbook.

In Retrospect

You should now have a general feeling of what statistics is about—an image that will grow and change as you work your way through this book. You know what a sample and a population are, the distinction between attribute and variable data, and the difference between continuous and discrete variables. You even know the difference between statistics and probability (although we will not study probability in detail until chapter 4). You should also have an appreciation for and a partial understanding of how sample data are obtained.

You should realize that the news articles at the beginning of the chapter represent various aspects of statistics. All of these articles, in one way or another, are the matter of statistics. The three graphic displays about women workers show annual results of record-keeping by the Bureau of Labor Statistics. The same three graphics also show predicted values or estimates for future years.

The "It's Your Life" article reports several statistics about the American teen. The intent in reporting the values is to estimate the level of concern all American teenagers feel, not just the sampled ones. The sign in the cartoon reports another statistic, "1.3 kids."

As members of a computer-age society, we are constantly bombarded by "statistical" information, by way of advertisements and public opinion polls. Information involving numbers can be found in nearly every daily newspaper. However, the fact that the information is numerical does not make it statistical, even though the terms data *and* statistics *are sometimes used in referring to such information. "Forty-three billion hamburgers" is not a piece of data as it is used here.*

After some exercises using the material in this chapter, we will consider the analysis of data in detail in the next chapters.

Chapter Exercises

1-13 How would you test Pickadilly Pete's claim that he is loved by everybody (see illustration 1-2 page 5)?

1-14 How would you arrive at a satisfactory answer to the question posed by the owners of Custom Sport Coupe (see illustration 1-3 page 5), "How tall are sports car drivers?"

1-15 We want to describe the so-called typical student at your college. Describe a variable that measures some characteristic of a student and results in

(a) attribute data,

(b) discrete data,

(c) continuous data.

1-16 Refer to the information shown in the news article "*It's Your Life*" at the beginning of this chapter.

(a) What is the population of interest?

(b) Name one variable discussed. (*Hint*: Information asked of each person sampled.)

(c) What type of data results from the variable named in part (b)?

(d) One of the numerical values reported is "eighty percent." Is this numerical value a piece of data or a statistic? Explain.

1-17 "*The Lasting Changes Brought by Women Workers*" news article reports several statistics. Suppose you had collected the information that led to the construction of these graphs. You might have asked each woman in your sample one question corresponding to each graph. From left to right, they might have been:

1. Are you employed or looking for work?

2. If married, are you a dual-earner?

3. What is your annual salary?

(a) What type of data (attribute or variable) results from the answer to each of these questions?

(b) Describe the statistic that is reported annually in each graph that could be obtained from the answers to the above questions.

1-18 Explain why "43 billion hamburgers" is not statistical.

1-19 It is impossible for any family to have "1.3 kids" as the sign in the cartoon states. Explain what "1.3 kids" means.

1-20 A researcher studying consumer buying habits questions every twentieth person entering Publix Supermarket. He asks, "How many times per week do you go grocery shopping?" He then records the answer as T.

(a) $T = 3$ is an example of (1) a sample, (2) a variable, (3) a statistic, (4) a parameter, or (5) a piece of data.

Suppose the researcher questions 427 shoppers during the survey.

(b) Give an example of a question that relates to this survey that the tools of descriptive statistics can be used to answer.

(c) Give an example of a question that relates to this survey that the tools of inferential statistics can be used to answer.

1–21 A researcher studying the attitudes of parents of preschool children interviews a random sample of 50 mothers. Each mother has one preschool child. She asks each mother, "How many times did you compliment your child yesterday?" She records the answer as C.

(a) C is a (1) piece of data, (2) statistic, (3) parameter, (4) variable, or (5) sample.

(b) Give an example of a question that relates to this survey that the tools of descriptive statistics can be used to answer.

(c) Give an example of a question that relates to this survey that the tools of inferential statistics can be used to answer.

1–22 Describe, in your own words, and give an example of the following terms. Your example should not be one given in class or in the textbook.

(a) variable　　(b) data　　(c) sample

(d) population　　(e) statistic　　(f) parameter

1–23 Find an article or an advertisement from a newspaper or a magazine that exemplifies the use of statistics.

(a) Describe and identify one population of interest (if there is more than one, identify just one for your answer).

(b) Describe and identify one variable used in the article.

(c) Describe and identify one statistic reported in the article.

Vocabulary List

Be able to define each term. In addition describe, in your own words, and give an example of each term. Your examples should not be ones given in class or in the textbook.

attribute data	random
census	random sample
cluster sample	representative sampling frame
continuous data	response variable
data	sample
descriptive statistics	sampling frame
discrete data	sampling plan
experiment	simple random sample
inferential statistics	statistic
judgment sample	statistics
parameter	strata
population	stratified sample
probability	systematic sample
proportional sample	variability
quota sample	variable data

Quiz

Answer "True" if the statement is always true. If the statement is not always true, replace the words shown in color with words that make the statement always true.

1–1   **Inferential** statistics is the "study" and description of data that result from an experiment.

1–2   **Descriptive** statistics is the "study" of a sample which enables us to make projections or estimates about the population from which the sample was drawn.

1–3   A **population** is typically a very large collection of individuals or measurements about which we desire information.

1–4   A statistic is the calculated measure of some characteristic of a **population**.

1–5   A parameter is the measure of some characteristic of a **sample**.

1–6   As a result of surveying 50 freshmen, it was found that 16 had participated in interscholastic sports, 23 had served as officers of classes and clubs, and 18 had been in school plays during their high school years. This is an example of **discrete variable data**.

1–7   The number of rotten apples in a shipping crate is an example of **continuous** variable data.

1–8   The thickness of the sheet metal that a company uses in its manufacturing process is an illustration of **attribute** data.

1–9   A **representative** sample is a sample obtained in such a way that all individuals had an equal chance to be selected.

1–10   The basic objective of **statistics** is that of obtaining a sample, inspecting this sample, and then making inferences about the unknown characteristics of the population from which the sample was drawn.

2 Descriptive Analysis and Presentation of Single-Variable Data

"Average" means different things

When it comes to convenience, few things can match that wonderful mathematical device called averaging.

How handy it is! With an average you can take a fistful of figures on any subject—temperatures, incomes, velocities, populations, light-years, hairbreadths, anything at all that can be measured—and compute one figure that will represent the whole fistful.

But there is one thing to remember. There are several kinds of measures ordinarily known as averages. And each gives a different picture of the figures it is called on to represent.

Take an example. Here are the annual incomes of ten families:

$54,000	$31,500
$39,000	$31,500
$37,500	$31,500
$36,750	$31,500
$35,250	$25,500

What would this group's "typical" income be? Averaging would provide the answer, so let's compute the typical income by the simpler and most frequently used kinds of averaging.

The arithmetic mean. When anyone cites an average without specifying which kind, you can probably assume that he has the arithmetic mean in mind.

It is the most common form of average, obtained by adding items in the series, then dividing by the number of items. In our example, the sum of the ten incomes divided by 10 is $35,400.

The mean is representative of the series in the sense that the sum of the amounts by which the higher figures exceed the mean is exactly the same as the sum of the amounts by which the lower figures fall short of the mean.

The median. As you may have observed, six families earn less than the mean, four earn more. You might very well wish to represent this varied group by the income of the family that is right smack dab in the middle of the whole bunch.

To do this, you need to find the median. It would be easy if there were 11 families in the group. The family sixth from highest (or sixth from lowest) would be in the middle and have the median income. But with ten families there is no middle family. So you add the two central incomes ($31,500 and $35,250 in this case) and divide by 2. The median works out to $33,375, less than the mean.

The midrange. The median, you will note, is the middle item in the series. Another number that might be used to represent the group is the midrange, computed by calculating the figure that lies halfway between the highest and lowest incomes. To find this figure, add the highest and lowest incomes ($54,000 and $25,500), divide by 2 and you have the amount that lies halfway between the extremes, $39,750.

The mode. So, three kinds of averages, and not one family actually has an income matching any of them.

Say you want to represent the group by stating the income that occurs most frequently. That kind of representativeness is called a mode. In this example $31,500 would be the modal income. More families earn that income than any other. If no two families earned the same income, there would be no mode.

Four different averages, each valid, correct and informative in its way. But how they differ!

arithmetic mean	$35,400
median	$33,375
midrange	$39,750
mode	$31,500

And they would differ still more if just one family in the group were a millionaire—or one were jobless!

So there are three lessons to take away from today's class in averages. First, when you see or hear an average, find out which average it is. Then you'll know what kind of picture you are being given.

Second, think about the figures being averaged so you can judge whether the average used is appropriate.

And third, don't assume that a literal mathematical quantification is intended every time somebody says "average." It isn't. All of us often say "the average person" with no thought of implying a mean, median or mode. All we intend to convey is the idea of other people who are in many ways a great deal like the rest of us.

Chapter Objectives

In this chapter we will learn how to present and describe single-variable data. **Single variable** *means that we are going to deal with only one numerical response value from a given source. In chapter 3 we will work with two values from a common source.*

The basic idea of descriptive statistics is to describe a set of data. Primarily we will try to describe a set of numerical values in a variety of abbreviated ways. Suppose your instructor just returned an exam paper to you. It carries a grade of 78, and you wish to interpret its value with respect to the rest of the class. Students usually are interested in three different things. What is the first thing you would ask your instructor?

Most students immediately ask: What was the average exam grade? The informqtion being sought here is the location of the "middle" for the set of all the grades. "Measures of central tendency" are used to describe this concept. Your instructor replies that the class average was 68. So you know that your paper was above average by 10 points. Now you would like an indication of the "spread" of the grades. Since 78 is 10 points above the average, you ask: "How close to the top is it?" Your instructor replies that the grades ranged from 42 to 87 points. Pictorially, the information we have so far looks like the accompanying figure.

The third question is usually less important, since by now you have a fairly good idea how your grade compares with those of your classmates. However, the third concept still has some importance. It answers the question, How are the data distributed? You may have heard about the "normal curve," but what is it? As you will see later, the idea of distribution describes the data by telling you whether the values are evenly distributed or clustered (bunched) around a certain value. For example, your instructor says that half the class had grades between 65 and 75.

When very large sets of data are involved, measures of "position" are useful. When college board exams are taken, there are thousands of scores that result. Your concern is the position of your score with respect to all the others. Averages, ranges, and so on, are not enough. We would like to say that your specific score is better than a certain percentage of scores. For example, your score is better than 75% of all the scores.

The four concepts necessary to describe sets of single-variable data are (1) measures of central tendency, (2) measures of dispersion (spread), (3) measures of position, and (4) types of distribution.

Unit 1 Graphic Presentation of Data

Section 2–1 Graphs and Stem-and-Leaf Displays

Once we collect the sample of data, we must "get acquainted" with it. One of the most helpful ways to become acquainted is to employ an initial exploratory technique that will result in a pictorial representation of the data.

The resulting displays visually reveal patterns of behavior of the variable being studied. There are several graphic (pictorial) ways to describe data. The method used is determined by the type of data and the idea to be presented.

NOTE: You should realize from the start that there is no single correct answer when constructing a graphic display. The analyst's judgment and the circumstances surrounding the problem will play a major role in the development of the graphic.

Circle and Bar Graphs

Circle graphs (pie diagrams) and bar graphs are often used to summarize attribute data. You have seen many examples of these before.

Table 2–1 lists the number of each type of operation performed at General Hospital last year. These data are displayed on a circle graph in figure 2–1 with each type being represented by its proportion and reported in percentages. Figure 2–2 displays the same set of type-of-operation data but in the form of a bar graph.

TABLE 2–1
Operations Performed at General Hospital

Type of Operation	Number of Cases
Thoracic	20
Bones and joints	45
Eye, ear, nose, and throat	58
General	98
Abdominal	115
Urologic	74
Proctologic	65
Neurosurgery	23
	498

FIGURE 2–1
Circle Graph

Operations Performed at General Hospital Last Year

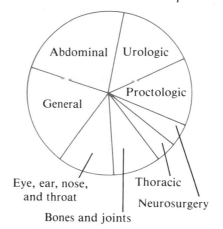

FIGURE 2–2
Bar Graph

Operations Performed at General Hospital Last Year

Stem-and-Leaf Displays

stem-and-leaf display   In recent years a technique known as the **stem-and-leaf display** has become very popular and these displays are well-suited for computer application. This technique, very simple to create and use, is a combination of a graphic technique and a sorting technique. (To **sort** data is to make a listing of the data in rank order according to numerical value.) The data values themselves are used to do this sorting. The **stem** is the leading digit(s) of the data, while the **leaf** is the trailing digit(s). For example, the numerical data value 458 might be split 45—8 as shown:

Leading Digits	Trailing Digits
45	8
Used in sorting	Shown in display

Illustration 2–1

Let's construct a stem-and-leaf display of the following set of 20 test scores.

82 74 88 66 58 74 78 84 96 76
62 68 72 92 86 76 52 76 82 78

At a quick glance we see that there are scores in the 50s, 60s, 70s, 80s, and 90s. Let's use the first digit of each score as the stem and the second digit as the leaf. Typically we will construct the display in a vertical position. Draw a vertical line and to the left of it locate the stems in order.

```
5 |
6 |
7 |
8 |
9 |
```

Next we place each leaf on its stem. This is accomplished by placing the trailing digit on the right side of the vertical line opposite its corresponding leading digit. Our first data value is 82, 8 is the stem and 2 is the leaf; thus we place a 2 opposite the stem 8.

8 | 2

We continue until each of the other 19 leaves are placed on the display. Figure 2–3 shows the resulting stem-and-leaf display.

FIGURE 2–3

Once the data has been entered into the computer, reworking it to create a display with different stems is usually very easy to accomplish.

```
            20 Exam Scores
         ──────────────────────────
     5 | 8 2
     6 | 6 2 8
     7 | 4 4 8 6 2 6 6 8
     8 | 2 8 4 6 2
     9 | 6 2
```

In figure 2–3 all scores with the same tens digit are placed on the same branch. This may not always be desired. Suppose we reconstruct the display; this time instead of grouping ten possible values on each stem, let's group the values so that only five possible values could fall on each stem. Do you notice a difference in appearance in Figure 2–4? The general shape,

FIGURE 2–4

```
                20 Exam Scores
            ──────────────────────────
   (50-54)  5 | 2
   (55-59)  5 | 8
   (60-64)  6 | 2
   (65-69)  6 | 6 8
   (70-74)  7 | 4 4 2
   (75-79)  7 | 8 6 6 6 8
   (80-84)  8 | 2 4 2
   (85-89)  8 | 8 6
   (90-94)  9 | 2
   (95-99)  9 | 6
```

approximately symmetrical about the 70s is very much the same. It is fairly typical of many variables to display a distribution that is concentrated (mounded) about a central value and then in some manner dispersed in both directions. □

Often a graphic display reveals something that the analyst may or may not have anticipated. The next illustration demonstrates what generally occurs when two populations are sampled together.

Illustration 2–2

A random sample of 50 college students was selected. Their weights were obtained from their medical records. The resulting data are listed in table 2–2.

TABLE 2–2

Student	1	2	3	4	5	6	7	8	9	10
Male/Female	F	M	F	M	M	F	F	M	M	F
Weight	98	150	108	158	162	112	118	167	170	120

Student	11	12	13	14	15	16	17	18	19	20
Male/Female	M	M	M	F	F	M	F	M	M	F
Weight	177	186	191	128	135	195	137	205	190	120

Student	21	22	23	24	25	26	27	28	29	30
Male/Female	M	M	F	M	F	F	M	M	M	M
Weight	188	176	118	168	115	115	162	157	154	148

Student	31	32	33	34	35	36	37	38	39	40
Male/Female	F	M	M	F	M	F	M	F	M	M
Weight	101	143	145	108	155	110	154	116	161	165

Student	41	42	43	44	45	46	47	48	49	50
Male/Female	F	M	F	M	M	F	F	M	M	M
Weight	142	184	120	170	195	132	129	215	176	183

Notice that the weights range from 98 to 215. Let's group the weights on stems of ten units using the hundreds and the tens digits as stems and the units digit as the leaf. (See figure 2–5.)

Close inspection of figure 2–5 suggests that there may be two overlapping distributions involved. That is exactly what we have: a distribution of female weights and a distribution of male weights. Figure 2–6, which shows a "back-to-back" stem-and-leaf display of this set of data, makes it quite obvious that we do in fact have two distinct distributions.

FIGURE 2–5

	Weights of 50 College Students (lbs.)
09	8
10	8 1 8
11	2 8 8 5 5 0 6
12	0 8 0 0 9
13	5 7 2
14	8 3 5 2
15	0 8 7 4 5 4
16	2 7 8 2 1 5
17	0 7 6 0 6
18	6 8 4 3
19	1 5 0 5
20	5
21	5

FIGURE 2–6

Weights of 50 College Students (lbs.)

Female		Male
8	09	
8 1 8	10	
6 0 5 5 8 8 2	11	
9 0 0 8 0	12	
2 7 5	13	
2	14	8 3 5
	15	0 8 7 4 5 4
	16	2 7 8 2 1 5
	17	0 7 6 0 6
	18	6 8 4 3
	19	1 5 0 5
	20	5
	21	5

Exercises

2–1 Sales revenues for Gulf Resources & Chemical generated in 1981 follow as shown below:

Product	Sales Revenue (in millions $)
Coal	$11.8
Lithium	7.5
Industrial explosives	6.6
Fertilizers and salt	1.5
Oil and gas	22.7
Other	0.1
Total	$50.2

Prepare a vertical bar graph of the sales revenue by product.

2–2 The Madison Gas & Electric Company uses various forms of energy for generating its electricity. The sources and the percentages of energy derived from each source for last year and ten years ago are shown below:

Source of Energy	10 Years Ago	Last Year
Coal	16.8%	62.6%
Nuclear	26.9%	34.1%
Gas	55.9%	2.9%
Other (including oil)	0.4%	0.4%
Total	100.0%	100.0%

Prepare a horizontal bar graph for this data. Put percentage along the horizontal axis and the sources of energy along the vertical axis.

2–3 The distribution of the number of Washington Gas Light Company shares owned by stockholders is as follows:

Number of Shares Owned	Proportion of Total
1– 99	0.43
100–249	0.41
250–999	0.14
1,000 or more	0.02
	1.00

Prepare a pie chart that shows the distribution of the number of shares owned.

2–4 Construct a stem-and-leaf display of the following data:

13.7 13.7 15.6 11.3 11.7 11.2 13.9
14.0 10.5 12.4 11.2 12.8 11.4 15.1

2–5 Construct a stem-and-leaf display of the following data:

39 52 39 56 87 40 73 38
48 55 44 65 51 63 60

2–6 An electronic Geiger counter was used to count the number of detectable emissions during a 10-second period. The experiment was repeated 22 times, and the following counts resulted:

8   23 18 22 22 15 21 23 25 18 24
22 21 37 19 22 22 12 27 16 26 32

Construct a stem-and-leaf display of these data.

2-7 On the first day of class last semester, 50 students were asked for their one-way travel times from home to college (to the nearest 5 minutes). The resulting data were as follows:

```
20  20  30  25  20  25  30  15  10  40
35  25  15  25  25  40  25  30   5  25
25  30  15  20  45  25  35  25  10  10
15  20  20  20  20  25  20  20  15  20
 5  20  20  10   5  20  30  10  25  15
```

Construct a stem-and-leaf display of these data.

Section 2–2 Frequency Distributions, Histograms, and Ogives

Listing a large set of data does not present much of a picture to the reader. Sometimes we want to condense the data into a more manageable form. This can be accomplished with the aid of a **frequency distribution**.

frequency distribution

Frequency Distributions

To demonstrate the concept of a frequency distribution, let's use the following set of data:

```
3  2  2  3  2
4  4  1  2  2
4  3  2  0  2
2  1  3  3  1
```

A frequency distribution is used to represent this set of data by listing the x values with their frequencies. For example, the value 1 occurs in the sample three times; therefore, the frequency for $x = 1$ is 3. The set of data is represented by the frequency distribution shown in table 2–3. The **frequency** f is the number of times the value x occurs in the sample. This is an **ungrouped frequency distribution**. We say "ungrouped" because each value of x in the distribution stands alone.

ungrouped frequency distribution

TABLE 2–3
Frequency Distribution

x	f
0	1
1	3
2	8
3	5
4	3

When a large set of data has many different x values instead of a few repeated values, as in the previous example, we can group the data into a set of **classes** and construct a frequency distribution. The stem-and-leaf display in figure 2–3 (page 29) shows, in picture form, a grouped frequency distribution. Each stem represents a class. The number of leaves on each stem is the same as the frequency for that same class. The data represented in figure 2–3 are listed as a frequency distribution in table 2–4:

TABLE 2–4

Class	Frequency
50–59	2
60–69	3
70–79	8
80–89	5
90–99	2
	20

TABLE 2–5
Ranked Data for Statistics Exam Score

27	68	79	91	107
43	71	80	91	108
43	71	81	93	108
44	71	82	94	116
47	73	82	94	120
49	73	84	94	120
50	74	84	96	122
54	75	86	97	123
58	76	88	103	127
65	77	88	106	128

The stem-and-leaf process can be used to construct a frequency distribution; however, the stem representation is not compatible with all class widths. For example, class widths of 3, 4, or 7 are awkward to use. Thus, sometimes we will find it advantageous to have a separate procedure for constructing a grouped frequency distribution. To illustrate this grouping (or classifying) procedure, let's use a sample of 50 final exam scores. Table 2–5 lists the 50 scores in ranked order.

The two basic guidelines that should be followed in constructing a grouped frequency distribution are:

1. Each class should be of the same width.
2. Classes should be set up so that they do not overlap and so that each piece of data belongs to exactly one class.

Three additional helpful (but not necessary) guidelines are these:

3. Five to 12 classes are most desirable for the exercises given in this textbook.
4. When it is convenient, an odd class width is often advantageous.
5. Use a system that takes advantage of a number pattern, to guarantee accuracy. (This will be demonstrated in the following example.)

PROCEDURE:

1. Identify the high and the low scores ($H = 128$, $L = 27$) and find the range. Range $= H - L = 128 - 27 = 101$.
2. Select a number of classes ($m = 10$) and a class with ($c = 11$) so that the product ($mc = 110$) is a bit larger than the range (range $= 101$).
3. Pick a starting point. This starting point should be a little smaller than the lowest score L. Suppose that we start at 22; counting from there by elevens (the class width), we get 22, 33, 44, 55, . . . , 132. These are called the lower class limits. (They are all multiples of eleven, an easily recognized pattern.)

lower and upper class limits

The **lower class limit** is the smallest piece of data that could go into each class. The **upper class limits** are the largest values fitting into each class; these are 32, 43, 54, and so on in our example. Our *classes* for this example are

22–32	77–87
33–43	88–98
44–54	99–109
55–65	110–120
66–76	121–131

NOTES:

1. At a glance you can check the number pattern to determine if the arithmetic used to form the classes was correct.

class width

2. The **class width** is the difference between a lower class limit and the next lower class limit. (It is *not* the difference between the lower and upper limit of the same class.)

class boundary

3. **Class boundaries** are numbers that do not occur in the sample data but are halfway between the upper limit of one class and the lower limit of the next class. In the example above the class boundaries are 21.5, 32.5, 43.5, 54.5, . . . , 120.5, and 131.5. The difference between the upper and lower class boundaries will also give you the class width.

When classifying data it helps to use a **standard chart** (see table 2–6).

TABLE 2–6
Standard Chart for Frequency Distribution

Class Number	Class Limits	Tallies	Frequency, f
1	22–32	|	1
2	33–43	||	2
3	44–54	⊬⊬	5
4	55–65	||	2
5	66–76	⊬⊬ ||||	9
6	77–87	⊬⊬ ||||	9
7	88–98	⊬⊬ ⊬⊬	10
8	99–109	⊬⊬	5
9	110–120	|||	3
10	121–131	||||	4
			50

Once the classes are set up, we need to tally the data (table 2–6). If the data have been ranked, this tallying is unnecessary; if not ranked, be careful as you tally them. The frequency f for each class is the number of pieces of data that belong in that class. The sum of the frequencies should be exactly equal to the number of pieces of data n ($n = \sum f$). This summation serves as a good check.

class mark

In table 2–7 there is a column for the class mark, x. The **class mark** is the numerical value that is exactly in the middle of each class. In table 2–7 the class marks are

$$x_1 = \frac{22 + 32}{2} = 27, \qquad x_2 = \frac{33 + 43}{2} = 38$$

and so on. As a check of your arithmetic, successive class marks should be a class width apart, which is 11 in this illustration.

TABLE 2–7

Class Number	Class Limits	f	Class Mark, x
1	22–32	1	27
2	33–43	2	38
3	44–54	5	49
4	55–65	2	60
5	66–76	9	71
6	77–87	9	82
7	88–98	10	93
8	99–109	5	104
9	110–120	3	115
10	121–131	4	126
		50	

NOTES:

1. Now you can see why it is helpful to have an odd class width; the class marks are whole numbers.

2. The occasional use of subscripts is for the purpose of identifying the class being discussed.

grouped frequency distribution   Once the class marks are determined, we have a **grouped frequency distribution**. You should note that when we classify data in this fashion, we lose some information. Only when we have all the raw data before us do we know the exact values that were actually observed in each class. For example, we put a 58 and a 65 into class number 4, with class limits of 55 and 65. Once they are placed in the class, their values are lost to us and we use the class mark, 60, as their representative value. This loss of identity costs some accuracy, but the computations are made easier.

Histograms

histogram   The **histogram** is a type of bar graph representing an entire set of data. The distribution of frequencies from table 2–7 appears in histogram form in figure 2–7. A histogram is made up of the following components:

1. A title, which identifies the population of concern.

2. A vertical scale, which identifies the frequencies in the various classes. Figure 2–7 is a frequency histogram.

3. A horizontal scale, which identifies the variable x. Values for the class boundaries, class limits, or class marks may be labeled along the x-axis. Use whichever *one* of these sets of class numbers best presents the variable.

(*Note*: Be sure to identify both scales so that the histogram tells the complete story.)

FIGURE 2–7
Frequency Histogram

50 Final Exam Scores in Elementary Statistics

By changing the vertical scale to $\frac{0}{50}$, $\frac{1}{50}$, $\frac{2}{50}$, and so on, the histogram

relative frequency histogram
relative frequency

would become a **relative frequency histogram**. The **relative frequency** is a proportional measure of the frequency of an occurrence. It is found by dividing the class frequency by the total number of observations. For example, in this illustration the frequency associated with the seventh class (88–98) is 10. The relative frequency is $\frac{10}{50}$ (or 0.20), since these values occurred 10 times in 50 observations. Relative frequency can often be useful in a presentation. Figure 2–8 is a relative frequency histogram of the sample of 50 scores. Compare and contrast figures 2–7 and 2–8.

FIGURE 2–8
Relative Frequency Histogram

50 Statistics Exam Scores

20 Exam Scores

FIGURE 2–9

A stem-and-leaf display contains all the information in a histogram. Figure 2–3 shows the stem-and-leaf display constructed in illustration 2–1. Figure 2–3 has been rotated 90° and labels have been added to form the histogram shown in figure 2–9.

Histograms are valuable tools. For example, the histogram of the sample should have a distribution shape that is very similar to that of the population from which the sample was drawn. If the reader of a histogram is at all familiar with the variable involved, he or she will usually be able to interpret several important facts from a histogram. Figure 2–10 presents histograms with descriptive labels resulting from their geometric shape. Can you think of populations whose samples might yield histograms like these?

Briefly, the terms used to describe histograms are as follows:

Symmetrical: Both sides of this distribution are identical.

Uniform (rectangular): Every value appears with equal frequency.

Skewed: One tail is stretched out longer than the other. The direction of skewness is on the side of the longer tail.

J-shaped: There is no tail on the side of the class with the highest frequency.

Bimodal: The two most populous classes are separated by one or more classes. This situation often implies that two populations are being sampled.

FIGURE 2–10

Shapes of Histograms

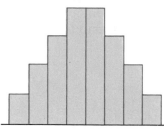

Normal or triangular
(distribution is symmetrical)

Uniform or rectangular
(distribution is symmetrical)

Skewed to right

Skewed to left

J-shaped

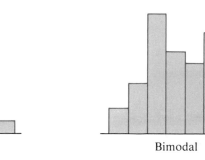

Bimodal

NOTES:

mode

1. The **mode** is the value of the piece of data that occurs with the greatest frequency. (Mode will be discussed in section 2–3.)

modal class

2. The **modal class** is the class with the highest frequency.

bimodal frequency
distribution

3. A **bimodal frequency distribution** has the two highest frequency classes separated by classes with lower frequencies.

Ogive

cumulative, and
cumulative relative
frequency distributions

A frequency distribution can very easily be converted to a **cumulative frequency distribution** by replacing the frequencies with cumulative frequencies. This is accomplished by placing a subtotal of the frequencies next to each class, as in table 2–8. The **cumulative frequency** for any given class is the sum of the frequency for that class and the frequencies of all classes of smaller values.

The same information can be presented by using a **cumulative relative frequency distribution** (see table 2–9). This combines the cumulative frequency idea and the relative frequency idea.

ogive

An **ogive** (pronounced o'jīv) is a cumulative frequency or cumulative relative frequency graph. An ogive has the following components:

1. A title, which identifies the population.

2. A vertical scale, which identifies either the cumulative frequencies or the cumulative relative frequencies. (Figure 2–11 shows an ogive with cumulative relative frequencies.)

3. A horizontal scale, which identifies the upper class boundaries. Until the upper boundary of a class has been reached, you cannot be sure you have accumulated all the data in that class. Therefore, **the horizontal scale for an ogive is always based on the upper class boundaries**

TABLE 2–8
A Cumulative
Frequency Distribution

Frequency Distribution		Cumulative Frequency Distribution	
Class Limits	Frequencies	Class Boundaries	Cumulative Frequencies
22–32	1	21.5–32.5	1
33–43	2	32.5–43.5	3 (1 + 2)
44–54	5	43.5–54.5	8 (1 + 2 + 5)
55–65	2	54.5–65.5	10 (1 + 2 + 5 + 2)
66–76	9	65.5–76.5	19 (1 + 2 + 5 + 2 + 9)
77–87	9	76.5–87.5	28
88–98	10	87.5–98.5	38
99–109	5	98.5–109.5	43
110–120	3	109.5–120.5	46
121–131	4	120.5–131.5	50
	50		

TABLE 2–9
A Cumulative Relative Frequency Distribution

Class Boundaries	Cumulative Relative Frequency
21.5–32.5	$\frac{1}{50} = 0.02$
32.5–43.5	$\frac{3}{50} = 0.06$
43.5–54.5	$\frac{8}{50} = 0.16$
54.5–65.5	$\frac{10}{50} = 0.20$
65.5–76.5	$\frac{19}{50} = 0.38$
76.5–87.5	$\frac{28}{50} = 0.56$
87.5–98.5	$\frac{38}{50} = 0.76$
98.5–109.5	$\frac{43}{50} = 0.86$
109.5–120.5	$\frac{46}{50} = 0.92$
120.5–131.5	$\frac{50}{50} = 1.00$

NOTE: Every ogive starts on the left with a relative frequency of 0 at the lower class boundary of the first class and ends on the right with a relative frequency of 100% at the upper class boundary of the last class.

All graphic representations of sets of data should be completely self-explanatory. That includes a descriptive and meaningful title and proper identification of the vertical and horizontal scales.

FIGURE 2–11
Ogive

50 Final Exam Scores in Statistics

Exercises

2–8 The monthly percentage changes in the Consumer Price Index for last year follow:

0.7   1.0   0.6   0.4   0.7   0.7   1.2   0.8   1.2   0.4   0.5   0.4

(a) Prepare a frequency distribution of these changes.

(b) Prepare a frequency histogram of these changes.

2-9   The ages of 50 dancers, who responded to a call to audition for a musical comedy, are as follows:

$$
\begin{array}{cccccccccc}
21 & 19 & 22 & 19 & 18 & 20 & 23 & 19 & 19 & 20 \\
19 & 20 & 21 & 22 & 21 & 20 & 22 & 20 & 21 & 20 \\
21 & 19 & 21 & 21 & 19 & 19 & 20 & 19 & 19 & 19 \\
20 & 20 & 19 & 21 & 21 & 22 & 19 & 19 & 21 & 19 \\
18 & 21 & 19 & 18 & 22 & 21 & 24 & 20 & 24 & 17 \\
\end{array}
$$

(a) Prepare an ungrouped frequency distribution of these ages.

(b) Prepare an ungrouped relative frequency distribution of the same data.

(c) Prepare a cumulative relative frequency distribution of the same data.

(d) Prepare a relative frequency histogram of this data.

(e) Prepare an ogive of this data.

2-10 The speed of 55 cars was measured by a radar device on a city street:

$$
\begin{array}{ccccccccccc}
27 & 23 & 22 & 38 & 43 & 24 & 35 & 26 & 28 & 18 & 20 \\
25 & 23 & 22 & 52 & 31 & 30 & 41 & 45 & 29 & 27 & 43 \\
29 & 28 & 27 & 25 & 29 & 28 & 24 & 37 & 28 & 29 & 18 \\
26 & 33 & 25 & 27 & 25 & 34 & 32 & 36 & 22 & 32 & 33 \\
21 & 23 & 24 & 18 & 48 & 23 & 16 & 38 & 26 & 21 & 23 \\
\end{array}
$$

(a) Classify these data into a grouped frequency distribution by using classes of 15–19, 20–24, ..., 50–54. (Retain this solution for use in answering exercises 2–21 and 2–23.)

(b) Find the class width.

(c) For the class 20–24, find these:
(1) the class mark
(2) the lower class limit
(3) the upper class boundary

(d) Construct a frequency histogram of these data.

2-11 The following 40 amounts are the fees that Fast Delivery charged for delivering small freight items last Thursday afternoon:

$$
\begin{array}{cccccccc}
2.03 & 1.56 & 1.10 & 4.04 & 3.62 & 1.16 & 0.93 & 1.82 \\
2.30 & 1.86 & 2.57 & 1.59 & 2.57 & 4.16 & 0.88 & 3.02 \\
3.46 & 1.87 & 4.81 & 2.91 & 1.62 & 1.62 & 1.80 & 1.70 \\
2.15 & 2.07 & 1.77 & 3.77 & 5.86 & 2.63 & 2.81 & 0.86 \\
3.02 & 3.24 & 2.02 & 3.44 & 2.65 & 1.89 & 2.00 & 0.99 \\
\end{array}
$$

(a) Classify these data into a grouped frequency distribution by using classes of 0.01–1.00, 1.01–2.00, ..., 5.01–6.00. (Retain this solution for use in answering exercises 2–22 and 2–34.)

(b) Find the class width.

(c) For the class 2.01–3.00, name the value of these:
(1) the class mark
(2) the class limits
(3) the class boundaries

(d) Construct a relative frequency histogram of these data.

2–12 A salesman made the following commissions (in dollars) on his last 42 commission sales:

```
180  256  687  566  443  102  277
384  314  460  145   32  654  255
384   21  276   87  320   85  142
296  546   61  141  503  690  413
372  469  110  259  389  396  120
561   81  657  407  440  319  281
```

(a) Construct the grouped frequency distribution of the data by using class marks of 49.5, 149.5, 249.5, ..., 649.5.

(b) Find the class width.

(c) Name the class limits for the classes whose class marks are 49.5 and 149.5.

(d) Name the boundary between the classes of 49.5 and 149.5.

(e) Construct a cumulative relative frequency distribution of these data.

(f) Prepare an ogive depicting these data.

Unit 2 Calculated Descriptive Statistics

Section 2–3 Measures of Central Tendency

measure of central tendency **Measures of central tendency** are numerical values that tend to locate in some sense the *middle* of a set of data. The term *average* is often associated with these measures. Each of the several measures of central tendency can be called the "average" value.

Mean

mean To find the **mean,** $\bar{x}$ (read "x bar"), the average with which you are probably most familiar, you will add all the values of the variable x (this sum of x values is symbolized by $\sum x$) and divide by the number of these values, n. We express this in formula form as

$$\text{mean} = \bar{x} = \frac{\sum x}{n} \qquad (2\text{–}1)$$

summation

1. See appendix A for information about the $\sum$ (read "**sum**") **notation**
2. No indexes are shown in the formulas in this chapter because the summations are over all the data.

Illustration 2–3

A set of data consists of the five values 6, 3, 8, 5, and 3. Find the mean.

Solution   Using formula (2–1), we find

$$\bar{x} = \frac{6 + 3 + 8 + 5 + 3}{5} = \frac{25}{5} = 5$$

Therefore, the mean of this sample is 5.   □

A physical representation of the mean can be constructed by thinking of a number line balanced on a fulcrum. A weight is placed on a number on the line corresponding to each number in the sample. In figure 2–12 there is one weight each on the 6, 8, and 5 and two weights on the 3 since there were two 3s in the sample of illustration 2–3. The mean is the value that balances the weights on the number line, in this case, 5.

FIGURE 2–12
Physical Representation of Mean

TABLE 2–10
Ungrouped Frequency Distribution

x	f
1	5
2	9
3	8
4	6
	28

Don't be misled by illustration 2–3 into believing that the mean value has to be a value in the data set. For example, consider the case where a salesperson sells $130 worth of merchandise one week and $110 in another. The mean of these values is $120, which is not a value in the data set.

When the sample data has the form of a frequency distribution, we will need to make a slight adaptation in order to find the mean. Consider the frequency distribution of table 2–10. This frequency distribution represents a sample of 28 values, five 1s, nine 2s, eight 3s, and six 4s. To calculate the mean $\bar{x}$ using formula (2–1), we need $\sum x$, the sum of the 28 x-values.

$$\sum x = \underbrace{1 + 1 + \cdots + 1}_{5 \text{ of them}} + \underbrace{2 + 2 + \cdots + 2}_{9 \text{ of them}} + \underbrace{3 + 3 + \cdots + 3}_{8 \text{ of them}} + \underbrace{4 + 4 + \cdots + 4}_{6 \text{ of them}}$$

$$\sum x = (5)(1) + (9)(2) + (8)(3) + (6)(4)$$

$$= 5 + 18 + 24 + 24 = 71$$

TABLE 2–11
Extension xf

x	f	xf
1	5	5
2	9	18
3	8	24
4	6	24
Total	28	71

We can also obtain this sum by direct use of the frequency distribution. The extensions xf (see table 2–11) are formed for each row and then totaled to obtain $\sum xf$. The $\sum xf$ is the sum of the data. Therefore, the mean of a frequency distribution may be found by dividing the sum of the data, $\sum xf$, by the sample size, $\sum f$. We can rewrite formula (2–1) for use with a frequency distribution as

$$\bar{x} = \frac{\sum xf}{\sum f} \qquad (2\text{–}2)$$

The mean, for the example is found by using formula (2–2).
Mean:

$$\bar{x} = \frac{\sum xf}{\sum f},$$

$$\bar{x} = \frac{71}{28} = 2.536 = \mathbf{2.5}$$

NOTES:

1. The extensions in the xf column are the same subtotals as we found previously.

2. The two column totals, $\sum f$ and $\sum xf$, are the same values as were previously known as n and $\sum x$, respectively. That is, $\sum f = n$, the sum of the frequencies is the number of pieces of data. The $\sum xf$ is the sum of all the pieces of data.

3. The f in the summation expression indicates that the sum was obtained with the use of a frequency distribution.

Let's return now to the sample of 50 exam scores. In table 2–7 (see page 36) you will find a grouped frequency distribution for the 50 scores. We will calculate the mean for the sample by using the frequencies and the class marks in the same manner as in the example above. Table 2–12 shows the extensions and totals. (*Note*: The class marks are now being used as representative values for the observed data. Hence the sum of the class marks is meaningless and is therefore not found.)
The mean: Using formula (2–2) we obtain

$$\bar{x} = \frac{\sum xf}{\sum f} = \frac{4{,}144}{50} = 82.88 = \mathbf{82.9}$$

To illustrate again what the mean is, consider a balance beam (a real number line) with a cube placed on the value for each response (class mark). That is, 9 cubes will be placed on 71, and so on. When the cubes are all placed, one cube for each piece of data, the beam will balance when the

TABLE 2-12
Calculations and Totals
for Exam Scores

Class Number	Class Limits	f	Class Marks x	xf
1	22–32	1	27	27
2	33–43	2	38	76
3	44–54	5	49	245
4	55–65	2	60	120
5	66–76	9	71	639
6	77–87	9	82	738
7	88–98	10	93	930
8	99–109	5	104	520
9	110–120	3	115	345
10	121–131	4	126	504
		50		4,144

fulcrum is at 82.9, the mean. See figure 2–13. Formula (2–2) gives an approximation of $\bar{x}$ when working with a grouped frequency distribution. Recall, the value of the class mark is used to represent the value of each piece of data that fell into that class interval.

FIGURE 2–13
Mean is 82.9

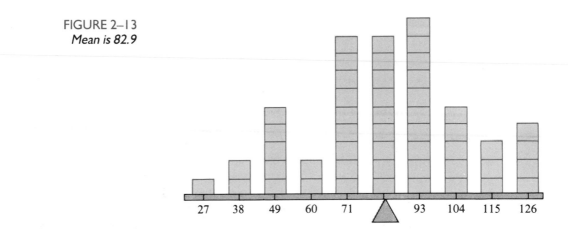

Median

median The **median** $\tilde{x}$ (read "x tilde"), is the middle value of the sample when the data are ranked in order according to size. The data in illustration 2–3 ranked in order of size are 3, 3, 5, 6, and 8. The 5 is the third or middle value of the five numbers and thus the median is 5.

The **position** of the median, i, is determined by the formula

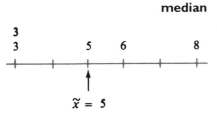

$$\text{position of the median} = i = \frac{n + 1}{2} \qquad (2\text{--}3)$$

where n represents the number of pieces of data and i is the position that the median occupies in the ranked data. To find the median $\tilde{x}$ you must first determine its position number i and then count the ranked data, starting from either end, finding the value in the ith position. The median will have the same value regardless of which end of the ranked data (high or low) you count from. In fact, counting from both ends will serve as an excellent check.

When n is an odd number, the median will be the exact middle piece of data. In our example, $n = 5$, and therefore the position of the median is

$$i = \frac{5 + 1}{2} = 3$$

That is, the median is the third number from either end in the ranked data, or $\tilde{x} = 5$.

However, if n is even, the position of the median will always be a half number. For example, let's look at a sample whose ranked data are 6, 7, 8, 9, 9, and 10. Here $n = 6$, and therefore the median's position is

$$i = \frac{6 + 1}{2} = 3.5$$

FIGURE 2–14
Median

This says that the median is halfway between the third and fourth pieces of data. To find the number halfway between any two values, add the two values together and divide by 2. In this case, add the 8 and the 9, then divide by 2. **The median is 8.5**, a number halfway between the "middle" two numbers. Notice in figure 2–14 that there are three numbers in the sample that are smaller in value than 8.5 and three that are larger.

Other Averages

mode   The **mode** is the value of x that occurs most frequently. In the examples above, the mode in the sample of size 5 is 3. The mode in the sample of size 6 is 9. In both cases these are the only numbers that occur more than once. If it happens that more than one of the values have the same frequency (number of occurrences), we say there is no mode. For example, in the sample 3, 3, 4, 5, 5, and 7, both the 3 and the 5 appear an equal number of times. There is no one value that appears most often. This sample has no mode.

midrange   Another measure of central tendency is the **midrange**. A set of data will always have a lowest value L and a highest value H. The midrange is a number exactly midway between them. It is found by averaging the low and the high values:

$$\text{midrange} = \frac{L + H}{2} \tag{2-4}$$

For the sample 6, 7, 8, 9, 9, and 10, $L = 6$ and $H = 10$. Therefore, the midrange is

$$\frac{6 + 10}{2} = 8$$

The midrange is the numerical value halfway between the two extreme values, the low L and the high H.

The four measures of central tendency represent four different methods of describing the middle. These four values may be the same but more likely they will result in different values. For the sample data shown in figure 2–14, the mean $\bar{x}$ is 8.2, the median $\tilde{x}$ is 8.5, the mode is 9, and the midrange is 8.

Exercises

2–13 Consider the sample 2, 4, 7, 8, and 9. Find the following:

(a) the mean $\bar{x}$ (b) the median $\tilde{x}$
(c) the mode   (d) the midrange

2–14 Consider the sample 6, 8, 7, 5, 3, and 7. Find the following:

(a) the mean $\bar{x}$ (b) the median $\tilde{x}$
(c) the mode   (d) the midrange

2–15 Consider the sample 7, 6, 10, 7, 5, 9, 3, 7, 5, and 13. Find the following:

(a) the mean $\bar{x}$ (b) the median $\tilde{x}$
(c) the mode   (d) the midrange

2–16 Fifteen randomly selected college students were asked to state the number of hours they slept last night. The resulting data are 5, 6, 6, 8, 7, 7, 9, 5, 4, 8, 11, 6, 7, 8, 7. Find the following:

(a) the mean $\bar{x}$ (b) the median $\tilde{x}$
(c) the mode   (d) the midrange

2–17 A class had the following set of quiz grades:

19	13	20	22	19	17
9	10	19	13	23	15
22	14	18	21	20	18
9	15	13	10	17	19

(a) Find the mean $\bar{x}$.   (b) Find the median $\tilde{x}$.
(c) Find the mode.   (d) Find the midrange.

2–18 Use formula (2–2) to find the mean for the following frequency distribution:

x	f
0	1
1	3
2	8
3	5
4	3

2–19 Find the mean for the following grouped frequency distribution:

Class Limits	f
3–5	2
6–8	10
9–11	12
12–14	9
15–17	7

2–20 Find the mean for the following grouped frequency distribution:

Class Limits	f
2–5	7
6–9	15
10–13	22
14–17	14
18–21	2

2–21 Find the mean for the set of speeds for the 55 cars given in exercise 2–10. Use the frequency distribution found in answering exercise 2–10. (Retain these solutions for use in answering exercise 2–33.)

2–22 Find the mean for the 40 small freight delivery charges given in exercise 2–11. Use the frequency distribution found in answering exercise 2–11. (Retain these solutions for use in answering exercise 2–34.)

Section 2–4 Measures of Dispersion

Once the "middle" of a set of data has been determined, our search for information immediately turns to the measures of dispersion (spread). The measures of dispersion include the range, variance, and standard deviation. These numerical values describe the amount of spread or variability that is found among the data. **Closely grouped data will have relatively small values and more widely spread-out data will have larger values for these various measures of dispersion.**

Range

range The **range** is the simplest measure of dispersion. It is the difference between the highest- (largest) valued (H) and the lowest- (smallest) valued (L) pieces of data:

$$\text{range} = H - L \tag{2–5}$$

For the sample 3, 3, 5, 6, and 8, the range is $8 - 3 = 5$. The range tells us that the five pieces of data all fall within a distance of 5 units on the number line.

The other two measures of dispersion, variance and standard deviation, are actually measures of dispersion about the mean. To develop a measure of dispersion about the mean, let's look first at the concept of **deviation from the mean**. An individual value x deviates, or is located, from the mean by an amount equal to $(x - \bar{x})$. This deviation $(x - \bar{x})$ is zero when x is equal to the mean. The deviation $(x - \bar{x})$ is positive if x is larger than $\bar{x}$ and negative if x is less than $\bar{x}$.

deviation from the mean

Consider the sample 6, 3, 8, 5, and 3. Using formula (2–1). $\bar{x} = (\sum x)/n$, we find that the mean is 5. Each deviation is then found by subtracting 5 from each x value.

x	6	3	8	5	3
$x - \bar{x}$	1	−2	3	0	−2

Figure 2–15 shows the deviation from the mean for each value. We might suspect that the sum of all these deviations, $\sum (x - \bar{x})$, would serve as a measure of dispersion about the mean. However, that sum is exactly zero. As a matter of fact, it will always be zero. Why? Think back to the definition of the mean (page 42) and see if you can justify this statement.

(page 42)

FIGURE 2–15

Deviations from the Mean

If the sum of the deviations, $\sum (x - \bar{x})$, is always zero, it is not going to be of any value in describing a particular set of data. However, we still want to be able to use the idea of deviation about the mean, since the mean is the most common average used. Yet there is a canceling effect between the deviations of x values smaller than the mean (negative) and those values larger than the mean (positive). This canceling effect could be removed if we make all the deviations positive. By using the absolute value of $x - \bar{x}$, $|x - \bar{x}|$, we can accomplish this. In the previous illustration we now obtain the following absolute deviations:

x	6	3	8	5	3
$|x - \bar{x}|$	1	2	3	0	2

mean absolute deviation The sum of these deviations is 8. Thus we define a value known as the **mean absolute deviation** :

$$\text{mean absolute deviation} = \frac{\sum |x - \bar{x}|}{n} \qquad (2\text{-}6)$$

For our example the mean absolute deviation then becomes 8/5, or **1.6**. Although this particular measure of spread is not used too frequently, it is a measure of dispersion. It tells us the average distance that a piece of data is from the mean.

There is another way to eliminate the canceling effect. Squaring the deviations will cause all these values to be nonnegative (positive or zero). When these values are totaled, the result is positive. The sum of the squares of the deviations from the mean, $\sum (x - \bar{x})^2$, is used to define the variance.

Variance

variance The **variance**, s^2, of a sample is the numerical value found by applying the following formula to the data:

$$\text{sample variance} = s^2 = \frac{\sum (x - \bar{x})^2}{n - 1} \qquad (2\text{-}7)$$

where n is the sample size, that is, the number of items in the sample. The variance of a sample is a measure of the spread of the data about the mean. The variance of our sample 6, 3, 8, 5, and 3 is found as shown in table 2–13.

Let's look at another sample: 1, 3, 5, 6, and 10. The calculations of the totals, the mean, and the variance are shown in table 2–14.

TABLE 2–13
Computing s^2 for 6, 3, 8, 5, 3 Using Formula (2–7)

Step 1. x	Step 3. $x - \bar{x}$	Step 4. $(x - \bar{x})^2$
6	1	1
3	−2	4
8	3	9
5	0	0
3	−2	4
25	0 (ck)	18

Step 2. $\bar{x} = \dfrac{\sum x}{n} = \dfrac{25}{5} = 5$

Step 5. $s^2 = \dfrac{\sum (x - \bar{x})^2}{n - 1} = \dfrac{18}{4} = 4.5$

TABLE 2–14
Computing s^2 for 1, 3, 5,
6, 10 Using Formula (2–7)

1. x	3. $x - \bar{x}$	4. $(x - \bar{x})^2$
1	−4	16
3	−2	4
5	0	0
6	1	1
10	5	25
25	0 ⓒⓚ	46

2. $\bar{x} = \dfrac{\sum x}{n} = \dfrac{25}{5} = 5$

5. $s^2 = \dfrac{\sum (x - \bar{x})^2}{n - 1} = \dfrac{46}{4} = 11.5$

NOTES:

1. The sum of all the x's is used to find $\bar{x}$.

2. The sum of the deviations, $\sum (x - \bar{x})$, is always zero. Use this as a check in your calculations, as we did in tables 2–13 and 2–14.

The last set of data is more dispersed than the previous set, and therefore its variance is larger. A comparison of these two samples is shown in figure 2–16.

FIGURE 2–16
Comparison of Data

First sample: 3 3 5 6 8 $s^2 = 4.5$
Second sample: 1 3 5 6 10 $s^2 = 11.5$

Standard Deviation

standard deviation The **standard deviation** of a sample, s, is the positive square root of the variance:

$$s = \sqrt{s^2} = \sqrt{\frac{\sum (x - \bar{x})^2}{n - 1}} \qquad (2-8)$$

For the samples shown in figure 2–16, the standard deviations are $\sqrt{4.5}$, or **2.1**, and $\sqrt{11.5}$, or **3.4**.

Round-off Rule

When rounding off an answer, let's agree to keep one more decimal place in our answer than was present in our original data. Round off only the

final answer, not the intermediate steps. That is, do not use a rounded variance to obtain a standard deviation. In our previous examples, the data were composed of whole numbers; therefore, those answers that have decimal values should be rounded to the nearest tenth. See appendix C for specific instructions on how to perform the rounding off.

NOTE: The numerator for sample variance, $\sum (x - \bar{x})^2$, is often called the "sum of squares for x" and symbolized by SS(x). Thus formula (2–7) can be expressed

$$s^2 = \frac{SS(x)}{n - 1}, \quad \text{where} \quad SS(x) = \sum (x - \bar{x})^2.$$

The formulas for variance can be modified into other forms for easier use in various situations. For example, suppose that we have the sample 6, 3, 8, 5, and 2. The variance for this sample is computed in table 2–15.

TABLE 2–15
Computing s^2 for 6, 3, 8, 5, 2 Using Formula (2–7)

x	x − x̄	(x − x̄)²
6	1.2	1.44
3	− 1.8	3.24
8	3.2	10.24
5	0.2	0.04
2	− 2.8	7.84
24	0 ⓒⓚ	22.80

Using formula (2–1): $\quad \bar{x} = \dfrac{\sum x}{n} = \dfrac{24}{5} = 4.8$

Using formula (2–7): $\quad s^2 = \dfrac{\sum (x - \bar{x})^2}{n - 1} = \dfrac{22.80}{4} = 5.7$

The arithmetic for this example has become more complicated because the mean is in decimal form. However, the sum of squares for x can be rewritten:

$$SS(x) = \sum (x - \bar{x})^2 = \sum x^2 - \frac{(\sum x)^2}{n} \tag{2–9}$$

Combining formulas (2–7) and (2–9) yields formula (2–10).

$$s^2 = \frac{SS(x)}{n - 1} = \frac{\sum x^2 - (\sum x)^2/n}{(n - 1)} \tag{2–10}$$

short-cut formula   Formula 2–10 is called the **short-cut formula** because it bypasses the calculation of $\bar{x}$.

This shortcut formula can be shown to be equivalent to the defining formula (exercise 2–36). The computations for s^2 using formula (2–9) are performed as shown in table 2–16.

TABLE 2–16
*Computing s^2 with
the Shortcut Formula*

x	x^2
6	36
3	9
8	64
5	25
2	4
24	138

Using formula (2–9):

$$SS(x) = 138 - \frac{(24)^2}{5} = 138 - 115.2 = 22.8$$

$$s^2 = \frac{22.8}{4} = \mathbf{5.7}$$

$$s = \sqrt{5.7} = \mathbf{2.4}$$

The units of measure for the standard deviation are the same as the units of measure for the data. For example, if our data are in pounds, then the standard deviation s will also be in pounds. The unit of measure for variance might then be thought of as units squared. In our example of pounds, this would be "pounds squared." As you can see, this unit has very little meaning.

The reason we have more than one formula to calculate variance is for convenience, not for confusion. As is so common in life, there is an easy way and a hard way to accomplish a goal. In statistics we have several formulas for variance, and if you use the appropriate formula your work will be kept to a minimum.

How do you decide which formula is applicable? With small samples there are only two formulas for variance,

$$\frac{\sum (x - \bar{x})^2}{n - 1} \quad \text{and} \quad \frac{\sum x^2 - (\sum x)^2/n}{n - 1}$$

Notice that the first formula involves the mean. If the mean is unknown or is a decimal (and thereby hard to work with), use the second formula. If the mean $\bar{x}$ is known and is a whole number (and thus convenient to work with), use the first formula. If you need to find both the mean and the variance of a sample, always find the mean first.

TABLE 2–17
Ungrouped Frequency Distribution

x	f
1	5
2	9
3	8
4	6
	28

TABLE 2–18
$\sum x$ and $\sum x^2$

x	x^2	x	x^2
1	1	3	9
1	1	3	9
1	1	3	9
1	1	3	9
1	1	3	9
2	4	3	9
2	4	3	9
2	4	3	9
2	4	4	16
2	4	4	16
2	4	4	16
2	4	4	16
2	4	4	16
2	4	4	16
		71	209

TABLE 2–19
Extensions xf and x^2f

x	f	xf	x^2f
1	5	5	5
2	9	18	36
3	8	24	72
4	6	24	96
Total	28	71	209

In order to use formula (2–10) to calculate the variance when the sample is in the form of a frequency distribution, we must determine the values of n, $\sum x$, and $\sum x^2$ using an extensions table. Recall that in section 2–3 the values of n and $\sum x$ were found on an extensions table as the values $\sum f$ and $\sum xf$. To obtain $\sum x^2f$ (it is the value of $\sum x^2$) we need only add a column to our extensions table. Let's look at table 2–17, which is the same ungrouped frequency distribution used in section 2–3 (table 2–10).

To calculate the sum $\sum x^2$ needed, we find the extension x^2f for each class. These extensions, (see tables 2–18 and 2–19), will have the same value as the sum of the squares of the like values of x. For example, $x = 3$ has a frequency of 8, meaning that the value 3 occurs 8 times in the sample. Thus, $x^2f = (3^2)(8) = 72$, which is the same value obtained by adding the squares of eight 3s. These extensions serve as subtotals. When added we obtain the sum $\sum x^2f$, the replacement for $\sum x^2$.

NOTES:

1. The extension x^2f is found by multiplying xf by x: $x \cdot xf = x^2f$. (See table 2–19.)

2. The three column totals, $\sum f$, $\sum xf$, and $\sum x^2f$, are the same values as were previously known as n, $\sum x$, and $\sum x^2$, respectively. That is, $\sum f = n$, the sum of the frequencies is the number of pieces of data. The $\sum xf = \sum x$ and $\sum x^2f = \sum x^2$. The f in the summation expression only indicates that the sum was obtained with the use of a frequency distribution. With these ideas in mind, the formula for the variance (2–10) becomes

$$s^2 = \frac{\sum x^2f - \frac{(\sum xf)^2}{\sum f}}{\sum f - 1} \qquad (2\text{–}11)$$

The standard deviation is the square root of variance.

The variance and standard deviation for the example are found by using formula (2–11).

Variance:

$$SS(x) = 209 - \frac{(71)^2}{28}$$

$$= 209 - 180.036 = 28.964$$

$$s^2 = \frac{28.964}{27} = 1.073 = \mathbf{1.1}$$

Standard deviation:

$$s = \sqrt{s^2} = \sqrt{1.073} = 1.036 = \mathbf{1.0}$$

Let's return now to the sample of 50 exam scores. In table 2–7 you will find a grouped frequency distribution for the 50 scores. We will calculate the variance and standard deviation for the sample by using the frequencies and the class marks in the same manner as in the example above. Table 2–20 shows the extensions and totals. (*Note:* The class marks are now being used as representative values for the observed data. Hence the sum of the class marks is meaningless and is therefore not found.)

TABLE 2–20
Calculations and Totals
for Exam Scores

Class Number	Class Limits	f	Class Marks, x	xf	x^2f
1	22–32	1	27	27	729
2	33–43	2	38	76	2,888
3	44–54	5	49	245	12,005
4	55–65	2	60	120	7,200
5	66–76	9	71	639	45,369
6	77–87	9	82	738	60,516
7	88–98	10	93	930	86,490
8	99–109	5	104	520	54,080
9	110–120	3	115	345	39,675
10	121–131	4	126	504	63,504
		50		4,144	372,456

The variance: Using formula (2–11) we have

$$s^2 = \frac{\sum x^2 f - \dfrac{(\sum xf)^2}{\sum f}}{\sum f - 1}$$

$$= \frac{372,456 - \dfrac{(4,144)^2}{50}}{49} = 591.86$$

The standard deviation is

$$s = \sqrt{s^2} = \sqrt{591.86} - 24.33 = 24.3$$

Exercises

2–23 Consider the sample 2, 4, 7, 8, and 9. Find the following:

(a) range

(b) variance s^2; use formula (2–7)

(c) standard deviation s

2-24   Consider the sample 6, 8, 7, 5, 3, and 7. Find the following:

(a) range

(b) variance s^2; use formula (2–7)

(c) standard deviation s

2-25   Given the sample 7, 6, 10, 7, 5, 9, 3, 7, 5, 13, find the following:

(a) variance s^2; use formula (2–7)

(b) variance s^2; use formula (2–10)

(c) standard deviation s

2-26   Consider the sample of number of hours slept from exercise 2–16. Find the following:

(a) variance s^2; use formula (2–7)

(b) variance s^2; use formula (2–10)

(c) standard deviation s

2-27   Calculate the standard deviation for the set of quiz grades given in exercise 2–17.

2-28   The average annual rainfall for selected cities in California is:

Bakersfield	5.72 in.	Los Angeles	14.05
San Diego	9.75	San Francisco	20.06
Sacramento	17.22	Fresno	10.24

Determine the mean average annual rainfall and standard deviation for these six cities.

2-29   During the last few years Portland General Electric Company has requested rate increases several times. As a result of the granted rate increases, the following revenues are the amounts to be earned:

$34.5 million	$62.3 million
13.3	83.8
22.0	58.3
41.5	10.8

What is the variance for the revenue amounts realized?

2-30   Use formula (2–11) to find the variance and the standard deviation for the following frequency distribution:

x	f
0	1
1	3
2	8
3	5
4	3

2-31   Find the variance and the standard deviation for the following grouped frequency distribution:

Class Limits	f
3–5	2
6–8	10
9–11	12
12–14	9
15–17	7

2-32   Find the mean and variance for the following grouped frequency distribution:

Class Limits	f
2–5	7
6–9	15
10–13	22
14–17	14
18–21	2

2-33   Find the standard deviation for the set of speeds for the 55 cars given in exercise 2–10. Use the frequency distribution found in answering exercise 2–10.

2-34   Find the standard deviation for the 40 small freight delivery charges given in exercise 2–11. Use the frequency distribution found in answering exercise 2–11.

2-35   A sheet metal firm employs several trouble shooters to make emergency repairs of furnaces. Typically, the trouble shooters take many short trips. For the purpose of estimating travel expenses for the coming year a sample of 20 travel expense vouchers related to trouble shooting was taken. The information below resulted:

Dollar Amount on Voucher	Number of Vouchers
$0.01–10.00	2
10.01–20.00	8
20.01–30.00	7
30.01–40.00	2
40.01–50.00	1
Total in Sample	20

Calculate the mean and the standard deviation for these travel account dollar amounts.

2-36   Show that the short-cut formula for the sum of squares for x, $SS(x)$, and $\sum x^2 - (\sum x)^2/n$ [formula (2-9)] is equivalent to $\sum (x - \bar{x})^2$, which is the sum of squares by definition.

Section 2–5 Measures of Position

measure of position

Measures of position are used to describe the location of a specific piece of data in relation to the rest of the sample. Quartiles and percentiles are the two most popular measures of position.

Quartiles

quartile

Quartiles are number values of the variable that divide the ranked data into quarters; each set of data has three quartiles. The **first quartile**, Q_1, is a number such that at most one-fourth of the data are smaller in value than Q_1 and at most three-fourths are larger. The **second quartile** is the median. The **third quartile**, Q_3, is a number such that at most three-fourths of the data are smaller in value than Q_3 and at most one-fourth are larger. See figure 2–17.

FIGURE 2–17

Quartiles

Ranked data, increasing order

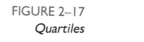

The procedure for determining the value of the quartiles is the same as for percentiles and is shown in the description of percentiles which follows.

Percentiles

percentile

Percentiles are number values of the variable that divide a set of ranked data into 100 equal parts; each set of data has 99 percentiles. See figure 2–18. The kth percentile, P_k, is a number value such that at most $k\%$ of the data are smaller in value than P_k and at most $(100 - k)\%$ of the data are larger. See figure 2–19.

Ranked data, increasing order

FIGURE 2–18

Percentiles

Ranked data, increasing order

FIGURE 2–19

kth Percentile

NOTES:

1. The 1st quartile and the 25th percentile are the same; that is, $Q_1 = P_{25}$. Also, $Q_3 = P_{75}$.

2. The median, the 2d quartile, and the 50th percentile are all the same: $\tilde{x} = Q_2 = P_{50}$. Therefore, when asked to find P_{50} or Q_2, use the procedure for finding the median.

The procedure for determining the value of any kth percentile (or quartile) involves three basic steps.

First: The data must be ranked: start at the lowest-valued piece of data and proceed toward the larger-valued.

Second: The position number i for the percentile in question must be determined. It is found by first calculating the value of $nk/100$ (n = sample size). If $nk/100$ is *not* an integer (that is, it contains a fraction), then i is equal to the next larger integer. For example, suppose $nk/100 = 17.2$; then $i = 18$. If $nk/100$ is an integer, then i will be equal to $(nk/100) + 0.5$. For example, suppose $nk/100 = 23$; then $i = 23.5$.

Third: Locate the value of P_k. To do this, count from the lowest-valued piece of data L until the ith value is found, if i is an integer. If i is not an integer, then it contains the fraction one-half. This means that the value of P_k is halfway between the $(nk/100)$th and the $[(nk/100) + 1]$th piece of data.

Illustration 2–4

A sample of 50 raw scores was taken from a population of raw scores for an elementary statistics final exam. These scores are listed in table 2–21 in the order of collection. Find the first quartile Q_1, the 56th percentile P_{56}, and the third quartile Q_3.

TABLE 2–21
Raw Scores for Elementary Statistics Exam

75	96	74	120	71
97	58	73	94	106
71	94	68	79	86
65	43	54	80	108
84	116	50	82	84
27	123	49	71	93
108	91	81	88	77
91	120	128	88	107
122	94	103	47	44
82	43	76	73	127

Solution The first step is to rank the data. The resulting ranked data are listed in table 2–22 in uniform columns (i.e., each column has the same number of items).

TABLE 2–22
Ranked Data

27	68	79	91	107
43	71	80	91	108
43	71	81	93	108
44	71	82	94	116
47	73	82	94	120
49	73	84	94	120
50	74	84	96	122
54	75	86	97	123
58	76	88	103	127
65	77	88	106	128

To find Q_1 we need to determine the location of Q_1 by using $nk/100$.

$$n = 50 \qquad \text{since there are 50 pieces of data}$$

$$k = 25 \qquad \text{since } Q_1 = P_{25}$$

$$\frac{nk}{100} = \frac{(50)(25)}{100} = 12.5$$

Thus $i = 13$. So Q_1 is the 13th value, counting from L. Thus $Q_1 = 71$.

To find P_{56} we need to determine its location, using $nk/100$.

$$n = 50 \qquad \text{since there are 50 pieces of data}$$

$$k = 56 \qquad \text{from } P_{56}$$

$$\frac{nk}{100} = \frac{(50)(56)}{100} = 28$$

Thus $i = 28.5$. So P_{56} is the value halfway between the 28th and the 29th pieces of data. Therefore,

$$P_{56} = \frac{86 + 88}{2} = 87$$

To find Q_3 we need to determine its location, using $nk/100$.

$$n = 50 \quad \text{and} \quad k = 75 \quad \text{since } Q_3 = P_{75}$$

$$\frac{nk}{100} = \frac{(50)(75)}{100} = 37.5$$

Thus $i = 38$. So Q_3 is the 38th value, counting from L. Thus $Q_3 = 97$.

Q_3 can also be found by counting from H. If the value of a kth percentile is going to be located by counting from H, the highest-(largest) valued data, use $(100 - k)$ in place of k in the formula $nk/100$. For Q_3, $k = 75$; therefore, you may use $k = 25$, and $nk/100$ then equals $(50)(25)/100 = 12.5$ and $i = 13$. Count, starting with $i = 1$ at $H = 128$, and you will find that $i = 13$ at the data whose value is 97, exactly the data whose position was $i = 38$ when counting from L. (*Note:* The same answer results using either method.) ☐

An additional measure of central tendency can now be defined, the **midquartile** midquartile. The **midquartile** is the numerical value midway between the first quartile and the third quartile.

$$\text{midquartile} = \frac{Q_1 + Q_3}{2} \qquad\qquad (2\text{--}12)$$

Illustration 2–5

Find the midquartile for the set of 50 exam grades given in illustration 2–4.

Solution $Q_1 = 71$ and $Q_3 = 97$, as found in illustration 2–4. Thus

$$\text{midquartile} = \frac{71 + 97}{2} = \frac{168}{2} = 84 \qquad \square$$

NOTE: The median, the midrange, and the midquartile are not necessarily the same value.

z Score

z or standard score   The z score, also called the **standard score**, is the position of a value x in terms of the number of standard deviations it is located from the mean. The z score is found by the formula

$$z = \frac{\text{piece of data} - \text{mean of all data}}{\text{standard deviation of all data}} = \frac{x - \bar{x}}{s} \qquad (2\text{–}13)$$

Illustration 2–6

Find the standard scores for (a) 92 and (b) 72 with respect to a sample of exam grades that have a mean score of 75.9 and a standard deviation of 11.1.

Solution

(a) $x = 92$, $\bar{x} = 75.9$, and $s = 11.1$. Thus

$$z = \frac{92 - 75.9}{11.1} = \frac{16.1}{11.1} = 1.45$$

(b) $x = 72$, $\bar{x} = 75.9$, and $s = 11.1$. Thus

$$z = \frac{72 - 75.9}{11.1} = \frac{-3.9}{11.1} = -0.35$$

This means that the score 92 is approximately one and one-half standard deviations above the mean, while the score 72 is approximately one-third of a standard deviation below the mean. $\qquad \square$

NOTES:

1. Typically the calculated value of z is rounded to the nearest hundredth.
2. z scores typically range in value from approximately -3.0 to $+3.0$.

The z score, being a measure of relative position with respect to the mean, can be used to help make a comparison of two raw scores that come from separate populations. For example, suppose that you want to compare a grade that you received on a test with the grade that a friend gets on a comparable exam in her course. You received a raw score of 45 points; she obtained 72 points. Is her grade better? We need more information before we may draw such a conclusion. Suppose that the mean on the exam you took was 38 and the mean on her exam was 65. You both have grades 7 points above the mean, so we still can't draw a definite conclusion. However, the standard deviation on the exam you took was 8 points, and it was 14 points on your friend's exam. That means that your score is $\frac{7}{8}$ of a standard deviation above the mean $(z = \frac{7}{8})$, whereas your friend's grade is only $\frac{1}{2}$ of a standard deviation above the mean $(z = \frac{1}{2})$. Since your score has the "better" relative position, we would conclude that your score was slightly higher than your friend's score. (Again, this is speaking from a relative point of view.)

Exercises

2-37   The following data are the yields in pounds of hops:

$$
\begin{array}{ccccc}
3.9 & 3.4 & 5.1 & 2.7 & 4.4 \\
7.0 & 5.6 & 2.6 & 4.8 & 5.6 \\
7.0 & 4.8 & 5.0 & 6.8 & 4.8 \\
3.7 & 5.8 & 3.6 & 4.0 & 5.6
\end{array}
$$

(a) Find the first and the third quartiles of the yield.

(b) Find the midquartile.

(c) Find the following percentiles:
   (1) P_{15} (2) P_{33} (3) P_{90}

2-38   (a) Find the first and third quartiles for the set of 15 data for hours slept last night given in exercise 2–16.

(b) Find the midquartile for the data.

2-39   (a) Find the first and third quartiles for the quiz scores given in exercise 2–17.

(b) Find the midquartile of the quiz scores.

2-40   Find the following percentiles for the set of 50 final exam scores shown in table 2–22.

(a) P_{35} (b) P_{60} (c) P_{95}

2-41   A sample has a mean of 50 and a standard deviation of 4.0. Find the z score for each value of x.

(a) $x = 54$ (b) $x = 50$ (c) $x = 59$ (d) $x = 45$

2-42   A sample has a mean of 120 and a standard deviation of 20.0. Find the value of x that corresponds to each of these standard scores:

(a) $z = 0.0$ (b) $z = 1.2$ (c) $z = -1.4$ (d) $z = 2.05$

2-43 (a) What does it mean to say that $x = 152$ has a standard score of $+1.5$?

(b) What does it mean to say that a particular value of x has a z score of -2.1?

(c) In general, the standard score is a measure of what?

2-44 Which x value has the higher value relative to the set of data from which it comes?

$A: x = 85$, where mean $= 72$ and standard deviation $= 8$

$B: x = 93$, where mean $= 87$ and standard deviation $= 5$

2-45 Which x value has the lower relative position with respect to the set of data from which it comes?

$A: x = 28.1$, where $\bar{x} = 25.7$ and $s = 1.8$

$B: x = 39.2$, where $\bar{x} = 34.1$ and $s = 4.3$

Section 2–6 Interpreting and Understanding Standard Deviation

Standard deviation is a measure of fluctuation (dispersion) in the data. It has been defined as a value calculated with the use of formulas. But you may wonder what it really is. It is a kind of yardstick by which we can compare one set of data with another. This particular "measure" can be understood further by examining two statements, Chebyshev's theorem and the empirical rule.

Chebyshev's theorem

Chebyshev's Theorem

The proportion of any distribution that lies within k standard deviations of the mean is at least $1 - (1/k^2)$, where k is any positive number larger than 1. This theorem applies to any distribution of data.

This theorem says that within two standard deviations of the mean ($k = 2$) you will always find at least 75% (that is, 75% *or more*) of the data.

$$1 - \frac{1}{k^2} = 1 - \frac{1}{2^2} = 1 - \frac{1}{4} = \frac{3}{4} = 0.75$$

Figure 2–20 shows a mounded distribution that illustrates this.

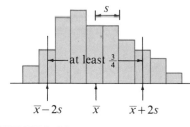

FIGURE 2–20
Chebyshev's Theorem with k = 2

If we were to consider the interval enclosed by three standard deviations on either side of the mean ($k = 3$), the theorem says that we will find at least 8/9, or 89%, of the data $[1 - (1/k^2) = 1 - (1/3^2) = 1 - (1/9) = 8/9]$, as shown in figure 2–21.

FIGURE 2–21
Chebyshev's Theorem with k = 3

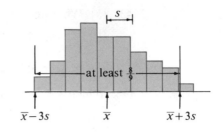

Empirical Rule

If a variable is normally distributed, then within one standard deviation of the mean there will be approximately 68% of the data. Within two standard deviations of the mean there will be approximately 95% of the data, and within three standard deviations of the mean there will be approximately 99.7% of the data. [This rule applies specifically to a normal (bell-shaped) distribution, but it is frequently applied as an interpretive guide to any mounded distribution.]

Figure 2–22 shows the intervals of one, two, and three standard deviations about the mean of an approximately normal distribution. Usually these proportions will not occur exactly in a sample, but your observed values will be close when a large sample is drawn from a normally distributed population.

Normality may be tested crudely with the empirical rule. Let's demonstrate this by working with the distribution of final exam scores that we have used previously throughout this chapter. The mean $\bar{x}$ was found to be 82.9, and the standard deviation s was 24.3. The interval from one standard

FIGURE 2–22
Empirical Rule

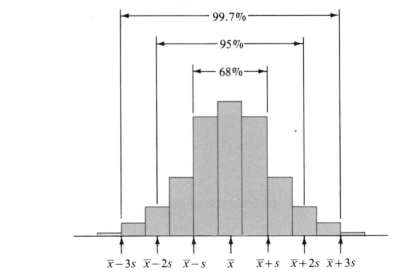

deviation below the mean, $\bar{x} - s$, to one standard deviation above the mean, $\bar{x} + s$, is $82.9 - 24.3 = \mathbf{58.6}$ to $82.9 + 24.3 = \mathbf{107.2}$. This interval includes $59, 60, 61, \ldots, 106, 107$. Upon inspection of the ranked data (table 2–5, p. 34), we see that 32 of the 50 data pieces, or 64%, lie within one standard deviation of the mean. Further, $\bar{x} - 2s = 82.9 - (2)(24.3) = 82.9 - 48.6 = \mathbf{34.3}$, $\bar{x} + 2s = 82.9 + 48.6 = \mathbf{131.5}$, and the interval from 35 to 131, two standard deviations about the mean, includes 49 of the 50 data pieces, or 98%. All 50 data, or 100%, are included within three standard deviations of the mean (from 10.0 to 155.8). This information can be placed in a table for comparison with the values given by the empirical rule (see table 2–23).

TABLE 2–23
Observed Percentages
Versus the Empirical Rule

Interval	Empirical Rule Percentage	Percentage Found
$\bar{x} - s$ to $\bar{x} + s$	≈ 68	64
$\bar{x} - 2s$ to $\bar{x} + 2s$	≈ 95	98
$\bar{x} - 3s$ to $\bar{x} + 3s$	≈ 99.7	100

These percentages are reasonably close to those found by using the empirical rule. By combining this evidence with the histogram of the data, we can safely say that the sample data are approximately normally distributed.

If a distribution is approximately normal, it will be nearly symmetrical and the mean (the mean and the median are the same) will divide the distribution in half. This allows us to refine the empirical rule. Figure 2–23 shows this refinement.

FIGURE 2–23
Refinement of
Empirical Rule

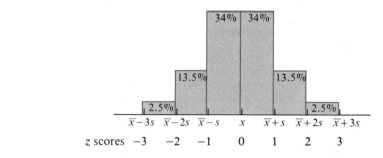

There is also a graphic way to test for normality. This is accomplished by drawing a relative frequency ogive of the grouped data on **probability paper** (which can be purchased at your college bookstore). On this paper the vertical scale is measured in percentages and is placed on the right side of the graph paper. All the directions and guidelines given on page 39 for drawing an ogive must be followed. An ogive of the statistics exam scores is drawn and labeled on a piece of probability paper in figure 2–24. The **test for normality** is to draw a straight line from the lower left corner to the

probability paper

test for normality

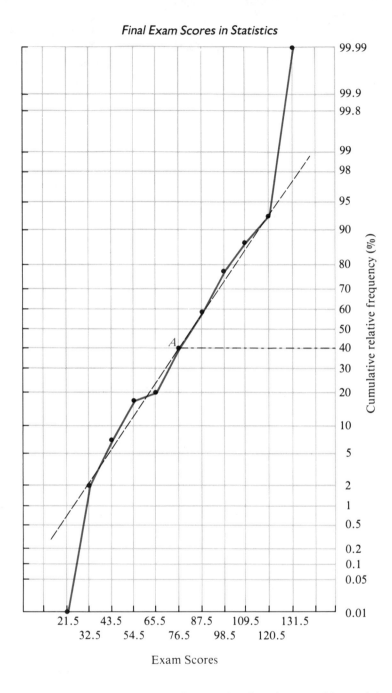

FIGURE 2–24
Ogive on Probability Paper

Final Exam Scores in Statistics

upper right corner of the graph. If the ogive lies close to this straight line, the distribution is said to be approximately normal. The ogive for an exactly normal distribution will trace the straight line.

The dashed line in figure 2–24 is the straight-line test for normality. The ogive suggests that the distribution of exam grades is approximately normal. (*Warning*: This graphic technique is very sensitive to the scale used along the horizontal axis.)

The ogive is a very handy tool to use for finding percentiles of a frequency distribution. If the frequency distribution is drawn on probability paper, as in figure 2–24, it is a very simple matter to determine any kth percentile. For example, let's find the value of P_{40} for the data shown on figure 2–24. On figure 2–24 locate the value of k (40) along the right-hand side of the graph. Follow the horizontal line at k (40) until it intersects the graph of the ogive. Then determine the value of x that corresponds to this point of intersection (point A on figure 2–24) by reading it from the scale on the x-axis: $P_{40} = 76.5$. This method can be used to find percentiles, quartiles, or the median. (*Note*: Because of the grouping into classes, the results may differ slightly from the answers obtained when using the ranked data.)

Exercises

2–46 Sixty college freshmen were asked for the number of children in their family (number of their brothers and sisters plus 1). The data collected follow:

```
1  6  3  5  5  3  4  1  2  7  3  2
3  4  5  3  1  3  2  1  4  4  2  2
3  9  4  3  3  5  3  5  7  3  1  1
3  5  2  6  4  3  3  3  3  3  2  3
4  3  5  7  3  2  1  2  3  2  4  3
```

(a) Construct an ungrouped frequency distribution of these data.

(b) Use the ungrouped frequency distribution found in answer (a) to find the mean and standard deviation of the "number of children" data.

(c) Find the values of $\bar{x} - s$ and $\bar{x} + s$.

(d) How many of the 60 pieces of data have values between $\bar{x} - s$ and $\bar{x} + s$? What percentage of the sample is this?

(e) Find the values of $\bar{x} - 2s$ and $\bar{x} + 2s$.

(f) How many of the 60 pieces of data have values between $\bar{x} - 2s$ and $\bar{x} + 2s$? What percentage of the sample is this?

(g) Find the values of $\bar{x} - 3s$ and $\bar{x} + 3s$.

(h) What percentage of the sample has values between $\bar{x} - 3s$ and $\bar{x} + 3s$?

(i) Compare the answers found in (f) and (h) to the results predicted by Chebyshev's theorem.

(j) Compare the answers found in (d), (f), and (h) to the results predicted by the empirical rule. Does the result suggest an approximately normal distribution?

2-47 The following information was generated from a study of stocks listed on the New York Stock Exchange and the American Stock Exchange:

	New York Exchange	American Exchange
Number of Stocks	24	25
Mean 52-week Price Difference	$5.57	$11.79
Standard Deviation	3.57	10.55

(a) At least 89% of the American Stock Exchange stocks had 52-week price differences within what range?

(b) At least 75% of the New York Stock Exchange stocks had 52-week price differences within what range?

2-48 The average clean-up time for a crew of a medium-sized firm is 84.0 hours and the standard deviation is 6.8 hours. Assuming that the empirical rule is appropriate;

(a) What proportion of the time will it take the clean-up crew 97.6 or more hours to clean the plant?

(b) The total clean-up time will fall within what interval ninety-five percent of the time?

2-49 On the first day of class last semester, 50 students were asked for the one-way distance from home to college (to the nearest mile). The resulting data were:

```
 6   5   3  24  15  15   6   2   1   3
 5  10   9  21   8  10   9  14  16  16
10  21  20  15   9   4  12  27  10  10
 3   9  17   6  11  10  12   5   7  11
 5   8  22  20  13   1   8  13   4  18
```

(a) Construct a grouped frequency distribution of the data by using 1–3 as the first class.

(b) On probability paper, draw and label carefully the ogive depicting the distribution of one-way distances.

(c) Does this distribution appear to have an approximately normal distribution? Explain.

(d) Estimate the values of P_{10}, Q_1, $\tilde{x}$, Q_3, and P_{90} from the ogive drawn in part (b).

Section 2–7 The Art of Statistical Deception

"There are three kinds of lies—lies, damned lies, and statistics." These remarkable words spoken by Disraeli (nineteenth-century British prime minister) represent the cynical view of statistics held by many people. Most people are on the consumer end of statistics and therefore have to "swallow" them.

Good Arithmetic, Bad Statistics

Let's explore an outright statistical lie. Suppose that a small business firm employs eight people who earn between $200 and $240 per week. The owner of the business pays himself $800 per week. He reports to the general public that the average wage paid to the employees of his firm is $280 per week. That may be an example of good arithmetic, but it is also an example of bad statistics. It is a misrepresentation of the situation, since only one employee, the owner, receives more than the mean salary. The public will think that most of the employees earn about $280 per week.

Tricky Graphs

Graphic representation can be tricky and misleading. The vertical scale (which is usually the frequency) should start at zero in order to present a true picture. Usually graphs that do not start at zero are used to save space. Nevertheless, this can be, and usually is, deceptive.

The graph in figure 2–25 was presented in an annual report of a small business in connection with the amount of sales made by the company's three salespersons. Without careful study the viewer of this graph would conclude that *B* sold only about half of what *C* sold. In reality, the graph should look as shown in figure 2–26. The graph is not really different, it only creates a different "impression."

FIGURE 2–25
A Graph That Doesn't Start at 0

FIGURE 2–26
The Same Graph, but Starting at 0

Figure 2–27 represents a daily report of air pollution as found in a local newspaper. Notice that the attribute scales (Low, Med., High) are approximately equal, but the numerical scale is not uniform. (The non-uniform scale is due to the pollutants' effect upon a person breathing.) Likewise, the arrow indicator points to the middle of the attribute level and is not in a position relative to the numerical scale. This seems to be a two-way deception.

FIGURE 2–27
Air Pollution Chart:
Two-way Deception?

Air Pollution

Reading represents average air-pollution levels for a 24-hour period ending at midnight this morning as measured at the Environmental Conservation Department's automatic monitoring station on Farmington Road near the Rochester-Irondequoit line.

Sulfur Dioxide		
Low	Med.	High

0   0.03 ↑ 0.09
Medium—0.04

(In parts per million): low, less than 0.03; medium, 0.03 to 0.08; high, 0.09 or above.

Carbon Monoxide		
Low	Med.	High

0   ↑ 4.9   15
Low—1.5

(In parts per million): low, less than 4.9; medium, 5.0 to 14.9; high, 15 or above.

Soiling or Dust		
Low	Med.	High

0   ↑ 0.4   0.8
Low—0.3

(In reflective units of dirt shade): low, less than 0.4; medium, 0.4 to 0.7; high, 0.8 or above.

From *The Times-Union*, Rochester, N.Y.; by permission.

The histogram in figure 2–28 shows the number of fatally injured drivers involved in automobile accidents in New York State in the year 1968. Notice the "normal" distribution of ages of fatally injured drivers. The reason for grouping the ages in this manner is completely unclear. Certainly the age of fatally injured automobile drivers is not expected to be normally distributed. In this histogram notice also that the class width varies considerably. The histogram seems to indicate that 30- to 39-year-olds are the "poorest drivers." This may or may not be the case. Perhaps this age group has more drivers than any of the other age groups. The group with the most drivers might be expected to have the highest number of driver fatalities. A vertical scale that showed the number of driver fatalities per thousand drivers might be more representative. An age grouping that created classifications with the same number of drivers in each class would also be more representative.

There are situations in which a change in the class width might be appropriate for the graphic representation. However, one must be extremely careful when calculating the values of the numerical measures. The annual salaries paid to the employees of a large firm is a situation in which a change in class width might be appropriate. A large firm has employees with salaries ranging from $8,000 to $80,000. Ten or 12 classes of equal width in this case

FIGURE 2–28
Histogram

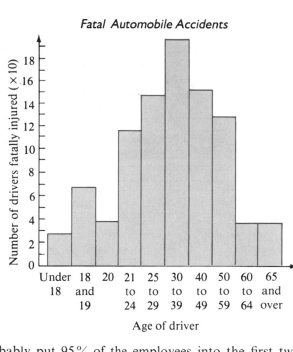

Fatal Automobile Accidents

would probably put 95% of the employees into the first two classes. To present a more accurate picture, one might consider a classification system similar to 8,000—9,000—10,000—12,000 . . . 15,000—18,000—24,000—30,000 —50,000 and up. This system should allow for more accurate description for the lower salaries and will also allow the higher salaries to be lumped together. There are, generally speaking, usually only a few very high salaries.

Insufficient Information

A realtor describes a mountainside piece of land as a perfect place for a nice summer home and tells you that the average high temperature during the months of July and August is 77 degrees. If the standard deviation is 5 degrees, this is not too bad. Assuming that these daily highs are approximately normally distributed, the temperature will range from three standard deviations below to three standard deviations above the mean temperature. The range would be from 62 to 92 degrees, with only a few days above 87 and a few days below 67. However, what she didn't tell you was that the standard deviation of temperature highs is 10 degrees, which means the daily high temperatures range from 47 degrees to 107 degrees. These extremes are not too pleasant!

The familiar advertisement that shows the "higher level of pain reliever" is a tricky graph with no statistics. It has a statistical appearance due only to the clever use of a graphic presentation. No variable, no units—just implications.

What it all comes down to is that statistics, like all languages, can be and is abused. In the hands of the careless, the unknowledgeable, or the unscrupulous, statistical information can be as false as "damned lies."

In Retrospect

You have been introduced to some of the more common techniques of descriptive statistics. There are far too many specific types of statistics used in nearly every specialized field of study for us to review here. We have outlined the uses of only the most universal statistics. Specifically, you have seen several basic graphic techniques (circle and bar graphs, stem-and-leaf displays, histograms, and ogives) that are used to present sample data in picture form. You have also been introduced to some of the more common measures of central tendency (mean, median, mode, midrange, and midquartile), measures of dispersion (range, variance, and standard deviation), and measures of position (quartiles, percentiles, and z score).

You should now be aware that an average can be any one of five different statistics, and you should understand the distinctions among the different types of averages. The "Average Means Different Things" article at the beginning of the chapter discusses four of the "averages" studied in this chapter. You might reread it now and find that it has more meaning and is of more interest. It will be time well spent!

You should also have a feeling for, and an understanding of, the concept of a standard deviation. You were introduced to Chebyshev's theorem and the empirical rule for this purpose.

The exercises in this chapter are extremely important; they will help you nail down the concepts studied before going on to learn how to use these ideas in later chapters.

Chapter Exercises

2-50   Samples A and B are shown in the table. Notice that the two samples are the same except that the 8 in A has been replaced by a 9 in B.

$$A: \quad 2 \quad . \quad 4 \quad \tfrac{5}{5} \quad . \quad 7 \quad 8$$

$$B: \quad 2 \quad . \quad 4 \quad \tfrac{5}{5} \quad . \quad 7 \quad . \quad 9$$

What effect does changing the 8 to a 9 have on each of the following statistics? Explain why.

(a) mean

(b) median

(c) mode

(d) midrange

(e) range

(f) variance

(g) standard deviation

2-51 The following figures are the "average family income needed to live moderately" in the selected cities:

Atlanta	$20,797	Dallas-Fort Worth	$20,219
Baltimore	22,439	Houston	21,028
Boston	27,029	Los Angeles	21,954
Buffalo	23,393	San Diego	22,185
Chicago	22,717	Seattle-Everett	23,392

(a) Determine the median income needed to live moderately for this sample of ten cities.

(b) List the cities that fall below the first quartile.

2-52 The average daily occupancy rate for the Metropolitan Hospital for each month last year is given below:

Month	Jan.	Feb.	Mar.	Apr.	May	June	July	Aug.	Sept.	Oct.	Nov.	Dec.
Patients	265	259	258	242	245	249	234	222	226	254	215	190

(a) Determine the median average daily occupancy rate for this information.

(b) Determine the third quartile for this data.

(c) What is the mean monthly occupancy rate?

2-53 A random sample of ten state government employees showed that each had the following number of months of service:

110 90 84 26 46 50 60 12 21 156

(a) What is the median length of service?

(b) Determine the 80th percentile for length of service.

2-54 The acreage of each offshore Louisiana lease acquired by Kerr-McGee last year is given below:

Location	Gross Acres
Main Pass	4,995
Ship Shoal	3,327
South Timbalier	3,772
Vermilion	5,000
West Cameron	5,000
West Cameron	2,500
West Cameron	5,000
West Delta	1,250

(a) What is the range of the acreage of these acquisitions?

(b) What is the mean?

(c) What is the median?

(d) What is the mode?

(e) Which measure do you think is most representative of the plots acquired? Explain.

2-55 The percent earned on average shareholders' equity (the rate of return on common stock) for the Pacific Lighting Corporation for the last several years has been:

$$8.6\% \quad 9.8\% \quad 10.9\% \quad 11.8\% \quad 13.7\% \quad 13.9\% \quad 14.6\%$$

Determine:

(a) the median rate of return for these years

(b) the mean rate of return for these years

(c) the mean absolute deviation for this data

2-56 Draw a sample of 12 single digits from the random number table and calculate $\bar{x}$ and s.

2-57 Ask one of your instructors for a list of exam grades (15 to 25 grades) from a class.

(a) Find five measures of central tendency.

(b) Find the three measures of dispersion.

(c) Construct a stem-and-leaf display. Does this diagram suggest that the grades are normally distributed?

(d) Find the following measures of location:

(1) Q_1 and Q_3
(2) P_{15} and P_{60}
(3) the standard score z for the highest grade

2-58 A survey of 32 workers at building 815 of Eastman Kodak Company was taken last May. Each worker was asked: "How many hours of television did you watch yesterday?" The results were as follows:

$$
\begin{array}{cccccccc}
0 & 0 & \frac{1}{2} & 1 & 2 & 0 & 3 & 2\frac{1}{2} \\
0 & 0 & 1 & 1\frac{1}{2} & 5 & 2\frac{1}{2} & 0 & 2 \\
2\frac{1}{2} & 1 & 0 & 2 & 0 & 2\frac{1}{2} & 4 & 0 \\
6 & 2\frac{1}{2} & 0 & \frac{1}{2} & 1 & 1\frac{1}{2} & 0 & 2
\end{array}
$$

(a) Construct a stem-and-leaf display.

(b) Find the mean.

(c) Find the median.

(d) Find the mode.

(e) Find the midrange.

(f) Which one of the measures of central tendency would "best" represent the average viewer if you were trying to portray the typical television viewer? Explain.

(g) Which measure of central tendency would best describe the amount of television watched? Explain.

(h) Find the range.

(i) Find the variance.

(j) Find the standard deviation.

2-59 Draw a sample of 25 two-digit numbers from the random number table and calculate its mean and standard deviation.

2-60 The Crab Run Gas Company drills many gas wells each year. The depths (in feet) of successful wells drilled last year are:

$$18,150 \quad 16,361 \quad 16,800 \quad 17,700$$
$$16,500 \quad 16,621 \quad 16,744$$

Determine the mean and the standard deviation for the depths of these wells.

2-61 A time-study analyst observed a packaging operation and collected the following times (in seconds) required for the operation to fill packages of a fixed volume:

$$10.8 \quad 14.4 \quad 19.6 \quad 18.0 \quad 8.4 \quad 15.2 \quad 11.0 \quad 13.3 \quad 23.1 \quad 17.2$$
$$17.1 \quad 16.1 \quad 12.6 \quad 14.6 \quad 9.1 \quad 12.0 \quad 11.6 \quad 14.7 \quad 12.4 \quad 17.3$$

(a) Find the mean.   (b) Find the variance.

(c) Find the standard deviation.

2-62 The following set of data gives the ages of the 118 known offenders who committed an auto theft last year in Garden City, Michigan:

11	14	15	15	16	16	17	18	19	21	25	36
12	14	15	15	16	16	17	18	19	21	25	39
13	14	15	15	16	17	17	18	20	22	26	43
13	14	15	15	16	17	17	18	20	22	26	46
13	14	15	16	16	17	17	18	20	22	27	50
13	14	15	16	16	17	17	19	20	23	27	54
13	14	15	16	16	17	18	19	20	23	29	59
13	15	15	16	16	17	18	19	20	23	30	67
14	15	15	16	16	17	18	19	21	24	31	
14	15	15	16	16	17	18	19	21	24	34	

(a) Find the mean.   (b) Find the median.

(c) Find the mode.   (d) Find Q_1 and Q_3.

(e) Find P_{10} and P_{95}.

2-63 The number of pigs born per litter last year at the Circle J Pig Farm are given in the following list:

11	8	13	14	11	8	14	7	11	13	10	8
7	4	9	11	10	12	6	12	5	10	9	10
12	10	3	6	9	10	10	13	12	9	12	7
9	5	12	7	11	9	4	8	13	12	11	
13	11	12	8	6	13	11	6	12	7	5	

(a) Construct an ungrouped frequency distribution of these data.

(b) How many litters of pigs were there last year?

(c) What is the meaning and value of $\sum f$?

2-64 You are given the frequency distribution shown in the accompanying table of ages, x, in years, of cars found in a parking lot.

x	1	2	3	4	5	6	7	8	9	10	11
f	16	20	18	12	9	7	6	3	5	3	1

(a) Draw a histogram of the data.

(b) Find the five measures of central tendency.

(c) Find Q_1 and Q_3.

(d) Find P_{15} and P_{12}.

(e) Find the three measures of dispersion (range, s^2, s).

2-65 Eighty-five of the newly admitted freshmen at a college were randomly selected and asked to perform a certain physical fitness task. The time required to complete this task was recorded and the results are shown in the grouped frequency distribution of the accompanying table.

Class Boundaries	Frequency
10.95–11.95	16
11.95–12.95	29
12.95–13.95	17
13.95–14.95	11
14.95–15.95	6
15.95–16.95	6

(a) Draw a histogram that depicts the distribution.

(b) Find the mean.

(c) Find the standard deviation.

2-66 The grouped frequency distribution in the accompanying table represents a sample of the ages of 120 randomly selected patients who were admitted to Memorial Hospital last June.

Class Limits	Frequency
0–8	17
9–17	14
18–26	10
27–35	14
36–44	10
45–53	16
54–62	9
63–71	11
72–80	8
81–89	11

(a) Construct a relative frequency histogram that shows the ages of the 120 patients.

(b) Construct an ogive that shows the cumulative distribution of the ages.

(c) Calculate the mean age of the patients.

(d) Calculate the standard deviation of the ages.

2-67 The following is a collection of IQ scores:

```
66  81  88  93  97  100  102  106  112  119
71  83  89  93  98  100  102  107  112  119
71  83  89  95  98  100  102  107  113  121
72  84  89  95  98  100  102  107  113  122
73  85  90  96  98  100  103  108  114  123
74  85  91  96  98  100  103  110  114  126
76  85  92  96  99  100  103  110  115  126
77  86  92  97  99  101  104  111  117  127
80  86  92  97  99  101  105  111  118  130
81  88  92  97  99  101  106  112  118  136
```

(a) Find the mean.　　　　　(b) Find the median.

(c) Find the mode.　　　　　(d) Find the midrange.

(e) Find the range.　　　　　(f) Find Q_1 and Q_3.

(g) Find the midquartile.　　(h) Find P_{35} and P_{64}.

(i) Construct a grouped frequency distribution that uses a class width of ten. Start with 60–69.

(j) Construct a histogram of the frequency distribution.

(k) Construct a cumulative relative frequency distribution.

(l) Construct an ogive of the cumulative relative frequency distribution.

(m) Find the mean of the frequency distribution.

(n) Find the standard deviation of the frequency distribution.

2-68 The lengths (in millimeters) of 100 brown trout in pond 2-B at Happy Acres
      Fish Hatchery on June 15 of last year were as follows:

15.0	15.3	14.4	10.4	10.2	11.5	15.4	11.7	15.0	10.9
13.6	10.5	13.8	15.0	13.8	14.5	13.7	13.9	12.5	15.2
10.7	13.1	10.6	12.1	14.9	14.1	12.7	14.0	10.1	14.1
10.3	15.2	15.0	12.9	10.7	10.3	10.8	15.3	14.9	14.8
14.9	11.8	10.4	11.0	11.4	14.3	15.1	11.5	10.2	10.1
14.7	15.1	12.8	14.8	15.0	10.4	13.5	14.5	14.9	13.9
10.1	14.8	13.7	10.9	10.6	12.4	14.5	10.5	15.1	15.8
12.0	15.5	10.8	14.4	15.4	14.8	11.4	15.1	10.3	15.4
15.0	14.0	15.0	15.1	13.7	14.7	10.7	14.5	13.9	11.7
15.1	10.9	11.3	10.5	15.3	14.0	14.6	12.6	15.3	10.4

Answer the questions asked in exercise 2–67. Use 10.0–10.5 as the first class
for your grouped distribution.

2-69 The length of life of 220 incandescent 60-watt lamps was obtained and
      yielded the frequency distribution shown in the accompanying table.

Class Limit	Frequency
500–599	3
600–699	7
700–799	14
800–899	28
900–999	64
1000–1099	57
1100–1199	23
1200–1299	13
1300–1399	7
1400–1499	4

(a) Construct a histogram of these data, using a vertical scale for relative
frequencies.

(b) Find the mean length of life.

(c) Find the standard deviation.

2-70 Earnings per share for 40 firms in the radio and transmitting equipment
      industry follow:

4.62	0.25	1.07	5.56	0.10	1.34	2.50	1.62
1.29	2.11	2.14	1.36	7.25	5.39	3.46	1.93
6.04	0.84	1.91	2.05	3.20	−0.19	7.05	2.75
9.56	3.72	5.10	3.58	4.90	2.27	1.80	0.44
4.22	2.08	0.91	3.15	3.71	1.12	0.50	1.93

(a) Prepare a frequency distribution and a frequency histogram for this data.

(b) Which class of your frequency distribution contains the median?

2-71　The accompanying table shows the age distribution of heads of families.

Age of Head of Family (years)	Number
20–24	23
25–29	38
30–34	51
35–39	55
40–44	53
45–49	50
50–54	48
55–59	39
60–64	31
65–69	26
70–74	20
75–79	16
	450

(a) Find the mean age of the heads of families.

(b) Find the standard deviation.

2-72　Bill and Bob are good friends, although they attend different high schools in their city. The city school system uses a battery of fitness tests to test all high school students. After completing the fitness tests, Bill and Bob are comparing their scores to see who did better on each item. They need help.

	Sit-ups	Pull-ups	Shuttle Run	50 yard dash	Softball Throw
Bill	$z = -1$	$z = -1.3$	$z = 0.0$	$z = 1.0$	$z = 0.5$
Bob	61	17	9.6	6.0	179 ft.
Mean	70	8	9.8	6.6	173 ft.
Standard Deviation	12	6	0.6	0.3	16 ft.

Bill received his test results in z-scores while Bob was given raw scores. Since both boys understand raw scores, convert Bill's z-scores to raw scores in order to make an accurate comparison.

2-73　Compare an individual's achievement in relation to the rest of his class on a midterm exam and a final exam. You have the following information:

	His score	Class Mean	Class Standard Deviation
Midterm	70	75	10
Final	50	55	15

On which exam did he perform better?

2-74　Mrs. Adam's twin daughters, Jean and Joan, are in 5th grade (different sections) and the class has been given a series of ability tests.

	Results	
Skill	Jean: z-score	Joan: percentile
Fitness	2.0	99
Posture	1.0	69
Agility	1.0	88
Flexibility	−1.0	35
Strength	0.0	50

If the scores for these ability tests are approximately normally distributed, which girl has the higher relative score on each of the skills listed? Explain your answers.

2-75　Given the information below, would a pitcher with a winning percentage of 0.400 and a strikeout-to-walk ratio of 1.6 have a higher relative standing in winning percentage or strikeout-to-walk ratio? [*Notes*: (1) In both cases, the higher the value the better. (2) "Strikeout-to-walk" ratio equals number of batters struck out divided by number of batters walked.]

	Winning Percentage	Strikeout-to-Walk Ratio
Mean	0.500	2.00
Standard Deviation	0.070	0.50
Median	0.502	2.06

2-76　The units of product sold by the salesmen of a firm fluctuates as the seasons change. For July, the number of units sold per salesman has had a mean of 900 and a standard deviation of 94; while for December the units sold had a mean of 330 and a standard deviation of 40. John Scranton sold 1,080 in July and 450 in December. In which month did John Scranton have the better relative performance?

2-77　Chebyshev's theorem guarantees what proportion of a distribution to be included between the following?

(a) $\bar{x} - 2s$ and $\bar{x} + 2s$　　(b) $\bar{x} - 3s$ and $\bar{x} + 3s$

2–78 The empirical rule indicates that we can expect to find what proportion of the sample to be included between the following?

(a) $\bar{x} - s$ and $\bar{x} + s$ (b) $\bar{x} - 2s$ and $\bar{x} + 2s$

(c) $\bar{x} - 3s$ and $\bar{x} + 3s$

2–79 Why is it that the z score for a value belonging to a normal distribution usually lies between -3 and 3?

2–80 If a distribution is mounded so as to be approximately normal, how wide will an interval centered at the mean have to be to span the middle 95% of the data?

Vocabulary List

Be able to define each term. In addition, describe, in your own words, and give an example of each term. Your examples should not be ones given in class or in the textbook.

bell-shaped distribution	modal class
bimodal distribution	mode
Chebyshev's theorem	normal distribution
class	ogive
class boundary	percentile
class limits (lower and upper)	probability paper
class mark	quartile
class width	range
cumulative frequency distribution	rectangular distribution
cumulative relative frequency distribution	relative frequency
	relative frequency distribution
deviation from mean	relative frequency histogram
distribution	sample
empirical rule	single-variable data
frequency	skewed distribution
frequency distribution	standard deviation
frequency histogram	standard score
grouped frequency distribution	stem-and-leaf display
histogram	summation
mean	tally
mean absolute deviation	variability
measure of central tendency	ungrouped frequency distribution
measure of dispersion	
measure of position	variance
median	x bar $(\bar{x})$
midquartile	z score
midrange	

Quiz

Answer "True" if the statement is always true. If the statement is not always true, replace the words shown in color with words that make the statement always true.

2–1   The **mean** of a sample always divides the data into two equal halves—half larger and half smaller in value than itself.

2–2   A measure of **central tendency** is a quantitative value that describes how widely the data are dispersed about a central value.

2–3   The sum of the squares of the deviations from the mean, $\sum (x - \bar{x})^2$, will **sometimes** be negative.

2–4   For any distribution, the sum of the deviations from the mean equals **zero**.

2–5   The standard deviation for the set of values 2, 2, 2, 2, and 2 is **2**.

2–6   On a test John scored at the 50th percentile and Jim scored at the 25th percentile; therefore, John's test score was **twice** Jim's test score.

2–7   The frequency of a class is the number of pieces of data whose values fall within the **boundaries** of that class.

2–8   **Frequency distributions** are used in statistics to present large quantities of repeating values in a concise form.

2–9   The unit of measure for the standard score is always in **standard deviations**.

2–10   For a bell-shaped distribution, the range will be approximately equal to **six standard deviations**.

3 Descriptive Analysis and Presentation of Bivariate Data

Points and Fouls in Basketball

ALBERT P. SHULTE

Sometimes you may think that two different sets of numbers are related in some way, although not perfectly. How can you decide if they are related? If they are related, is there any reason why they should be, or is it purely accidental? Does a change in one of the variables cause a change in the other? This example is designed to help answer the first question, but usually the second and third ones are much harder, and people often disagree about the answers. . . .

The data come from the records of a junior-varsity basketball team . . . for two separate years. . . . A quick look at Table 2 makes one feel that a strong relationship exists between the number points . . . and the number of personal fouls. . . . In Figure 1, the data for the first year have been plotted. . . .

Now let's think a bit more about the relationship that seems to exist. . . . The table and the figure both indicate that the more fouls he commits the more points he scores. . . . Surely no coach would coach a player to make lots of fouls in the hope that he would therefore score more points. The crucial word in the previous sentence is "therefore." Is the fact that a player commits more fouls the reason that he scores more points? Of course not. But maybe both of these . . . are the results of some other act. Perhaps . . . more game time . . . Or maybe . . . .

A graph such as that in Figure 1 can show that a high degree of correlation exists, but it does not tell us why.

TABLE 2 *Second Year*

Player	Total points	Personal fouls
Brummett	2	1
Cooper	75	24
Felice	0	1
Hook	59	18
Hurd	9	9
Kampsen	7	3
McPartlin	35	5
Pointer	46	20
Schuback	0	1
Wilson	2	3
Zuelch	57	22

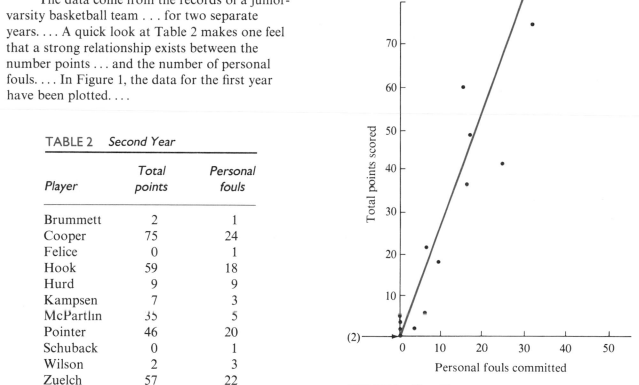

FIGURE I *First Year*

Chapter Objectives

In the field of statistics there are many problems that require the analysis of more than one variable. In business, in education, and in many other fields, we often try to obtain answers to the following questions: Are these two variables related? If so, how are they related? Do these variables correlate? The relationships discussed here are not cause-and-effect relationships. They are only mathematical relationships, which predict the behavior of one variable from knowledge about a second variable.

Let's look at a few specific illustrations.

Illustration 3–1

As a person grows taller he usually gains weight. Someone might ask, "Is there a relationship between height and weight?"

Illustration 3–2

Students are constantly spending their time studying and taking examinations while going to school. We could ask, "Is it true that the more you study, the higher your grade will be?"

Illustration 3–3

Research doctors test new drugs (old ones, too) by prescribing different amounts and observing the responses of their patients. One question we could ask here is this: "Does the amount of drug prescribed determine the amount of recovery time needed by the patient?"

Illustration 3–4

A high school guidance counselor would like to predict the academic success that students graduating from her school will have in college. In cases like this the predicted value (grade point average at college) depends on many traits of the students: (1) how well they did in high school, (2) their intelligence, (3) their desire to succeed at college, and so on.

These illustrations all require correlation or regression analysis to obtain answers. In this chapter we will take a first look at the very simplest form of correlation and regression analysis. The basic objectives of this chapter are (1) to gain an understanding of the distinction between the basic purposes of correlation analysis and regression analysis, and (2) to become familiar with the ideas of descriptive presentation. With these objectives in mind, we will restrict our discussion to the simplest and most basic form of correlation and regression analysis—the bivariate linear case.

Section 3–1 Bivariate Data

bivariate data

ordered pair

The term **bivariate data** is used to describe two pieces of data that are variable (that is, differ from each other) but are related in some way. Expressed mathematically, bivariate data comprise **ordered pairs** —let's call them x

and y, where x is the value of the first variable and y is the value of the second variable. We write an ordered pair as (x, y). The data are said to be "ordered" because one value, x in this case, is always written first. They are said to be "paired" because for each x value there is a corresponding y value. For example, if x is height and y is weight, a height and corresponding weight are recorded for each person.

input and output variables
It is customary to call the **input variable** (sometimes called "independent" variable) x and the **output variable** (sometimes called the "dependent" variable) y. The input variable x is measured or controlled in order to predict the output variable y. For example, in illustration 3–3 the researcher can control the amount of drug prescribed. Therefore, the amount of drug would be referred to as x. In the case of height and weight, either variable could be treated as input, the other as output, depending on the question being asked. However, different results will be obtained from the regression analysis, depending on the choice made.

scatter diagram
In problems that deal with both correlation and regression analysis, we will present our sample data pictorially on a scatter diagram. A **scatter diagram** is a plot of all the ordered pairs of bivariate data on a coordinate axis system. The input variable x is plotted on the horizontal axis and the output variable y is plotted on the vertical axis.

NOTE: When constructing a scatter diagram, it is convenient to construct scales so that the range of data along the vertical axis is equal to or slightly shorter than the range along the horizontal axis.

Illustration 3–5

In Mr. Chamberlain's physical fitness course, several fitness scores were taken. The following sample is the number of push-ups and sit-ups done by 10 randomly selected students:

$$(27, 30), (22, 26), (15, 25), (35, 42), (30, 38),$$
$$(52, 40), (35, 32), (55, 54), (40, 50), (40, 43)$$

Table 3–1 shows these sample data, and figure 3–1 shows a scatter diagram of the data.

TABLE 3–1
Data for Push-ups and Sit-ups

	Student									
	1	*2*	*3*	*4*	*5*	*6*	*7*	*8*	*9*	*10*
Push-ups, x	27	22	15	35	30	52	35	55	40	40
Sit-ups, y	30	26	25	42	38	40	32	54	50	43

FIGURE 3–1
Scatter Diagram

Push-ups Versus Sit-ups for 10 Students

Exercises

3–1 The accompanying data show the number x of hours studied for an exam and the grade y received on the exam. (y is measured in 10s; that is, $y = 8$ means that the grade, rounded to the nearest 10 points, is 80.) Draw the scatter diagram. (Retain this solution for use in answering exercises 3–6, and 3–14.)

x	2	3	3	4	4	5	5	6	6	6	7	7	7	8	8
y	5	5	7	5	7	7	8	6	9	8	7	9	10	8	9

3–2 An experimental psychologist asserts that the older a child is, the fewer irrelevant answers he or she will give during a controlled experiment. To investigate this claim, the accompanying data were collected. Draw a scatter diagram. (Retain this solution for use in answering exercises 3–7 and 3–15.)

Age, x	2	4	5	6	6	7	9	9	10	12
Number of Irrelevant Answers, y	12	13	9	7	12	8	6	9	7	5

3–3 Draw a scatter diagram for these data:

x	2	12	4	6	9	4	11	3	10	11	3	1	13	12	14	7	2	8
y	4	8	10	9	10	8	8	5	10	9	8	3	9	8	8	11	6	9

(Retain this solution for use in answering exercises 3–8 and 3–12.)

3-4 A sample of 15 upper classmen who commute to classes was selected at registration. They were asked to estimate the distance (x) and the time (y) required to commute each day to class; see the accompanying data. Construct a scatter diagram depicting this data. (Retain this solution for use in answering exercises 3–9 and 3–16.)

Distance, x (nearest mile)	Time, y (nearest 5 minutes)	Distance, x (nearest mile)	Time, y (nearest 5 minutes)
18	20	2	5
8	15	15	25
20	25	16	30
5	20	9	20
5	15	21	30
11	25	5	10
9	20	15	20
10	25		

3-5 The rate of increased sales (in dollars) and the rate of increased earnings after taxes for local gas utilities for the last seven years are presented in the following table.

Year	Rate of Increased Sales	Rate of Increased Earnings after Taxes
1	0.128	0.216
2	0.117	0.139
3	0.085	0.175
4	0.200	0.018
5	0.227	0.086
6	0.222	−0.058
7	0.222	0.117

Prepare a scatter diagram displaying this data.

Section 3–2 Linear Correlation

correlation analysis

independent and dependent variables

linear correlation

The primary purpose of linear correlation analysis is to measure the strength of a linear relationship between two variables. Let's examine some scatter diagrams demonstrating different relationships between input or **independent variables** x and output or **dependent variables** y. If as x increases there is no definite shift in the values of y, we say there is **no correlation**, or no relationship, between x and y. If as x increases there is a shift in the values of y, there is a **correlation**. The correlation is positive when y tends to increase and negative when y tends to decrease. If the values of x and y tend to follow a straight-line path, there is a **linear** correlation. The preciseness of the shift in y as x increases determines the strength of the linear correlation. The scatter diagrams in figure 3–2 demonstrate these ideas.

FIGURE 3–2
Scatter Diagrams and Correlation

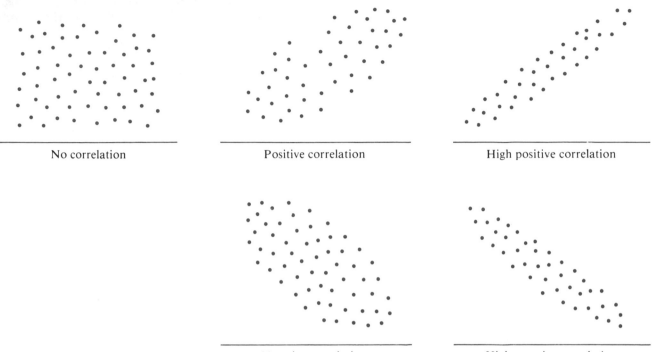

No correlation   Positive correlation   High positive correlation

Negative correlation   High negative correlation

FIGURE 3–3
Perfect Positive Correlation

Perfect linear correlation occurs when all the points fall exactly along a straight line, as shown in figure 3–3. This can be either positive or negative, depending on whether y increases or decreases as x increases. If the data form a straight horizontal or vertical line, there is no correlation, since one variable has no effect on the other.

Scatter diagrams do not always appear in one of the five forms shown in figure 3–2. Sometimes they will suggest relationships other than linear. Such is the case in figure 3–4. There appears to be a definite pattern; however, the two variables are not related linearly, and therefore they are not linearly correlated.

coefficient of linear correlation

The **coefficient of linear correlation**, r, is the measure of the strength of the linear relationship between two variables. The coefficient reflects the consistency of the effect that a change in one variable has on the other.

The value of linear correlation coefficient helps us to answer the question: Is there a linear correlation between the two variables under consideration? The linear correlation coefficient r always has a value between -1 and $+1$. A value of $+1$ signifies a perfect **positive correlation**, and a value of -1 shows a perfect **negative correlation**

positive and negative correlation

If as x increases there is a general increase in the value of y, then r will indicate a positive linear correlation. For example, a positive value of r

FIGURE 3–4
No Linear Correlation

would be expected for age and height of children, because as a child grows older, he grows taller. Also, consider the age x and relative value y of an automobile. As the car ages, its value decreases. Since as x increases, y decreases, their inverse relationship is denoted by a negative value.

The value of r for a sample is obtained from the formula

$$r = \frac{\sum (x - \bar{x})(y - \bar{y})}{(n - 1)s_x s_y} \qquad (3\text{–}1)$$

NOTE: s_x and s_y are the standard deviations of the x and y variables.

Pearson's product moment

This formula is called **Pearson's product moment** r The development of this formula is discussed in chapter 13.

To calculate r, we will use an alternative formula, which is equivalent to (3–1). We will calculate three separate sums of squares and then substitute them into formula (3–2) to obtain r.

$$r = \frac{\text{SS}(xy)}{\sqrt{\text{SS}(x)\text{SS}(y)}} \qquad (3\text{–}2)$$

where $\text{SS}(x) = \sum x^2 - ((\sum x)^2/n)$, from formula (2–9) on page 52,

$$\text{SS}(y) = \sum y^2 - \frac{(\sum y)^2}{n}, \qquad (3\text{–}3)$$

and

$$\text{SS}(xy) = \sum xy - \frac{(\sum x)(\sum y)}{n} \qquad (3\text{–}4)$$

Let's return to illustration 3–5 and find the linear correlation coefficient. First we need to construct a table (table 3–2) listing all the pairs of values

TABLE 3–2
Computations Needed to Calculate r

Student	Push-ups, x	x^2	Sit-ups, y	y^2	xy
1	27	729	30	900	810
2	22	484	26	676	572
3	15	225	25	625	375
4	35	1,225	42	1,764	1,470
5	30	900	38	1,444	1,140
6	52	2,704	40	1,600	2,080
7	35	1,225	32	1,024	1,120
8	55	3,025	54	2,916	2,970
9	40	1,600	50	2,500	2,000
10	40	1,600	43	1,849	1,720
Total	351	13,717	380	15,298	14,257

(x, y) and the extensions for x^2, xy, and y^2, and the column totals. Then we use the totals from the table and formulas (3–2), (2–9), (3–3), and (3–4) to calculate r.

Using formula (2–9),

$$SS(x) = 13{,}717 - \frac{(351)^2}{10} = 1396.9$$

Using formula (3–3),

$$SS(y) = 15{,}298 - \frac{(380)^2}{10} = 858.0$$

Using formula (3–4),

$$SS(xy) = 14{,}257 - \frac{(351)(380)}{10} = 919.0$$

Using formula (3–2), we find r to be

$$r = \frac{919.0}{\sqrt{(1396.9)(858.0)}} = 0.8394 = \mathbf{0.84}$$

NOTE: Typically r is rounded to the nearest hundredth.

The value of the calculated linear correlation coefficient is supposed to help us answer the question: Is there a linear correlation between the two variables under consideration? When the calculated value of r is close to 0, we conclude that there is no linear correlation. When the calculated value of r is close to $+1$ or -1, we suspect that there is a linear correlation between the two variables.

Between 0 and 1.0 there is a value that marks the division point between concluding that there is or there is not a linear correlation. This point is called a **decision point**. There is a similar point between 0 and -1.0; see figure 3–5. The value of the decision point is determined by the size of the sample. Table 3–3 presents the positive decision points of various sample sizes from 5 to 100. Notice that the values for the decision points decrease as n, the sample size, increases. If r lies between the negative and the positive value of a decision point, we can assume there is no evidence of a linear relationship between the two variables. If r is either less than the negative value or more than the positive value of a decision point, we can conclude there is evidence of a linear relationship between the two variables.

decision point

FIGURE 3–5

Decision Points for r

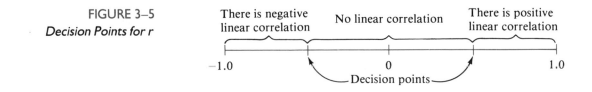

n	Decision Point	n	Decision Point	n	Decision Point	n	Decision Point
5	0.878	12	0.576	19	0.456	30	0.361
6	0.811	13	0.553	20	0.444	40	0.312
7	0.754	14	0.532	22	0.423	50	0.279
8	0.707	15	0.514	24	0.404	60	0.254
9	0.666	16	0.497	26	0.388	80	0.220
10	0.632	17	0.482	28	0.374	100	0.196
11	0.602	18	0.468				

Illustration 3–6

Is there evidence of linear correlation between the two variables "number of push-ups" and "number of sit-ups" in illustration 3–5?

Solution   Since $n = 10$, the values for the decision points are ± 0.632. Since the calculated value of r is 0.84 (see page 92), which falls between $+0.632$ and $+1.0$, we can conclude that yes, **there is evidence of linear correlation** See the accompanying figure.

(*indicates location of the calculated $r = 0.84$)

To conclude that there is a relationship between two variables does not mean that there is a cause-and-effect relationship. This must be determined by your understanding of the situation. For example, let x be police officers' salaries over the last 10 years and let y be the amount of alcoholic beverages sold in the United States over the same 10 years. A sample of these data would show a strong positive correlation. Certainly the increase in the salaries of police officers has not caused the increase in the sales of alcoholic beverages. Rather, both of these have increased. They both have reacted in a predictable pattern. One did not cause the other.

The decision-making process discussed above will reoccur in different contexts throughout the rest of the book. This process involves the following components: (1) probability, which determines the decision points and develops the theory of statistical decision making; and (2) inference procedures, which are used in testing a hypothesis. This testing is done by comparing a statistic to a decision point or critical value. Chapters 4 through 7 discuss probability while chapter 8 and later chapters present the methods of statistical inference.

With a formula as complex as formula (3–2), it would be very convenient to be able to inspect the scatter diagram of the data and be able to estimate the calculated value of r from the diagram. This would serve as a check on the calculations. The following **method for estimating r** is quick and generally yields a reasonable estimate.

1. Draw an oval around the plot of points; see figure 3-6.

FIGURE 3-6

An Oval Is Drawn Around the Plotted Points

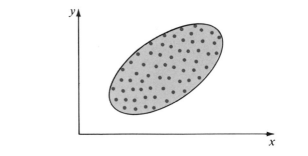

2. Measure the length of the maximum diameter D; see figure 3-7, where $D = 35$ millimeters.

FIGURE 3-7

The Maximum Diameter D

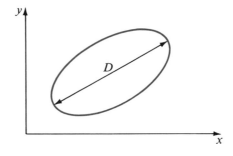

3. Measure the length of the minimum diameter d; see figure 3-8, where $d = 18$ millimeters.

FIGURE 3-8

The Minimum Diameter d

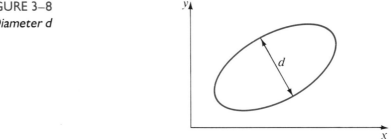

The value of r may be estimated as

$$\pm\left(1 - \frac{d}{D}\right)$$

The sign assigned to r may be determined by observing the general position of the maximum diameter. If it lies in an increasing position (see figure 3–9), r will be positive; if it lies in a decreasing position, r will be negative. If the diameters are in horizontal or vertical positions, r will be zero.

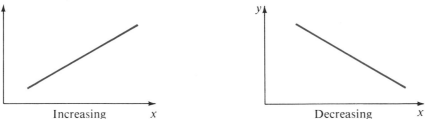

It should be noted that this estimating technique is very crude. It is very sensitive to the "spread" of the diagram. However, if the range of the x values and the range of the y values are approximately equal, the approximation will be helpful. **This technique should be used only as a mental check.**

Let's estimate the value of the linear correlation coefficient r for the relationship between the number of push-ups and sit-ups. From figure 3–10 we find that $d = 0.95$ inches and $D = 1.9$ inches. Therefore, our estimate for r is 0.5. It is positive because of the increasing position of the maximum diameter. The value is 0.5 because $1 - \frac{1}{2} = \frac{1}{2} = 0.5$.

FIGURE 3–10
Estimating r for the Push-up and Sit-up Data

Exercises

 3–6   (a) Use the scatter diagram drawn in answering exercise 3–1 to estimate r for the sample data relating number of hours studied and exam grade.

(b) Calculate r.

(c) Does there seem to be a linear correlation? Explain.

3–7 (a) Use the scatter diagram drawn in answering exercise 3–2 to estimate r for the sample data involving age and irrelevant answers.

(b) Calculate r.

(c) Does there seem to be a linear correlation? Explain. (Retain these solutions for use in answering exercise 3–15.)

3–8 Estimate the value of the correlation coefficient for the data on the scatter diagram drawn in answering exercise 3–3.

3–9 (a) Use the scatter diagram drawn in answering exercise 3–4 to estimate r for the sample data.

(b) Calculate r.

(c) Does there seem to be a linear correlation? Explain. (Retain these solutions for use in answering exercise 3–16.)

3–10 A marketing firm wished to determine whether or not the number of television commercials broadcast were linearly correlated to the sales of its product. The data obtained from each of several cities were as shown in the accompanying table.

City	A	B	C	D	E	F	G	H	I	J
No. TV Commercials, x	12	6	9	15	11	15	8	16	12	6
Sales Units, y	7	5	10	14	12	9	6	11	11	8

(a) Draw a scatter diagram.

(b) Estimate r.

(c) Calculate r.

(d) Is there evidence that the number of commercials is linearly correlated to the sales? Explain.

(Retain these solutions for use in answering exercise 3–17.)

3–11 Verify, algebraically, that formula (3–2) for calculating r is equivalent to the definition formula (3–1).

Section 3–3 Linear Regression

While correlation tells us the strength of a linear relationship, it does not tell us the exact numerical relationship. For example, the correlation coefficient calculated for the data in table 3–2 implies that there is a linear relationship between the number of push-ups and the number of sit-ups done by a student. That means we should be able to use the number of push-ups to predict the number of sit-ups. But correlation analysis does not show us how to determine a y value given an x value. This is done by regression analysis. **Regression analysis calculates an equation that provides values of y for given**

regression analysis

values of x. One of the primary objectives of regression analysis is to **make predictions**—for example, predicting the success a student will have in college based on results of high school, or predicting the distance required to stop a car based on its speed.

Generally the exact value of y is not predicted. We are usually satisfied if the predictions are reasonably close. The statistician seeks an equation to express the relationship between the two variables. The equation he or she chooses is the one that **best fits** the scatter diagram.

prediction equation

Here are some examples of various relationships; they are called **prediction equations**:

$$y = b_0 + b_1 x \quad \text{(linear)}$$

$$y = a + bx + cx^2 \quad \text{(quadratic)}$$

$$y = a(b^x) \quad \text{(exponential)}$$

$$y = a \log_b x \quad \text{(logarithmic)}$$

Figures 3–11 through 3–14 illustrate some of these relationships. These graphs might result if we were to sample the following bivariate populations:

The hourly wages earned by store clerks x and their weekly pay y.

FIGURE 3–11

Linear Regression with Positive Slope

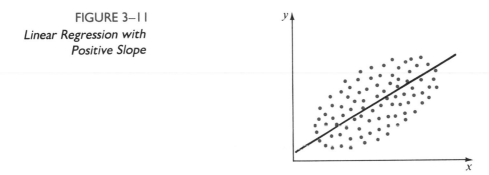

The age of a used American economy car x and its resale value as a used car y.

FIGURE 3–12

Linear Regression with Negative Slope

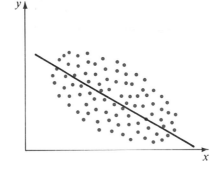

The week of the semester x and the number of hours spent studying statistics y.

FIGURE 3–13
Curvilinear Regression
(Looks Quadratic)

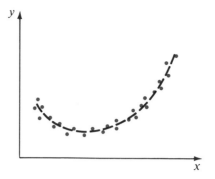

The number of the row that a student sat in for a large lecture course x and the final grade received for the course y.

FIGURE 3–14
No Relationship

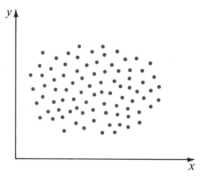

The relationship between these two variables will be an algebraic expression describing the mathematical relationship between x and y. In figures 3–11, 3–12, and 3–13 there does appear to be a relationship. In figure 3–14, however, the values of x and y do not seem to be related.

Given a set of bivariate (x, y) data, how do we find the line of best fit? If a straight-line relationship seems appropriate, the best-fitting straight line is found by using the **method of least squares**. Suppose that $\hat{y} = b_0 + b_1 x$ is the equation of a straight line, where $\hat{y}$ (read "y hat") represents the **predicted value** of y that corresponds to a particular value of x. The **least squares criterion** requires that we find the constants b_0 and b_1 such that the **sum $\sum (y - \hat{y})^2$ is as small as possible**

Figure 3–15 shows the distance of an observed value of y from a predicted value of y. The length of this distance represents the value $(y - \hat{y})$. Note that $(y - \hat{y})$ is positive when the point (x, y) is above the line and negative when (x, y) is below the line. Figure 3–16 shows a scatter diagram with what might appear to be the line of best fit, along with ten individual $(y - \hat{y})$'s. The sum of the squares of these differences is

method of
least squares

predicted value

least squares
criterion

minimized (made as small as possible) if the line is indeed the line of best fit. Figure 3–17 shows the same data points as figure 3–16 with the 10 individual $(y - \hat{y})$'s associated with a line that is definitely *not* the line of best fit.

FIGURE 3–15
Observed and Predicted Values of y

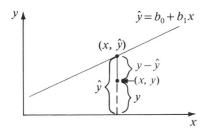

FIGURE 3–16
The Line of Best Fit

FIGURE 3–17
Not the Line of Best Fit

$$\Sigma \, (y - \hat{y})^2 = (-1)^2 + (+1)^2 + \cdots + (+1)^2 = 23.0$$

$$\Sigma \, (y - \hat{y})^2 = (-6)^2 + (-4)^2 + \cdots + (+6)^2 = 149.0$$

line of best fit

The **equation of the line of best fit** is determined by its slope (b_1) and its y-intercept (b_0). (See the study guide for a review of the concepts of slope and intercept of a straight line.) The values of the constants that satisfy the least squares criterion are found by using these formulas:

slope

$$\text{Slope:} \quad b_1 = \frac{\Sigma \, (x - \bar{x})(y - \bar{y})}{\Sigma \, (x - \bar{x})^2} \qquad (3\text{–}5)$$

intercept

$$\text{Intercept:} \quad b_0 = \frac{1}{n} \left(\Sigma \, y - b_1 \cdot \Sigma \, x \right) \qquad (3\text{–}6)$$

(The derivation of these formulas is beyond the scope of this text.)

We will use a mathematical equivalent of formula (3–5) that is easier to apply. We will calculate the slope b_1 by use of the formula

$$b_1 = \frac{\text{SS}(xy)}{\text{SS}(x)} \qquad (3\text{–}7)$$

and formulas (2–9), on page 52 and (3–4), on page 91. Notice that the numerator of formula (3–7) is the same as the numerator of formula (3–2) for the linear correlation coefficient. Notice also that the denominator is the first radicand (expression under a radical sign) in the denominator of formula (3–2). Thus if you calculate the linear correlation coefficient by using formula (3–2), you can easily find the slope of the line of best fit. If you did not calculate r, set up a table similar to table 3–2 and find the necessary values.

Now let's consider the data in illustration 3–5 and the question of predicting a student's sit-ups based on his or her number of push-ups. We want to find the line of best fit, $\hat{y} = b_0 + b_1 x$. The necessary calculations and totals for determining b_1 have already been obtained in table 3–2 (page 91). Recall that $SS(xy) = 919.0$ and $SS(x) = 1396.9$ (see page 92). Then, using formula (3–7) the slope is calculated to be

$$b_1 = \frac{919.0}{1396.9} = 0.6579 = \mathbf{0.66}$$

Using formula (3–6) the y-intercept is calculated to be

$$b_0 = \frac{1}{10}[380 - (0.6579)(351)]$$

$$= \frac{1}{10}[380 - 230.9229] = 14.9077 = \mathbf{14.9}$$

Remember to keep the extra decimal places in the calculations to ensure an accurate answer.

Thus the equation of the line of best fit is

$$\hat{y} = \mathbf{14.9 + 0.66}x$$

NOTE: When rounding off the calculated values of b_0 and b_1, always keep at least 2 significant digits.

Figure 3–18 shows a scatter diagram of the data and the line of best fit. There are some additional facts about the least squares method that we need to discuss:

1. The slope b_1 represents the predicted change in y per unit increase in x. In our example $b_1 = 0.66$; thus if a student had done an additional 10 push-ups (x), we would predict that he or she would have done an additional 7 (0.66×10) sit-ups (y).

2. The y-intercept is the value of y where the line of best fit intersects the y-axis, that is, where $x = 0$. However, in interpreting b_0 you first must consider whether $x = 0$ is a reasonable value before you conclude that you would predict $y = b_0$ if $x = 0$. To predict that if a student did no push-ups, he or she would still do approximately 15 sit-ups ($b_0 = 14.9$) is probably incorrect. An x value of 0 is outside the domain of the data on which the regression line is based. **In predicting y based on an x value, you should check to be sure that the x value is within the domain of the x values observed.**

3. The line of best fit will always pass through the point ($\bar{x}, \bar{y}$). When drawing the line of best fit on your scatter diagram, you should use this point as a check. It must lie on the line of best fit.

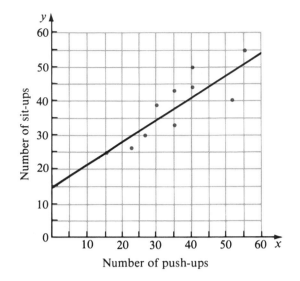

FIGURE 3–18
*Line of Best Fit for
Push-ups Versus Sit-ups*

Illustration 3–7

Each coed in a random sample of eight college women was asked for her height (to the nearest inch) and her weight (to the nearest 5 pounds). The data obtained are shown in table 3–4. Find an equation to predict the weight of a college woman based on height (the equation of the line of best fit) and draw it on the scatter diagram.

TABLE 3–4
*Data for College
Women's Heights and Weights*

	Woman							
	1	*2*	*3*	*4*	*5*	*6*	*7*	*8*
Height, x	65	65	62	67	69	65	61	67
Weight, y	105	125	110	120	140	135	95	130

Solution   Before we find the equation for the line of best fit, we need to decide whether or not the two variables appear to be linearly related. There are two ways to check on this: (1) draw the scatter diagram and see if the plot of points suggests a linear relationship, or (2) calculate the linear correlation coefficient and compare the calculated value to the decision point from table 3–3. If the scatter diagram or the calculated correlation coefficient suggests linear correlation, you are justified in calculating the equation for the line of best fit. The scatter diagram for the data on the height and weight of college women is shown in figure 3–19. (The calculation of r is left as an exercise.) The scatter diagram suggests that a linear line of best fit is appropriate.

FIGURE 3–19
Scatter Diagram

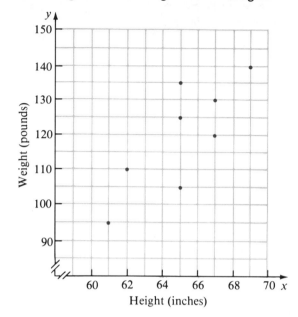

College Women's Heights Versus Weights

Let's calculate the equation for the line of best fit, using the summations in table 3–5. By formulas (3–7), (3–4), and (2–9) we find

$$SS(xy) = 62{,}750 - \frac{(521)(960)}{8} = 230.0$$

$$SS(x) = 33{,}979 - \frac{(521)(521)}{8} = 48.875$$

$$b_1 = \frac{230.0}{48.875} = 4.706 = \mathbf{4.71}$$

TABLE 3–5
Calculations Needed to Find b_1 and b_0

Coed	Height, x	x^2	Weight, y	xy
1	65	4.225	105	6,825
2	65	4.225	125	8,125
3	62	3,844	110	6,820
4	67	4,489	120	8,040
5	69	4,761	140	9,660
6	65	4,225	135	8,775
7	61	3,721	95	5,795
8	67	4,489	130	8,710
Total	521	33,979	960	62,750

By formula (3–6) we find

$$b_0 = \frac{1}{n} \left[\sum y - b_1 \cdot \sum x \right]$$

$$= \frac{1}{8} \left[960 - (4.706)(521) \right]$$

$$= -186.478 = -186.5$$

Therefore,

$$\hat{y} = -186.5 + 4.71x$$

To draw the line of best fit on the scatter diagram, we need to locate two points. Substitute two values for x, for example, 60 and 70, into the regression equation and obtain two corresponding values for y:

$$\hat{y} = -186.5 + (4.71)(60) = -186.5 + 282.6 = 96.1 = 96$$

and

$$\hat{y} = -186.5 + (4.71)(70) = -186.5 + 329.7 = 143.2 = 143$$

The values (60, 96) and (70, 143) represent two points (shown by + on figure 3–20) that enable us to draw the line of best fit; see the figure. □

NOTE: In figure 3–20 $(\bar{x}, \bar{y}) = (65.1, 120)$ is also on the line of best fit. It is the color circle. Use $(\bar{x}, \bar{y})$ as a check on your work.

FIGURE 3–20
*Line of Best Fit for College
Women's Heights Versus Weights*

One of the main purposes for obtaining a regression equation is for making predictions. Once a linear relationship has been established and the value of the input variable x is known, we can predict a value of y ($\hat{y}$). For example, in the physical fitness illustration, the equation was found to be $\hat{y} = 14.9 + 0.66x$. Student A can do 25 push-ups. How many sit-ups do you therefore predict that A will be able to do? (This predicted value will be the average number of sit-ups that you would expect from a sample taken of students who could do exactly 25 push-ups.) The predicted value is

$$\hat{y} = 14.9 + (0.66)(25) = 14.9 + 16.5 = 31.4 = \mathbf{31}$$

When making predictions based on the line of best fit, you must observe a few restrictions:

1. The equation should be used to make predictions only about the population from which the sample was drawn. For example, the use of the relationship between the height and weight of college women to predict the weight of professional athletes given their height would be questionable.

2. The equation should be used only over the sample domain of the input variable. For example, for illustration 3–7 the prediction that a college woman of height zero weighs -186.5 pounds is nonsense. You should not use a height outside the sample domain of 60 to 70 inches to predict weight.

3. If the sample was taken in 1983, do not expect the results to have been valid in 1929 or to hold in 1989. The women of today may be different from the women of 1929 and the women of 1989. On occasion you might wish to use the line of best fit to estimate values outside the domain interval of the sample. This can be done; however, you should do it with caution and only for values close to the domain interval.

Estimating the Line of Best Fit

It is possible to estimate the line of best fit and its equation. As with the approximation of r, it should be used mostly as a mental check of your work. On the scatter diagram of the data, draw the straight line that appears to be the line of best fit. (*Hint*: If you draw an oval around the points on the scatter diagram, the long diameter will be a reasonable approximation in most cases.) Then use this line to approximate the equation of the line of best fit.

Figure 3–21 shows the oval and the estimated line of best fit (drawn by eye) for the data of illustration 3–7. This line can now be used to approximate the equation of the line of best fit. Notice that the line passes through the two points (59, 85) and (66, 125). These points are circled in figure 3–21. An estimate for the equation can now be found by using the

FIGURE 3–21

*Estimating the Line of Best Fit
for the College Coed Data*

point-intercept form for a straight line (see the study guide). For the points (59, 85) and (66, 125), we have

$$m = \frac{\Delta y}{\Delta x} = \frac{125 - 85}{66 - 59} = \frac{40}{7} = 5.714 = 5.7$$

Now using the slope-intercept form for a straight line, we have

$$y = mx + b = 5.7x + b$$

Substituting the values $x = 59$ and $y = 85$ into this equation, we can determine the value of b:

$$85 = (5.7)(59) + b$$
$$b = 85 - (5.7)(59) = 85 - 336.3 = -251.3 = -250$$

Thus

$$y = 5.7x - 250$$

Or we estimate that $\hat{y} = -250 + 5.7x$. This should serve as a crude estimate. (*Note:* The graph's y-intercept is -250, not approximately 80, as might be read. Why?)

The techniques presented in this section presume that the scatter diagram suggests a linear relationship.

Exercises

3-12   Would you be justified in using the techniques of linear regression on the data given in exercise 3-3 to find the line of best fit? Explain.

3-13   Explain why the y-intercept is not 80 for the line of best fit in figure 3-21.

3-14   (a) Draw the line of best fit by eye and find an estimate of its equation by using the scatter diagram found in answering exercise 3-1.

(b) Calculate the equation of the line of best fit.

(c) Compare your answers for parts (a) and (b).

(d) Using the answer to part (b), what grade would you expect to receive if you had studied 6 hours?

(e) What exactly does your answer to (d) mean?

3-15   (a) Calculate the equation of the line of best fit for the psychologist's problem in exercise 3-2.

(b) Using the evidence found [answers to exercises 3-2, 3-7, and part (a) above], does the psychologist's claim seem to be valid?

3-16   (a) Draw the line of best fit by eye and estimate its equation for the data given in exercise 3-4.

(b) Calculate the line of best fit for the same data.

(c) Compare your answers.

3-17   (a) Calculate the line of best fit for the relationship between "number of TV commercials x" and "sales units y" for the data given in exercise 3-10.

(b) What does the value calculated for r in exercise 3-10 tell you about the answer you found in part (a)?

3-18   A record of maintenance costs is kept for each of several cash registers throughout a department store chain. A sample of 14 registers gave the accompanying data.

Age, x (years)	Maintenance Cost, y (dollars)	Age, x (years)	Maintenance Cost, y (dollars)
6	92	2	49
7	181	1	64
1	23	9	141
3	40	3	110
6	126	8	105
4	82	9	181
5	117	8	152

(a) Draw a scatter diagram that shows these data.

(b) Calculate the equation of the line of best fit.

(c) A particular cash register is 8 years old. How much maintenance (cost) do you predict it will require this year?

(d) Interpret your answer to part (c).

(e) Calculate r.

(f) What meaning does r have? How does it affect your answer in part (d)?

3-19 Verify, algebraically, that formula (3-7) is equivalent to formula (3-5).

In Retrospect

To sum up what we have just learned, there is a distinct difference between the purpose of regression analysis and the purpose of correlation. In regression analysis we seek a relationship between the variables. The equation that represents this relationship may be the answer desired or it may be the means to the prediction that is desired. In correlation analysis, on the other hand, we simply ask: Is there a significant linear relationship between the two variables?

The article at the beginning of the chapter relating the number of fouls committed and the number of points scored by basketball players summarizes the basic concepts of correlation and regression. The article is worth reading again. The data appear to fall along a straight line, which strongly suggests a linear relationship. But they do not in any way imply that if a player commits more fouls, he will score more points.

The linear methods we have studied thus far are the basic concepts presented for the purpose of a first, descriptive look. More details must, by necessity, wait until additional developmental work has been completed. After reading this chapter you should have a basic understanding of bivariate data and how to present it, of regression and correlation analysis, and a knowledge of when each is used. You also should be able to find the equation for the line of best fit and the correlation coefficient, which serve to describe the sample data. But before we delve more deeply into these techniques, we will study some probability.

Chapter Exercises

3-20 Calculate the linear correlation coefficient for the data on college women in illustration 3-7.

3-21 Determine whether each of these questions will require correlation analysis or regression analysis to obtain an answer.

(a) Is there a correlation between the grades a student attained in high school and the grades attained in college?

(b) What is the relationship between weight of a package and the cost of mailing it first class?

(c) Is there a linear relationship between a person's height and shoe size?

(d) What is the relationship between the number of worker hours and the number of units of production completed?

(e) Is the score obtained on a certain aptitude test linearly related to a person's ability to perform a certain job?

3-22 An automobile owner records the gallons of gasoline (x) required to fill the gasoline tank and the number of miles traveled (y) between fill-ups.

(a) If he does a correlation analysis on the data, what would be his purpose and what would be the nature of his result?

(b) If he does a regression analysis on the data, what would be his purpose and what would be the nature of his results?

3-23 Estimate the value of the correlation coefficient for each of the following pairs of variables.

(a) number of worker hours and number of units produced

(b) high school academic average and college grade point average

(c) a person's height and shoe size

(d) age of a car and its resale value

(e) number of fouls committed and the points scored by basketball players, as indicated in the article at the beginning of this chapter

3-24 A study of beef prices showed that as the demand for beef increased by 1000 units, the price tended to increase by 10¢ a pound.

(a) What type of analysis would produce this result?

(b) Is the correlation coefficient for the demand for beef and the price of beef positive or negative?

3-25 A regression line for estimating the yearly food cost $\hat{y}$ for a family of five based on yearly income x is given by $\hat{y} = 350 + 0.3x$.

(a) What is the average yearly expenditure for families with incomes of $6,000? $12,000?

(b) Why would this formula not be appropriate for use with incomes of $0 or $100,000?

3-26 Use the second-year data from the article at the beginning of this chapter.

(a) Construct the scatter diagram.

(b) Estimate the correlation coefficient.

(c) Estimate the equation for the line of best fit.

(d) Does there seem to be a relationship between the two variables?

(e) Does this mean that an increase in the number of fouls (x) will cause an increase in number of points (y)? Explain.

3-27 The following data were obtained from the weather records of 6 California cities.

	Number of Days of Measurable Rainfall	Annual Rainfall in Inches
Bakersfield	35	5.72
Los Angeles	35	14.05
San Diego	42	9.75
Fresno	44	10.24
Sacramento	57	17.22
San Francisco	66	20.66

(a) Plot the information using number of days as the input variable.

(b) Does it appear that there might be a linear relationship between the number of days of measurable rainfall and the annual average amount of rainfall?

(c) Compute the linear correlation coefficient.

(d) Using the decision rule presented in this chapter, decide whether or not there appears to be a relationship.

3-28 The accompanying data are the ages and the asking prices for 19 used small foreign cars.

Age, x (years)	Price, y (× $100)	Age, x (years)	Price, y (× $100)
3	34	6	21
5	26	8	11
3	31.5	5	25
6	12	6	18
4	30	5	23
4	30	7	18
6	14	4	24
7	18	7	10
2	34	5	18
2	32		

(a) Draw a scatter diagram.

(b) Calculate the linear correlation coefficient r.

(c) Is there evidence of correlation? Explain.

(d) Calculate the equation for the line of best fit.

(e) Graph the line of best fit on the scatter diagram.

(f) Predict the average asking price for all such foreign cars that are 5 years old.

3-29   The accompanying data resulted from a psychological experiment.

x	y	x	y
310	3.2	250	2.5
280	3.0	360	3.4
120	2.6	215	2.6
230	3.5	365	3.8
175	2.8	230	3.0
295	3.8	325	3.8
135	2.2	280	3.3

(a) Draw a scatter diagram.

(b) Calculate the linear correlation coefficient r.

(c) Is there evidence of correlation? Explain.

(d) Calculate the equation for the line of best fit.

(e) Graph the line of best fit on the scatter diagram.

(f) If x equals 300, what value can be predicted for y?

3-30   The number of one-family building permits issued and the number of one-family housing starts for the last 10 months in our town are recorded below:

Month	1	2	3	4	5	6	7	8	9	10
Permits Issued	528	494	453	398	413	454	450	436	460	444
Starts	696	614	623	507	554	550	592	568	627	564

Determine if there is a relationship between x, the number of permits issued, and y, the number of starts. The best relationship will probably be obtained by relating permits to starts one month later, since it takes time to get construction started after a permit is issued. *Hint*: There will be only 9 data points; (528, 614), (494, 623), ...

3-31   The following data are based upon an analysis of the Miss America Pageant participants:

Year	Mean Height in inches	Mean Weight in pounds	Mean Years of Education
1951	66.40	121.00	13.63
1956	66.10	120.00	13.72
1961	66.04	119.49	14.02
1966	66.34	119.48	14.49
1971	66.20	118.80	14.84
1976	66.32	116.02	15.06
1981	67.08	116.70	15.32

(a) Compute the correlation coefficient for the mean weight and mean years of education.

(b) Evaluate your result using the decision rule presented in this chapter. Should one conclude that this is a meaningful correlation because your decision rule indicates it is?

(c) Determine the line of best fit between year and mean weight. Use year as the input variable.

(d) What mean weight can be expected for 1986? Does it appear to be reasonable? Explain.

(e) Answer questions (c) and (d) using year as the input variable and mean height as the output variable.

(f) Answer questions (c) and (d) using year as the input variable and mean number of years of education as the output variable.

Vocabulary List

Be able to define each term. In addition, describe, in your own words, and give an example of each term. Your examples should not be ones given in class or in the textbook.

bivariate data	ordered pair
coefficient of linear correlation	output variable
correlation analysis	paired data
decision point	Pearson's product moment r
dependent variable	positive correlation
independent variable	predicted value
input variable	prediction equation
least squares criterion	regression analysis
linear	scatter diagram
line of best fit	slope, b_1
method of least squares	y-intercept, b_0
negative correlation	zero correlation

Quiz

Answer "True" if the statement is always true. If the statement is not always true, replace the words shown in color with words that make the statement always true.

3–1 Correlation analysis is a method of obtaining the equational relationship between two variables.

3–2 The linear correlation coefficient is used to determine the **equational** relationship between two variables.

3–3 A correlation coefficient of **zero** means that the two variables are perfectly correlated.

3 4 Whenever the slope of the regression line is zero, the **correlation coefficient** will also be zero.

3-5   When r is positive, b_1 will always be **negative**.

3-6   The **slope** of the regression line represents the amount of change that is expected to take place in y when x increases by one unit.

3-7   When the calculated value of r is between the table value of r and 1 (positive or negative), we **have** established a cause-and-effect relationship between the two variables.

3-8   Correlation coefficients range between **0 and + 1**.

3-9   The value that is being predicted is called the **input variable**.

3-10   The line of best fit is used to predict the **average value of** y that can be expected to occur at a given x value.

Working with Your Own Data

Each semester new students enter your college environment. You may have wondered, "What will the student body be like this semester?"

As a beginning statistics student, you have just finished studying three chapters of basic descriptive statistical techniques. Let's use some of these techniques to describe some characteristic of your college's student body.

A Single Variable Data

1. Define the population to be studied.

2. Choose a variable to define. (You may define your own variable or you may use one of the variables on the accompanying table of data if you are not able to collect your own data. Ask your instructor for guidance.)

3. Collect 35 pieces of data for your variable.

4. Construct a stem-and-leaf display of your data. Be sure to label it.

5. Calculate the value of the measure of central tendency that you believe best answers the question, "What is the average value of your variable?". Explain why you chose this measure.

6. Calculate the sample mean for your data (unless you used the mean in question 5).

7. Calculate the sample standard deviation for your data.

8. Find the value of the eighty-fifth percentile, P_{85}.

9. Construct a graphic display (other than a stem-and-leaf) that you believe "best" displays your data. Explain why the graph best presents your data.

B Bivariate Data

1. Define the population to be studied.

2. Choose and define two variables that will produce bivariate data. (You may define your own variables or you may use two of the variables on the accompanying table of data if you are not able to collect your own data. Ask your instructor for guidance.)

3. Collect 15 ordered pairs of data.

4. Construct a scatter diagram of your data. (Be sure to label it completely.)

5. Using a table to assist with the organization, calculate the extensions x^2, xy, and y^2, and the summations of x, y, x^2, xy, and y^2.

6. Calculate the linear correlation coefficient, r.

7. Calculate the equation of the line of best fit.

8. Draw the line of best fit on your scatter diagram.

The following table of data was collected on the first day of class last semester. You may use it as a source for your data if you are not able to collect your own data.

Variable A = student's sex (Male/Female)

Variable B = student's age at last birthday

Variable C = number of completed credit hours towards degree

Variable D = "Do you have a job (full/part-time)?" (Yes/No)

Variable E = number of hours worked last week, if D = yes

Variable F = wages (before taxes) earned last week, if D = yes

Student	A	B	C	D	E	F	Student	A	B	C	D	E	F
1	M	21	16	No			31	F	24	45	No		
2	M	18	0	Yes	10	34	32	M	34	4	No		
3	F	23	18	Yes	46	206	33	M	29	48	No		
4	M	17	0	No			34	M	22	80	Yes	40	336
5	M	17	0	Yes	40	157	35	M	21	12	Yes	26	143
6	M	40	17	No			36	F	18	0	No		
7	M	20	16	Yes	40	300	37	M	18	0	Yes	13	65
8	M	18	0	No			38	F	42	34	Yes	40	244
9	F	18	0	Yes	20	70	39	M	25	60	Yes	60	503
10	M	29	9	Yes	8	32	40	M	39	32	Yes	40	500
11	M	20	22	Yes	38	146	41	M	29	13	Yes	39	375
12	M	34	0	Yes	40	340	42	M	19	18	Yes	51	201
13	M	19	31	Yes	29	105	43	M	25	0	Yes	48	500
14	M	18	0	No			44	F	18	0	No		
15	M	20	0	Yes	48	350	45	M	32	68	Yes	44	473
16	F	27	3	Yes	40	130	46	F	21	0	No		
17	M	19	10	Yes	40	202	47	F	26	0	Yes	40	320
18	F	18	16	Yes	40	140	48	M	24	11	Yes	45	330
19	M	19	4	Yes	6	22	49	F	19	0	Yes	40	220
20	F	29	9	No			50	M	19	0	Yes	10	33
21	F	21	0	Yes	20	80	51	F	35	59	Yes	25	88
22	F	39	6	No			52	F	24	6	Yes	40	300
23	M	23	34	Yes	42	415	53	F	20	33	Yes	40	170
24	F	31	0	Yes	48	325	54	F	26	0	Yes	52	300
25	F	22	7	Yes	40	195	55	F	17	0	Yes	27	100
26	F	27	75	Yes	20	130	56	M	25	18	Yes	41	355
27	F	19	0	No			57	M	24	0	No		
28	M	22	20	Yes	40	470	58	M	21	0	Yes	30	150
29	F	60	0	Yes	40	390	59	M	30	12	Yes	48	555
30	M	25	14	No			60	F	19	0	Yes	38	169

Student	A	B	C	D	E	F	Student	A	B	C	D	E	F
61	M	32	45	Yes	40	385	81	F	27	0	Yes	40	550
62	M	26	90	Yes	40	340	82	F	42	47	Yes	37	300
63	M	20	64	Yes	10	45	83	F	41	21	Yes	40	250
64	M	24	0	Yes	30	150	84	M	36	0	Yes	40	400
65	M	20	14	No			85	M	25	16	Yes	40	480
66	M	21	70	Yes	40	340	86	F	18	0	Yes	45	189
67	F	20	13	Yes	40	206	87	M	22	0	Yes	40	385
68	F	33	3	Yes	32	246	88	F	27	9	Yes	40	260
69	F	25	68	Yes	40	330	89	F	26	3	Yes	40	240
70	F	29	48	Yes	40	525	90	F	23	9	Yes	40	330
71	F	40	0	Yes	40	400	91	M	41	3	Yes	23	253
72	F	36	3	Yes	40	300	92	M	39	0	Yes	40	110
73	F	35	0	Yes	40	280	93	M	21	0	Yes	40	246
74	F	28	0	Yes	40	350	94	F	32	0	Yes	40	350
75	M	40	64	Yes	40	390	95	F	48	58	Yes	40	714
76	F	31	0	Yes	40	200	96	F	26	0	Yes	32	200
77	F	32	0	Yes	40	270	97	F	27	0	Yes	40	350
78	F	37	0	Yes	24	150	98	F	52	56	Yes	40	390
79	F	35	0	Yes	40	350	99	F	34	27	Yes	8	77
80	M	21	72	Yes	45	470	100	F	49	3	Yes	24	260

The required calculations are accomplished most easily with the assistance of an electronic calculator or a canned program on a computer. Your local computer center will have a list of the available canned programs and will also provide you with the necessary information and assistance to run your data on the computer.

PART 2

Probability

Before continuing our study of statistics, we must make a slight detour and study some basic probability. Probability is often called the "vehicle" of statistics; that is, the probability associated with chance occurrences is the underlying theory for statistics. Recall that in chapter 1 we described probability as the science of making statements about what will occur when samples are drawn from known populations. Statistics was described as the science of selecting a sample and making inferences about the unknown population from which it is drawn. To make these inferences, we need to study sample results in situations where the population is known, so that we will be able to understand the behavior of chance occurrences.

In part 2 we will study the basic theory of probability (chapter 4), probability distributions of discrete variables (chapter 5), and probability distributions for continuous random variables (chapter 6). Following this brief study of probability, we will study the techniques of inferential statistics in part 3.

4 Probability

Chapter Outline

New Yorkers Would Restrict Public Smoking

The New York Poll was conducted Dec. 15–18 by the National Center for Telephone Research of New York for Gannett News Service. Trained interviewers surveyed 1,000 registered voters by telephone. The calls were placed at random, but in proportion to past voter turnouts around the state. The following report was written by John Omicinski, chief of the Albany GNS Bureau.

New Yorkers definitely favor restrictions on smoking in public.

By margins that may surprise smokers and non-smokers alike, they endorse changes in state laws to limit smoking in public places such as restaurants.

The New York Poll found 62 percent of those surveyed backed limits on smoking, while 36 percent opposed them. Only 2 percent weren't sure.

Backing for smoking limits ranged as high as 65 percent in New York City, and as low as 58 percent in Upstate counties.

In the 1978 legislative session, a bill to limit smoking in public to specified areas failed twice to win approval. It wasn't taken up in the Senate. That there is a legislative split over the issue is evident in the rules of the two houses: You can smoke in the Senate, but not in the Assembly.

Here are the figures, on the question, "Do you favor or oppose a proposed state law that forbids smoking in all public places unless designated as smoking areas?"

	Favor	Oppose	Not Sure
Total	62	36	2
Upstate	58	40	2
Western NY	58	41	1
NYC	65	34	1
NYC suburbs	63	34	3

Women were a bit more enthusiastic for a smoking-limit bill than were men. Females backed it, 63–35, and males were behind it, 60–38.

Low wage-earners—those making $10,000 or less—and those whose education extended to college and beyond gave the smoking limits their heaviest backing: 66 percent in favor. Less-than-high school graduates also favored the smoking controls, but by a lower margin of 58–38 percent.

Highest levels of opposition were in the 35-to-49-year age group: 41 percent opposed with 58 percent in favor, and among union members and those who call themselves "independent" voters: both 42 percent opposed.

The difference showed up between city dwellers and country residents. Urban people liked the idea, 63–36, and this dropped to 59–38 among town and rural residents.

Chapter Objectives

You may already be familiar with some ideas of probability, because probability is part of our everyday culture. We constantly hear people making probability-oriented statements:

"Our team will probably win the game tonight."

"There is a forty percent chance of rain this afternoon."

"I will most likely have a date for the winter weekend."

"If I park in the faculty parking area, I will probably get a ticket."

"I have a 50–50 chance of passing today's chemistry exam."

Everyone has made or heard these kinds of statements. What exactly do they mean? Do they, in fact, mean what they say? Some statements may be based on scientific information and others on subjective prejudice. Whatever the case may be, they are probabilistic inferences—not fact, but conjectures.

In this chapter we will study the basic concept of probability and the rules that apply to the probability of both simple and compound events.

Unit 1 Concepts of Probability

Section 4–1 The Nature of Probability

Let's consider an experiment in which we toss two coins simultaneously and record the number of heads that occur. 0H (zero heads), 1H (one head), and 2H (two heads) are the only possible outcomes. Let's toss the two coins 10 times and record our findings.

2H, 1H, 1H, 2H, 1H, 0H, 1H, 1H, 1H, 2H

Totals: 2H, 3   1H, 6   0H, 1

Suppose that we repeat this experiment 19 times. Table 4–1 shows the totals for 20 sets of 10 tosses. (Trial 1 shows the totals from above.) The total of 200 tosses of the pair of coins resulted in 2H on 53 occasions, 1H on 104 occasions, and 0H on 43 occasions. We can express these results in terms of relative frequencies and show the results using a histogram, as shown in figure 4–1.

What would happen if this experiment were repeated or continued? Would the relative frequencies change? If so, by how much? If we look at the individual sets of 10 tosses, we notice a large variation in the number of times each of the events (2H, 1H, and 0H) occurred. In both the 0H and the 2H categories, there were as many as 6 occurrences and as few as 0 occurrences in a set of 10 tosses. In the 1H category there were as few as 1 occurrence and as many as 9 occurrences.

TABLE 4–1
Experimental Results
of Tossing Two Coins

Outcome	\multicolumn{20}{c}{Trial}	Total

Outcome	1	2	3	4	5	6	7	8	9	10	11	12	13	14	15	16	17	18	19	20	Total
2H	3	3	5	1	4	2	4	3	1	1	2	5	6	3	1	4	1	0	3	1	53
1H	6	5	5	5	5	7	5	5	5	5	8	4	3	7	5	1	5	4	5	9	104
0H	1	2	0	4	1	1	1	2	4	4	0	1	1	0	4	5	4	6	2	0	43

FIGURE 4–1
Relative Frequency Histogram for
Coin-tossing Experiment

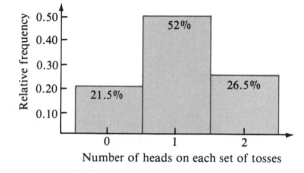

If we were to continue this experiment for several hundred more tosses, what would you expect to happen in terms of the relative frequencies of these three events? It looks as if we have approximately a 1:2:1 ratio in the totals of table 4–1. We might therefore expect to find the relative frequency for 0H to be approximately $\frac{1}{4}$ or 25%, the relative frequency for 1H to be approximately $\frac{1}{2}$ or 50%, and the relative frequency for 2H to be approximately $\frac{1}{4}$ or 25%. These relative frequencies accurately reflect the concept of probability.

Section 4–2 Probability of Events

We are now ready to define what is meant by probability. Specifically, we talk about "the probability that a certain event will occur."

probability of an event

Probability That an Event Will Occur

The relative frequency with which that event can be expected to occur.

**experimental or
empirical probability**

The probability of an event may be obtained in three different ways: (1) empirically, (2) theoretically, and (3) subjectively. The first method was illustrated in the experiment in section 4–1 and might be called **experimental** or **empirical probability**. This is nothing more than the **observed relative** frequency with which an event occurs. In our coin-tossing illustration we observed exactly one head (1H) on 104 of the 200 tosses of the pair of coins. The observed empirical probability for the occurrence of 1H was 104/200, or 0.52.

When the value assigned to the probability of an event results from experimental data, we will identify the probability of the event with the symbol $P'(\ )$. The name of the event is placed in parentheses:

$$P'(1\text{H}) = 0.52,\ P'(0\text{H}) = 0.215,\ \text{and}\ P'(2\text{H}) = 0.265$$

Thus the prime notation is used to denote empirical probabilities.

The value assigned to the probability of event A as a result of experimentation can be found by means of the formula

$$P'(A) = \frac{n(A)}{n} \tag{4-1}$$

when $n(A)$ is the number of times that event A actually is observed and n is the number of times the experiment is attempted.

$P'(A)$ represents the relative frequency with which event A occurred. This may or may not be the relative frequency with which that event "can be expected" to occur. $P'(A)$, however, represents our best *estimate* of what "can be expected" to occur. But what is meant by the relative frequency that "can be expected" to occur?

Consider the rolling of a single die. Define event A as the occurrence of a 1. In a single roll of a die, there are six possible outcomes. Assuming the die is symmetrical, each number should have an equal likelihood of occurring. Intuitively we see that the probability of A, or the expected relative frequency of a 1, is $\frac{1}{6}$. (Later we will formalize this calculation.)

What does this mean? Does it mean that once in every six rolls a 1 will occur? No, it does not. Saying that the probability of a 1, $P(1)$, is $\frac{1}{6}$ means that in the long run the proportion of times that a 1 occurs is approximately $\frac{1}{6}$. How close to $\frac{1}{6}$ can we expect the observed relative frequency to be?

Table 4–2 shows the number of 1s observed in each set of six rolls of a die [column (1)], an observed relative frequency for each set of six rolls [column (2)], and a cumulative relative frequency [column (3)]. Each trial is a set of six rolls.

Figure 4–2(a) (page 124) shows the fluctuation of the observed probability for event A on each of the 20 trials [column (2), table 4–2]. Figure 4–2(b) shows the fluctuation of the cumulative relative frequency [column (3), table 4–2]. Notice that the observed relative frequency on each trial of six rolls of a die tends to fluctuate about $\frac{1}{6}$. Notice also that the observed values on the cumulative graph seem to become more stable; in fact, they become relatively close to the expected $\frac{1}{6}$.

A cumulative graph such as figure 4–2(b) demonstrates the idea of long-term average. When only a few rolls were observed (as on each trial), the probability $P'(A)$ fluctuated between 0 and $\frac{1}{2}$ [see the relative frequency

Trial	(1) Numbers of 1s Observed	(2) Relative Frequency	(3) Cumulative Relative Frequency
1	1	1/6	1/6 = 0.17
2	2	2/6	3/12 = 0.25
3	0	0/6	3/18 = 0.17
4	2	2/6	5/24 = 0.21
5	1	1/6	6/30 = 0.20
6	1	1/6	7/36 = 0.19
7	2	2/6	9/42 = 0.21
8	2	2/6	11/48 = 0.23
9	0	0/6	11/54 = 0.20
10	0	0/6	11/60 = 0.18
11	2	2/6	13/66 = 0.20
12	0	0/6	13/72 = 0.18
13	2	2/6	15/78 = 0.19
14	1	1/6	16/84 = 0.19
15	1	1/6	17/90 = 0.19
16	3	3/6	20/96 = 0.21
17	0	0/6	20/102 = 0.20
18	1	1/6	21/108 = 0.19
19	0	0/6	21/114 = 0.18
20	1	1/6	22/120 = 0.18

of each trial, column (2), table 4–2]. As the experiment was repeated, however, the cumulative graph suggests a stabilizing effect on the observed cumulative probability. This stabilizing effect, or **long-term average** value, is often referred to as the "law of large numbers."

long-term average

Law of Large Numbers

Law of Large Numbers

If the numbers of times that an experiment is repeated is increased, the ratio of the number of successful occurrences to the number of trials will tend to approach the theoretical probability of the outcome for an individual trial.

The law of large numbers can be applied in a different manner. Consider the following experiment. A thumbtack is tossed and the result is observed and recorded as a point up ($\triangle$) or a point down ($\mathcal{A}$). What is the probability that the tack will land point up? That is, what is $P(\triangle)$? When we worked with the die in the preceding example, we could easily see the true probability. With the tack we have no idea. Is $P(\triangle)$ equal to $\frac{1}{2}$? Since the tack is

FIGURE 4–2

*Fluctuations Found in the
Die-tossing Experiment*

not symmetrical, we would suspect not. More than $\frac{1}{2}$? Less than $\frac{1}{2}$? About the only way we can tell is to conduct an experiment. And the law of large numbers can help us here. After many trials of the experiment, $P'(⏚)$ should indicate a value that is very close to the true value of $P(⏚)$, even though the true value cannot be proven mathematically.

Table 4–3 lists the data recorded after repeated tosses of a thumbtack. After each 10 tosses the number of times the tack landed point up was recorded. The cumulative value of the number of times it landed point up was computed and reported as a fraction of the total tosses. Data on 500 tosses are recorded. Figure 4–3 illustrates the value of $P(⏚)$ that we are attempting to estimate. The figure illustrates the tendency for the observed probability $P'(⏚)$ to be zeroing in on the value 0.605. We can conclude that $P(⏚)$ is very close to 0.605.

TABLE 4-3
Experimental Results of Tossing a Thumbtack

Multiples of 10 Tosses	n (☝) in Each Set of 10 Tosses	Cumulative P' (☝)	Multiples of 10 Tosses	n (☝) in Each Set of 10 Tosses	Cumulative P' (☝)
1	5	$5/10 = 0.500$	26	3	$156/260 = 0.600$
2	7	$12/20 = 0.600$	27	5	$161/270 = 0.596$
3	10	$22/30 = 0.733$	28	7	$168/280 = 0.600$
4	6	$28/40 = 0.700$	29	7	$175/290 = 0.603$
5	9	$37/50 = 0.740$	30	9	$184/300 = 0.613$
6	5	$42/60 = 0.700$	31	7	$191/310 = 0.616$
7	6	$48/70 = 0.686$	32	7	$198/320 = 0.619$
8	6	$54/80 = 0.675$	33	7	$205/330 = 0.621$
9	2	$56/90 = 0.622$	34	6	$211/340 = 0.621$
10	4	$60/100 = 0.600$	35	7	$218/350 = 0.623$
11	4	$64/110 = 0.582$	36	6	$224/360 = 0.622$
12	7	$71/120 = 0.592$	37	4	$228/370 = 0.616$
13	6	$77/130 = 0.592$	38	6	$234/380 = 0.616$
14	8	$85/140 = 0.607$	39	4	$238/390 = 0.610$
15	6	$91/150 = 0.607$	40	5	$243/400 = 0.608$
16	3	$94/160 = 0.588$	41	5	$248/410 = 0.605$
17	7	$101/170 = 0.594$	42	7	$255/420 = 0.607$
18	9	$110/180 = 0.611$	43	7	$262/430 = 0.609$
19	7	$117/190 = 0.616$	44	6	$268/440 = 0.609$
20	3	$120/200 = 0.600$	45	4	$272/450 = 0.604$
21	5	$125/210 = 0.595$	46	7	$279/460 = 0.607$
22	8	$133/220 = 0.605$	47	5	$284/470 = 0.604$
23	8	$141/230 = 0.613$	48	8	$292/480 = 0.608$
24	6	$147/240 = 0.613$	49	5	$297/490 = 0.606$
25	6	$153/250 = 0.612$	50	6	$303/500 = 0.606$

FIGURE 4-3
Experimental Probability of Thumbtack Pointing Up

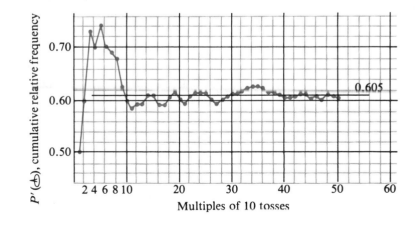

An important point is illustrated in figure 4–3 and table 4–3.

The larger the number of experimental trials n, the closer the experimental probability $P'(A)$ is expected to be to the true probability $P(A)$.

Many situations are analogous to the thumbtack experiment. We'll look at one example in illustration 4–1.

Illustration 4–1

The key to establishing proper life insurance rates is using the probability that the insureds will live one, two, or three years, and so forth, from the time they purchase policies. These probabilities are derived from actual life and death statistics and hence are experimental probabilities. They are published by the government and are extremely important to the life insurance industry.

□

Exercises

4–1 Explain what is meant by the statement, "The probability of a *one* when a single die is rolled is $\frac{1}{6}$."

4–2 Toss a single coin 20 times and record H (head) or T (tail) after each toss. Using your results, find the observed probabilities (a) $P'(H)$ and (b) $P'(T)$.

4–3 Roll a single die 20 times recording a 1, 2, 3, 4, 5, or 6 after each roll. Using your results, find the observed probabilities (a) $P'(1)$, (b) $P'(2)$, (c) $P'(3)$, (d) $P'(4)$, (e) $P'(5)$, and (f) $P'(6)$.

4–4 Place three coins in a cup, shake and dump them out, and observe the number of heads showing. Record 0H, 1H, 2H, or 3H after each trial. Repeat the process 25 times. Using your results find (a) $P'(0H)$, (b) $P'(1H)$, (c) $P'(2H)$, (d) $P'(3H)$.

4–5 Using a coin, perform the die and tack experiment discussed on pages 123–125. Toss a single penny 10 times, observing the number of heads (or use a cup containing 10 pennies and use one toss for each block of 10), and record the results. Repeat until you have accumulated 200 tosses. Chart and graph the data. Do your data tend to support the claim that $P(\text{head}) = \frac{1}{2}$? If not, what do you think caused the difference?

4–6 Take two dice (one white and one black) and roll them 50 times, recording the results as ordered pairs [for example, (3, 5) represents 3 on the white die and 5 on the black die]. Then calculate the following observed probabilities.
(a) P'(white die is an odd number)
(b) P'(sum is 6)
(c) P'(both dice show odd numbers)
(d) P'(number on black die is larger than number on white die)

Section 4–3 Simple Sample Spaces

Let's return to an earlier question: What values might we expect to be assigned to the three events (0H, 1H, 2H) associated with our coin-tossing experiment? As we inspect these three events, we see that they do not tend to happen with the same relative frequency. Why? Suppose the experiment of tossing two pennies and observing the number of heads had actually been carried out using a penny and a nickel—two distinct coins. Would this have changed our results? No, it would have had no effect on the experiment. However, it does show that there are more than three possible outcomes.

When a penny is tossed, it may land as heads or tails. When a nickel is tossed, it may also land as heads or tails. If we toss them simultaneously, we see that there are actually four different possible outcomes of each toss. Each observation would be one of the following possibilities: (1) heads on the penny and heads on the nickel, (2) heads on the penny and tails on the nickel, (3) tails on the penny and heads on the nickel, or (4) tails on the penny and tails on the nickel. Notice that the previous events (0H, 1H, and 2H) pair up with these four in the following manner: event (1) is the same as 2H, event (4) is the same as 0H, and events (2) and (3) together make up what was previously called 1H.

ordered pair

In this experiment with the penny and the nickel, let's use an **ordered pair** notation. The first listing will correspond to the penny and the second will correspond to the nickel. Thus (H, T) represents the event that a head occurs on the penny and a tail occurs on the nickel. Our listing of events for the tossing of a penny and a nickel looks like this:

$$(H, H), (H, T), (T, H), (T, T)$$

What we have accomplished here is a listing of what is known as the "sample space" for this experiment.

---

experiment

Experiment
Any process that yields a result or an observation.

outcome

Outcome
A particular result of an experiment.

sample space

Sample Space
The set of all possible outcomes of an experiment. The sample space is typically called S and may take any number of forms: a list, a tree diagram, a lattice grid system, and so on. The individual outcomes in a

sample points

sample space are called **sample points**. $n(S)$ is the number of sample points in sample space S.

event

Event
Any subset of the sample space. If A is an event, then $n(A)$ is the number of sample points that belong to event A.

Regardless of the form in which they are presented, the outcomes in a sample space can never overlap. Also, all possible outcomes must be represented. These characteristics are called **mutually exclusive** and **all inclusive**, respectively. A more detailed explanation of these characteristics will be presented later; for the moment, however, an intuitive grasp of their meaning is sufficient.

Now let's look at some illustrations of probability experiments and their associated sample spaces.

Experiment 4–1 A single coin is tossed once and the outcome—a head (H) or a tail (T)—is recorded.

$$\text{Sample space: } S = \{H, T\} \qquad \square$$

Experiment 4–2 Two coins, one penny and one nickel, are tossed simultaneously and the outcome for each coin is recorded using ordered pair notation: (penny, nickel).

The sample space is shown here in two different ways.

tree diagram   Tree diagram representation:*

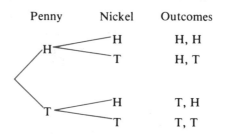

(four branches, each branch
shows a possible outcome)

Listing:

$$S = \{(H, H), (H, T), (T, H), (T, T)\}$$

Notice that both representations show the same four possible outcomes. For example, the top branch on the tree diagram shows heads on both coins, as does the first ordered pair in the listing. $\square$

Experiment 4–3 A die is rolled one time and the number of spots on the top face observed.

The sample space is

$$S = \{1, 2, 3, 4, 5, 6\} \qquad \square$$

* See the *Study Guide* for information about tree diagrams

Experiment 4–4 A box contains three poker chips (one red, one blue, one white), and two are drawn with replacement. This means that one chip is selected, its color is observed, and then the chip is replaced in the box. The chips are scrambled before a second chip is selected and its color observed.

The sample space is shown in two different ways.

Tree diagram representation:

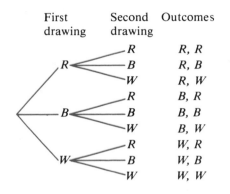

Listing:

$$S = \{(R, R), (R, B), (R, W), (B, R), (B, B),$$
$$(B, W), (W, R), (W, B), (W, W)\} \quad \square$$

Experiment 4–5 This experiment is the same as experiment 4–4 except that the first chip is not replaced before the second selection is made.

The sample space is shown in two ways.

Tree diagram representation:

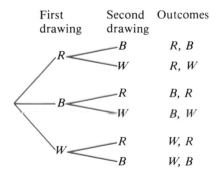

Listing:

$$S = \{(R, B), (R, W), (B, R), (B, W), (W, R), (W, B)\} \quad \square$$

Experiment 4–6   A white die and a black die are each rolled one time and the number of dots showing on each die is observed.

The sample space is shown by a chart representation.

Chart representation:

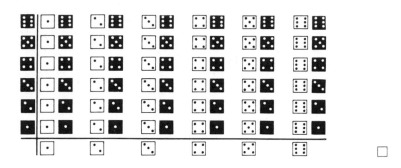

Experiment 4–7   Two dice are rolled and the sum of their dots is observed.

The sample space is

$$S = \{2, 3, 4, 5, 6, 7, 8, 9, 10, 11, 12\}$$

(or the 36-point sample space listed in experiment 4–6).   □

You will notice that two different sample spaces are suggested for experiment 4–7. Both of these sets satisfy the definition of a sample space and thus either could be used. We will learn later why the 36-point sample space is more useful than the other.

Experiment 4–8   A single marble is drawn from a box that contains 10 red and 90 blue marbles and its color is recorded.

The sample space is

$$S = \{R, B\}$$ □

Experiment 4–9   A thumbtack is tossed and whether it lands point up (⌁) or point down (⌁) is recorded.

The sample space is

$$S = \{⌁, ⌁\}$$ □

Experiment 4–10   A weather forecaster predicts that there will be a measurable amount of precipitation or no precipitation on a given day.

The sample space is

$$S = \{\text{precipitation, no precipitation}\}$$ □

Special attention should always be given to the sample space. Like the statistical population, the sample space must be well defined. Once the sample space is defined, you will find the remaining work much easier.

Exercises

4-7 An experiment consists of drawing one marble from a box that contains a mixture of red, yellow, and green marbles.

(a) List the sample space.

(b) Can we be sure that each outcome in the sample space is equally likely?

4-8 Two marbles are drawn from the box in exercise 4-7. List the sample space.

4-9 A bakery bakes six large birthday cakes each day. If a cake is not sold, it is discarded. List the sample space representing the number of cakes sold each day.

4-10 (a) A balanced coin is tossed twice. List a sample space showing the possible outcomes.

(b) A biased coin (it favors heads in a ratio of 3 to 1) is tossed twice. List a sample space showing the possible outcomes.

4-11 An experiment consists of tossing four coins and observing heads (H) or tails (T). Construct two alternative sample spaces showing the results of tossing the four coins.

4-12 An experiment consists of two trials. The first is tossing a penny and observing heads or tails; the second is rolling a die and observing a 1, 2, 3, 4, 5, or 6. Construct the sample space.

4-13 A coin is tossed and a head or a tail observed. If a head results, the coin is tossed a second time. If a tail results on the first toss, a die is rolled.

(a) Construct the sample space for this experiment.

(b) What is the probability that a die is rolled in the second stage of this experiment?

Section 4-4 Rules of Probability

Let's return now to the concept of probability and relate the sample space to it. Recall that the probability of an event was defined as the relative frequency with which the event could be expected to occur.

In the sample space associated with experiment 4-1, the tossing of one coin, we find two possible outcomes: heads (H) and tails (T). We have an "intuitive feeling" that these two events will occur with approximately the same frequency. The coin is a symmetrical object and therefore would not be expected to favor either of the two outcomes. We would expect heads to occur $\frac{1}{2}$ of the time. Thus the probability that a head will occur on a single toss of a coin is thought to be $\frac{1}{2}$.

The description above is the basis for the second technique for assigning the probability of an event. In a sample space containing **sample points** **equally likely events** that are **equally likely** to occur, the probability $P(A)$ of an event A is the ratio of the number $n(A)$ of points that satisfy the definition of event A to the number $n(S)$ of sample points in the entire sample space. That is,

$$P(A) = \frac{n(A)}{n(S)} \qquad (4\text{--}2)$$

theoretical probability This formula gives a **theoretical probability** value of event A's occurrence. The *prime symbol* of formula (4–1) *is not used* with theoretical probabilities.

The probability of event A is still a relative frequency, but it is now based on all the possible outcomes regardless of whether they actually occur. We can now see that the probability of obtaining two heads, $P(2H)$, when the penny and nickel are tossed is $\frac{1}{4}$. The probability of exactly one head, $P(1H)$ [(H, T) or (T, H)], is $\frac{2}{4}$, or $\frac{1}{2}$. And $P(0H) = \frac{1}{4}$. These are the values that were indicated by the experiment.

The use of formula (4–2) requires the existence of a sample space in which each outcome is equally likely. Thus when dealing with experiments that have more than one possible sample space, it is helpful to construct a sample space in which the sample points are equally likely.

Consider experiment 4–7, where two dice were rolled. If you list the sample space as the 11 sums, the sample points are not equally likely. If you use the 36-point sample space, all the sample points are equally likely. The 11 sums in the first sample space represent combinations of the 36 equally likely sample points. For example, the sum of 2 represents $\{(1, 1)\}$; the sum of 3 represents $\{(2, 1), (1, 2)\}$; and the sum of 4 represents $\{(1, 3), (3, 1), (2, 2)\}$. Thus we can use formula (4–2) and the 36-point sample space to obtain the probabilities for the 11 sums.

$$P(2) = \tfrac{1}{36}, \quad P(3) = \tfrac{2}{36}, \qquad P(4) = \tfrac{3}{36}$$

and so forth.

In many cases the assumption of equally likely events does not make sense. Experiments 4–1 through 4–6 are examples where the sample space elements are all equally likely. The sample points in experiment 4–7 are not equally likely, and there is no reason to believe that the sample points in experiments 4–8, 4–9, and 4–10 are equally likely.

What do we do when the sample space elements are not equally likely or not a combination of equally likely events? We could use empirical probabilities. But what do we do when no experiment has been done or can be performed?

Let's look again at experiment 4–10. The weather forecaster often assigns a probability to the event "precipitation." For example, "there is a 20% chance of rain today," or "there is a 70% chance of snow tomorrow." In such cases the only method available for assigning probabilities is personal
subjective probability judgment. These probability assignments are called **subjective probabilities**. The accuracy of subjective probabilities depends on the individual's ability to correctly assess the situation.

Often personal judgment of the probability of the possible outcomes of an experiment is expressed by comparing the likelihood between the various outcomes. For example, the weather forecaster's personal assessment might be that "it is five times more likely to rain (R) tomorrow than not rain (NR)"; $P(R) = 5 \cdot P(NR)$. If this is the case, what values should be assigned to $P(R)$ and $P(NR)$? To answer this question, we need to review some of the ideas about probability that we've already discussed.

1. Probability represents a relative frequency.

2. $P(A)$ is the ratio of the number of times an event can be expected to occur divided by the number of trials.

3. The numerator of the probability ratio must be a positive number or zero.

4. The denominator of the probability ratio must be a positive number (greater than zero).

5. The number of times an event can be expected to occur in n trials is always less than or equal to the total number of trials.

Thus it is reasonable to conclude that a probability is always a numerical value between 0 and 1.

$$\textbf{Property 1} \quad 0 \le P(A) \le 1$$

NOTES:

1. The probability is 0 if the event cannot occur.

2. The probability is 1 if the event occurs every time.

$$\textbf{Property 2} \quad \sum_{\text{all outcomes}} P(A) = 1$$

Property 2 states that if we add up the probabilities of each of the sample points in the sample space, the total probability must equal 1. This makes sense because when we sum up all the probabilities, we are really asking, "What is the probability the experiment will yield an outcome?" and this will happen every time.

Now we are ready to assign probabilities to $P(R)$ and $P(NR)$. The events R and NR cover the sample space, and the weather forecaster's personal judgment was

$$P(R) = 5 \cdot P(NR)$$

From property 2 we know

$$P(R) + P(NR) = 1$$

By substituting $5 \cdot P(NR)$ for $P(R)$, we get

$$5 \cdot P(NR) + P(NR) = 1$$
$$6 \cdot P(NR) = 1$$
$$P(NR) = \tfrac{1}{6}$$
$$P(R) = 1 - P(NR) = 1 - \tfrac{1}{6} = \tfrac{5}{6}$$

odds

The statement "It is five times more likely to rain tomorrow (R) than not rain (NR)" is often expressed as "The odds are 5 to 1 in favor of rain tomorrow" (also written as 5:1). **Odds** is simply another way of expressing probabilities. The relationships among **odds for an event**, **odds against an event**, and the **probability of an event** are expressed in the following rules.

If the **odds in favor of an event** A are a to b, then

1. The **odds against event** A are b to a

2. The **probability of event** A is

$$P(A) = \frac{a}{a + b}$$

3. The probability that event A will **not** occur is

$$P(A \text{ does not occur}) = \frac{b}{a + b}$$

To illustrate these rules, consider the above statement, "The odds favoring rain tomorrow are 5 to 1." The $P(R)$ is found using rule 2,

$$P(R) = \frac{5}{5 + 1} = \frac{5}{6}$$

The odds against it raining tomorrow are 1 to 5, and the probability that it does not rain tomorrow is

$$P(NR) = \frac{1}{5 + 1} = \frac{1}{6}$$

Exercises

4–14   Suppose a box of marbles contains an equal number of red marbles and yellow marbles but twice as many green marbles as red marbles. Draw one marble from the box and observe its color. Assign probabilities to the elements in the sample space.

4-15   Three coins are tossed and the number of heads observed is recorded. Find the probability for each of the possible results, 0H, 1H, 2H, and 3H.

4-16   Events A, B, and C are defined on sample space S. Their corresponding sets of sample points do not intersect and their union is S. Further, event B is twice as likely to occur as event A, and event C is twice as likely to occur as event B. Determine the probability of each of these three events.

4-17   Let x be the success rating of a new television show. The accompanying table lists the subjective probabilities assigned to each x for a particular new

	Judge		
Success Rating, x	A	B	C
Highly successful	0.5	0.6	0.3
Successful	0.4	0.5	0.3
Not successful	0.3	−0.1	0.3

show as assigned by three different media people. Which of these sets of probabilities are inappropriate because they violate a basic rule of probability? Explain.

4-18   Two dice are rolled (experiment 4-6). Find the probabilities in parts (b) through (e). Use the sample space given on page 130.

(a) Why is the set {2, 3, 4, ..., 12} not a useful sample space?

(b) P(white die is an odd number)

(c) P(sum is 6)

(d) P(both dice show odd numbers)

(e) P(number on black die is larger than number on white die)

4-19   Find the probability that event A will happen if the odds are:

(a) 7 to 3 in favor of A

(b) 1 to 3 in favor of A

(c) 3 to 2 against event A

(d) 3 to 5 against A

4-20   The oddsmakers claim that the odds are 2 to 5 in favor of the Dodgers to win the National League West pennant next year.

(a) What is the probability that the Dodgers will win the pennant next year, based on this information?

(b) What is the probability that they will not win the pennant?

Unit 2 Calculating Probabilities of Compound Events

compound events Compound events are formed by combining several simple events. We will study the following four compound events in the remainder of this chapter:

1. the probability of the complementary event, $P(\bar{A})$
2. the probability that either event A or event B will occur, $P(A \text{ or } B)$
3. the probability that both events A and B will occur, $P(A \text{ and } B)$
4. the probability that event A will occur given that event B has occurred, $P(A|B)$

NOTE: In determining which compound probability we are seeking, it is not enough to look for words "not," "either/or," "and," or "given" in the question. We must carefully examine the question asked to determine what combination of events is called for.

Section 4–5 Complementary, Mutually Exclusive Events, and the Addition Rule

Complementary Events

---

complementary event **Complement of an Event**

The set of all sample points in the sample space that do not belong to event A. The complement of event A is denoted by $\bar{A}$ (read "A complement").

For example, the complement of the event "success" is "failure"; the complement of "heads" is "tails" for the tossing of one coin; the complement of "at least one head" on 10 tosses of a coin is "no heads."

By combining the information in the definition of complement with property 2 (page 133), we can say

$$P(A) + P(\bar{A}) = 1.0, \qquad \text{for any event } A$$

It then follows that

$$P(\bar{A}) = 1 - P(A) \tag{4-3}$$

NOTE: Every event A has a complementary event $\bar{A}$. Complementary probabilities are very useful when the question asks for the probability of "at least one." Generally this represents a combination of several events, but the complementary event "none" is a single outcome. It is easier to solve for the complementary event and get the answer by using formula (4–3).

Illustration 4–2

Two coins are tossed. What is the probability that at least one head appears?

Solution   Let event A be the occurrence of no heads; then $\bar{A}$ represents the occurrence of one or more heads, that is, at least one head. The sample space is {(HH), (HT), (TH), (TT)}.

$$P(A) = \tfrac{1}{4} = 0.25 \quad \text{[using formula (4–2)]}$$

$$P(\bar{A}) = 1 - P(A) = 1 - \tfrac{1}{4} = \tfrac{3}{4} = 0.75 \quad \text{[using formula (4–3)]} \quad \Box$$

Mutually Exclusive Events

mutually exclusive events

___

Mutually Exclusive Events

Events defined in such a way that the occurrence of one event precludes the occurrence of any of the other events. (In short, if one of them happens, the others cannot happen.)

___

The following illustrations give examples of events that are mutually exclusive.

Illustration 4–3

In the experiment where 2 coins were tossed and the events of no heads (0H), one head (1H), and two heads (2H), were observed, the events 1H and 2H are examples of mutually exclusive events.

$$S = \{(H, H), (H, T), (T, H), (T, T)\}$$

$$1H = \{(H, T), (T, H)\}$$

and

$$2H = \{(H, H)\}$$

The events 1H and 2H have no common sample points (i.e., no intersection); therefore, they are mutually exclusive events. $\quad \Box$

Illustration 4-4

Consider an experiment in which two dice are rolled. Three events are defined:

A: The sum of the numbers on the two dice is 7.

B: The sum of the numbers on the two dice is 10.

C: Each of the two dice shows the same number.

Let's determine if these three events are mutually exclusive.

Are events A and B mutually exclusive? Yes, they are, since the sum on the two dice cannot be both 7 and 10 at the same time. If a sum of 7 occurs, it is impossible for the sum to be 10.

Figure 4-4 presents the sample space for this experiment. This is the same sample space shown in experiment 4-6, except that ordered pairs are used in place of the pictures. The ovals, diamonds, and rectangles show the ordered pairs that are in events A, B, and C, respectively. We **intersection** can see that events A and B do not **intersect** at a common sample point. Therefore, they are mutually exclusive. Point $(5, 5)$ satisfies both events B and C. Therefore, B and C are not mutually exclusive. Two dice can each show a 5, which satisfies C; and the total of the two 5s satisfies B.

FIGURE 4-4

Sample Space for the Roll of Two Dice

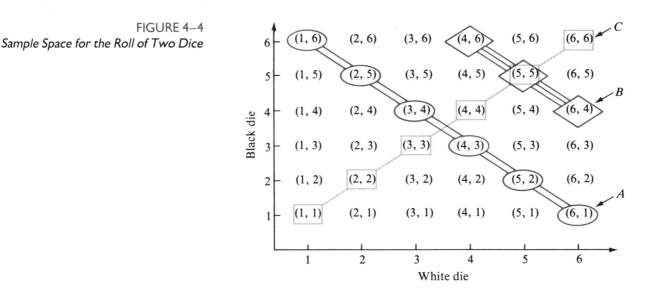

Venn diagram   Figures 4-5, 4-6, and 4-7 are **Venn diagrams** that show the relationships between the events defined in illustration 4-4. Venn diagrams are simple "area" or "region" diagrams. The basic concept of a Venn diagram is that those points (and only those points) belonging to a given event are shown inside a circle. (For further background information, see the study guide.)

Figure 4-5 shows that events A and B are mutually exclusive. There are six sample points that belong to event A and there are three sample

FIGURE 4–5
*Venn Diagram: Events A and B Are
Mutually Exclusive*

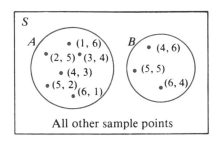

All other sample points

FIGURE 4–6
*Events A and C Are Mutually
Exclusive*

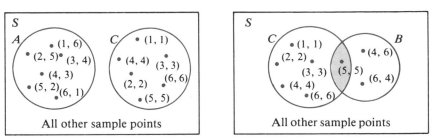

All other sample points

FIGURE 4–7
*Events B and C Are Not Mutually
Exclusive*

All other sample points

points in event B. The two events do not have a point in common. Therefore, the sets represented in the Venn diagram do not intersect; that is, they are mutually exclusive.

Figure 4–6 presents a similar situation. Events A and C do not intersect; therefore, they are mutually exclusive.

In figure 4–7 events C and B share a common sample point, $(5, 5)$. Notice that the point $(5, 5)$ is in the intersection of the two Venn diagram circles. Consequently, events C and B are not mutually exclusive. ☐

Addition Rule

Let us now consider the compound probability $P(A \text{ or } B)$, where A and B are mutually exclusive events.

Illustration 4–5

Earlier in this chapter we considered an experiment where two coins were tossed and we observed either no heads (0H), one head (1H), or two heads (2H). We found the probabilities of these events to be $P(0H) = \frac{1}{4}$, $P(1H) = \frac{1}{2}$, and $P(2H) = \frac{1}{4}$, as shown in the histogram of figure 4–8. The area in this histogram represents probability.

Let's find the probability of at least one head. This event, "at least one head," can be restated as "either one head or two heads occurred." These two events are represented on the histogram by the two shaded bars. It appears very reasonable to say that the probability of at least one head is $\frac{3}{4}$, the sum of $P(1H) = \frac{1}{2}$ and $P(2H) = \frac{1}{4}$. This value can be justified by looking at the sample space, where three of the four sample points satisfy the event. ☐

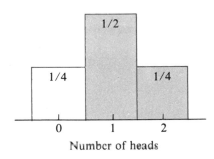

Number of heads

FIGURE 4–8
*Histogram for the Toss of Two Coins;
the Area Represents Probability*

When events are **not mutually exclusive**, we cannot find the probability that one or the other occurs by simply adding the individual probabilities, as shown above. Why not? Let's look at an illustration and see what happens when events are not mutually exclusive.

Illustration 4–6

Find the probability that a sum of 10 or a pair of double numbers are observed when a pair of dice are rolled. [Double numbers are (1, 1), (2, 2), (3, 3), and so on; this is the same as event C defined in illustration 4–4.]

Solution $P(10) = \frac{1}{12}$ and $P(\text{double}) = \frac{1}{6}$. If we add the two probabilities, we obtain $P(10 \text{ or double}) = \frac{1}{12} + \frac{1}{6} = \frac{1}{4}$. However, if we go to the sample space (figure 4–4) and count the number of points that satisfy the statement "a sum of 10 or a double was observed," we find eight points. Using an adaptation of formula (4–2), we obtain

$$P(10 \text{ or double}) = \frac{n(10 \text{ or double})}{n(S)} = \frac{8}{36} = \frac{2}{9}$$

But $\frac{2}{9}$ does not equal $\frac{1}{4}$. The probability of an event cannot have two values. What caused the difference?

Look at the sample space and you will find that the sample point (5, 5) satisfies both "sum is 10" and "double." In adding the two probabilities, we "double counted" the point (5, 5). However, this is not allowed.   □

We can add probabilities to find the probability of an "or" compound event, but we must make an adjustment in situations such as the previous example.

Addition Rule

General Addition Rule

Let A and B be two events defined in a sample space S.

$$P(A \text{ or } B) = P(A) + P(B) - P(A \text{ and } B) \qquad (4\text{–}4a)$$

Special Addition Rule

Let A and B be two events defined in a sample space. If A and B are **mutually exclusive** events, then

$$P(A \text{ or } B) = P(A) + P(B) \qquad (4\text{–}4b)$$

This can be expanded to consider **more than two** mutually exclusive events:

$P(A \text{ or } B \text{ or } C \text{ or } \ldots \text{ or } E)$

$$= P(A) + P(B) + P(C) + \cdots + P(E) \quad (4\text{–}4c)$$

The key to this formula is the property "mutually exclusive." If two events are mutually exclusive, there is no double counting of sample points. If events

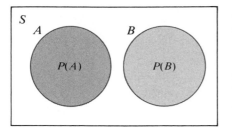

FIGURE 4–9
Mutually Exclusive Events

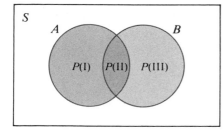

FIGURE 4–10
Nonmutually Exclusive Events

are not mutually exclusive, then when probabilities are added, the double counting will occur.

Let's look at some examples.

In figure 4–9 events A and B are mutually exclusive. Simple addition is justified, since the total probability of the shaded regions is sought. [In a Venn diagram, probabilities are often represented by enclosed areas. In figure 4–9, for example, we've shaded the areas representing the probabilities $P(A)$ and $P(B)$.]

In figure 4–10 events A and B are not mutually exclusive. The probability of the event "A and B," $P(A$ and $B)$, is represented by the area contained in region II. The probability $P(A)$ is represented by the area of circle A. That is, $P(A) = P(\text{region I}) + P(\text{region II})$. Also, $P(B) = P(\text{region II}) + P(\text{region III})$. And $P(A$ or $B)$ is the sum of the probabilities associated with the three regions:

$$P(A \text{ or } B) = P(\text{I}) + P(\text{II}) + P(\text{III})$$

However, if $P(A)$ is added to $P(B)$, we have

$$P(A) + P(B) = [P(\text{I}) + P(\text{II})] + [P(\text{II}) + P(\text{III})]$$
$$= P(\text{I}) + 2P(\text{II}) + P(\text{III})$$

This is the double count previously mentioned. However, if we subtract one measure of region II from this total, we will be left with the correct value.

The addition formula (4–4b) is a special case of the more general rule stated in formula (4–4a). If A and B are mutually exclusive events, $P(A$ and $B)$ = 0. (They cannot both happen at the same time.) Thus the last term in formula (4–4a) is zero when events are mutually exclusive.

Illustration 4–7

One white die and one black die are rolled. Find the probability that the white die shows a number smaller than 3 or the sum of the dice is greater than 9.

Solution I A = white die shows a 1 or a 2; B = sum of both dice is 10, 11, or 12.

$$P(A) = \frac{12}{36} = \frac{1}{3} \quad \text{and} \quad P(B) = \frac{6}{36} = \frac{1}{6}$$

$$P(A \text{ or } B) = P(A) + P(B) - P(A \text{ and } B)$$
$$= \frac{1}{3} + \frac{1}{6} - 0 = \frac{1}{2}$$

[$P(A$ and $B)$ = 0 since the events do not intersect.]   □

Solution 2

$$P(A \text{ or } B) = \frac{n(A \text{ or } B)}{n(S)} = \frac{18}{36} = \frac{1}{2}$$

(Look at the sample space, in figure 4–4, and count.) ☐

Exercises

4–21   Determine whether or not each of the following pairs of events is mutually exclusive.

(a) Five coins are tossed: "one head is observed," "at least one head is observed."

(b) A salesperson calls on a client and makes a sale: "the sale exceeds $100," "the sale exceeds $1000."

(c) One student is selected at random from a student body: the person selected is "male," the person selected is "over 21 years of age."

(d) Two dice are rolled: the total showing is "less than 7," the total showing is "more than 9."

4–22   Determine whether or not each of the following sets of events is mutually exclusive.

(a) Five coins are tossed: "no more than one head is observed," "two heads are observed," "three or more heads are observed."

(b) A salesperson calls on a client and makes a sale: the amount of the sale is "less than $100," is "between $100 and $1000," is "more than $500."

(c) One student is selected at random from the student body: the person selected is "female," is "male" or "over 21."

(d) Two dice are rolled: the numbers of dots showing on each die are "both odd," are "both even," are "total seven" or "total eleven."

4–23   If $P(A) = 0.3$ and $P(B) = 0.4$, and if A and B are mutually exclusive events, find the following:

(a) $P(\bar{A})$ (b) $P(\bar{B})$ (c) $P(A \text{ or } B)$ (d) $P(A \text{ and } B)$

4–24   Explain why $P(A \text{ and } B) = 0$ when events A and B are mutually exclusive.

4–25   One student is selected at random from the student body. Suppose the probability that this student is female is 0.5 and that the probability that this student is working a part-time job is 0.6. Are the two events "female" and "working" mutually exclusive events? Explain.

4–26   If $P(A) = 0.4$, $P(B) = 0.5$, and $P(A \text{ and } B) = 0.1$, find $P(A \text{ or } B)$.

4–27   Two cards are randomly drawn from a well-shuffled deck, with replacement. What is the probability that at least one is a heart? (Use two different methods to solve this problem.)

4-28 At our college it is known that 0.6 of the students are under 21 years of age. Also, 0.5 of the students are female, and 0.3 of all students are both female and under 21. If one student is selected at random, what is the probability that this student is

(a) female and under 21?

(b) male and 21 or over?

Section 4–6 Independence, Multiplication Rule, and Conditional Probability

Next we will study the compound event that A and B both occur. For example, what is the probability that two heads occur when two coins (one penny and one nickel) are tossed? Let A represent the occurrence of a head on the penny and B represent the occurrence of a head on the nickel. What is $P(A \text{ and } B)$? $P(A \text{ and } B)$ is found by using the definition $n(A \text{ and } B)/n(S)$. The sample space for experiment 4–2 suggests that this value should be $\frac{1}{4}$. How can we obtain this value by using $P(A)$ and $P(B)$? Both $P(A)$ and $P(B)$ have values of $\frac{1}{2}$. If $P(A)$ is multiplied by $P(B)$, we obtain $\frac{1}{4}$. Thus we might suspect that $P(A \text{ and } B)$ equals $P(A) \cdot P(B)$.

Consider this example. A is the event that 2 shows on a white die, and B is the event that 2 shows on a black die. If both dice are rolled once, what is the probability that two 2s occur?

$$P(A) = \frac{1}{6} \quad \text{and} \quad P(B) = \frac{1}{6}$$

$$P(A \text{ and } B) = \frac{n(A \text{ and } B)}{n(S)} = \frac{1}{36}$$

However, note that $\frac{1}{6}$ multiplied by $\frac{1}{6}$ is also $\frac{1}{36}$. In this case multiplication yields the correct answer. Multiplication does not always work, however. For example, $P(\text{sum of 7 and double})$ when two dice are rolled is 0 (as seen in figure 4–4). However, if $P(7)$ is multiplied by $P(\text{double})$, we obtain $(\frac{1}{6})(\frac{1}{6}) = \frac{1}{36}$.

Multiplication does not work for $P(\text{sum of 10 and double})$, either. By definition and by inspection of the sample space, we know that $P(10 \text{ and double}) = \frac{1}{36}$ [the point (5, 5) is the only element]. However, if we multiply $P(10)$ by $P(\text{double})$, we obtain $(\frac{3}{36})(\frac{6}{36}) = \frac{1}{72}$. The probability of this event cannot be both values.

independence The property that is required for multiplying probabilities is **independence**. Multiplication worked in the two examples cited above because the events were independent. In the other two cases the events were not independent.

Independence and Conditional Probabilities

independent events

Independent Events

Two events A and B are **independent events** if the occurrence (or nonoccurrence) of one does not affect the probability assigned to the occurrence of the other.

Consider again the experiment of tossing one penny and one nickel. Intuitively it appears reasonable that a result of a head or a tail on a nickel has no effect on the experiment of tossing a penny. The two events can be thought of as separate events. The same is true with rolling two dice and looking for 2s on both.

dependent events

Lack of independence, called dependence, is demonstrated by the following illustration. Reconsider the experiment of rolling two dice and observing the two events "sum of 10" and "double." As stated previously, $P(10) = \frac{3}{36} = \frac{1}{12}$ and $P(\text{double}) = \frac{6}{36} = \frac{1}{6}$. Does the occurrence of 10 affect the probability of a double? Think of it this way. A sum of 10 has occurred; it must be one of the following: $\{(4, 6,), (5, 5), 6, 4)\}$. One of these three possibilities is a double. Therefore, we must conclude that the P (double, *knowing* 10 has occurred), written P (double | 10), is $\frac{1}{3}$. Since $\frac{1}{3}$ does not equal the original probability of a double, $\frac{1}{6}$, we can conclude that the event "10" has an effect on the probability of a double. Therefore, "double" and "10" are dependent events.

Whether events are independent often becomes clear by examining the events in question. Rolling one die does not affect the outcome of a second roll. However, in many cases independence is not self-evident, and the question of independence itself may be of special interest. Consider the events "having a checking account at a bank" and "having a loan account at the same bank." Having a checking account at a bank may increase the probability that the same person has a loan account. This has practical implications. For example, it would make sense to advertise loan programs to checking account clients if they are more likely to apply for loans than are people who are not customers of the bank.

One approach to the problem is to *assume* independence or dependence. The correctness of the probability analysis depends on the truth of the assumption. In practice, we often assume independence and compare *calculated* probabilities with *actual frequencies* of outcomes in order to infer whether the assumption of independence is warranted.

conditional probability

The symbol $P(A|B)$ represents the probability that A will occur given that B has occurred. This is called a **conditional probability**.

The previous definition of independent events can now be written in a more formal manner.

Independent Events

Two events A and B are independent events if $P(A|B) = P(A)$ or if $P(B|A) = P(B)$.

(4–5)

Let's consider conditional probability. Take the experiment where a single die is rolled; $S = \{1, 2, 3, 4, 5, 6\}$. Two events that can be defined for this experiment are $B =$ an even number occurs and $A =$ a 4 occurs. Then $P(A) = \frac{1}{6}$. Event A is satisfied by exactly one of the six equally likely sample points in S. The conditional probability of A given B, $P(A|B)$, is found in a similar manner, but S is no longer the sample space. Think of it this way. A die is rolled out of your sight, and you are told that the number showing is even, that is, event B has occurred. That is the given condition. Knowing this condition, you are asked to assign a probability to the event that the even number is a 4. There are only three possibilities in the new (or reduced) sample space, $\{2, 4, 6\}$. Each of the three outcomes is equally likely; thus $P(A|B) = \frac{1}{3}$.

We can write this as

$$P(A|B) = \frac{P(A \text{ and } B)}{P(B)}$$

(4–6)

Thus for our example,

$$P(A|B) = \frac{\frac{1}{6}}{\frac{1}{2}} = \frac{1}{3}$$

Illustration 4–8

Each person, in a sample of 150 residents, was asked if he or she favored the concept of having a single countywide police agency. The county is composed of one large city and many suburban townships. The residence (city or outside the city) and the responses of the residents are summarized in table 4–4. If one of these residents was to be selected at random, what is the

TABLE 4–4
Sample Results for Illustration 1–8

Residence	Opinion		
	Favor, F	Oppose, O	Total
In city, C	80	40	120
Outside of city, $\bar{C}$	20	10	30
Total	100	50	150

probability that the person will (a) Favor the concept? (b) Favor the concept if the person selected is a city resident? (c) Favor the concept if the person selected is a resident from outside the city? (d) Are the events F (favor the concept) and C (reside in city) independent?

Solution

(a) $P(F)$ is the proportion of the total sample that favor the concept. Therefore,

$$P(F) = \frac{n(F)}{n(S)} = \frac{100}{150} = \frac{2}{3}$$

(b) $P(F|C)$ is the probability that the person selected favors the concept given that he or she lives in the city. The sample space is reduced to the 120 city residents in the sample. Of these, 80 favored the concept; therefore,

$$P(F|C) = \frac{n(F \text{ and } C)}{n(C)} = \frac{80}{120} = \frac{2}{3}$$

(c) $P(F|\bar{C})$ is the probability that the person selected favors the concept, knowing that the person lives outside the city. The sample space is reduced to the 30 noncity residents; therefore,

$$P(F|\bar{C}) = \frac{n(F \text{ and } \bar{C})}{n(\bar{C})} = \frac{20}{30} = \frac{2}{3}$$

(d) All three probabilities have the same value, $\frac{2}{3}$. Therefore, we can say that the events F (favor) and C (reside in city) are independent. The location of residence did not affect $P(F)$. □

Multiplication Rule

General Multiplication Rule

General Multiplication Rule

Let A and B be two events defined in sample space S. Then

$$P(A \text{ and } B) = P(A) \cdot P(B|A) \qquad (4\text{--}7a)$$

or

$$P(A \text{ and } B) = P(B) \cdot P(A|B) \qquad (4\text{--}7b)$$

If events A and B are independent, then the general multiplication rule (formula 4–7) reduces to the special multiplication rule, which is shown below as formula 4–8.

Special Multiplication Rule

Special Multiplication Rule

Let A and B be two events defined in sample space S. If A and B are **independent events**, then

$$P(A \text{ and } B) = P(A) \cdot P(B) \tag{4–8a}$$

This formula can be expanded. If $A, B, C, \ldots, G$ are independent events, then

$$P(A \text{ and } B \text{ and } C \text{ and} \ldots \text{and } G) = P(A) \cdot P(B) \cdot P(C) \cdots P(G)$$

$$\tag{4–8b}$$

Illustration 4–9

One white and one black die are rolled. Find the probability that the sum of their numbers is 7 and that the number on the black die is larger than the number on the white die.

Solution $A = $ sum is 7; $B = $ black number larger than white number. The "and" requires the use of the multiplication rule. However, we do not yet know if events A and B are independent. Refer to figure 4–4 for the sample space of this experiment. We see that $P(A) = \frac{6}{36} = \frac{1}{6}$. Also, $P(A|B)$ is obtained from the reduced sample space, which includes 15 points above the gray diagonal line. Of the 15 equally likely points, 3 of them—(1, 6), (2, 5), and (3, 4)—satisfy event A. Therefore, $P(A|B) = \frac{3}{15} = \frac{1}{5}$. Since this is a different value than $P(A)$, the events are dependent. So we must use formula (4–7b) to obtain $P(A \text{ and } B)$.

$$P(A \text{ and } B) = P(B) \cdot P(A|B) = \frac{15}{36} \cdot \frac{3}{15} = \frac{3}{36} = \frac{1}{12} \qquad \square$$

NOTES:

1. Independence and mutual exclusiveness are two completely separate concepts.
2. The term "mutually exclusive" describes whether the events can occur together, while "independence" describes the effect that one event has on the probability of the other event's occurrence.
3. If two events are *not* mutually exclusive, then their corresponding sets of sample points intersect.

(continued)

4. The relationship between independence (or dependence) and mutual exclusiveness (or nonmutual exclusiveness) is summarized by the following four statements:

(a) If two events are known to be mutually exclusive, then they are also dependent events.

For example: If $P(A) = 0.3$ and $P(B) = 0.4$ and if A and B are mutually exclusive, then $P(A|B) = 0$. Since $P(A) \neq P(A|B)$, A and B are dependent events.

(b) If two events are known to be independent, then they are *not* mutually exclusive events.

For example: If $P(C) = 0.2$ and $P(D) = 0.5$ and if C and D are independent, then $P(C \text{ and } D) = 0.2 \cdot 0.5 = 0.10$. Since $P(C \text{ and } D)$ is greater than zero, C and D intersect. Therefore C and D are not mutually exclusive events.

(c) If two events are known to be nonmutually exclusive, then they may be either independent *or* dependent events.

For example: (1) If $P(C) = 0.2$, $P(D) = 0.5$, and $P(C \text{ and } D) = 0.1$ (C and D intersect, that is, they are nonmutually exclusive), then $P(C|D) = 0.1/0.5 = 0.2$. Since $P(C) = P(C|D)$, C and D are independent events.

(2) If $P(E) = 0.2$, $P(F) = 0.5$, and $P(E \text{ and } F) = 0.05$ (E and F intersect, that is, they are nonmutually exclusive), then $P(E|F) = 0.05/0.5 = 0.1$. Since $P(E) \neq P(E|F)$, E and F are dependent events.

(d) If two events are known to be dependent, then they may be *either* mutually exclusive *or* nonmutually exclusive.

For example: (1) If $P(E) = 0.2$, $P(F) = 0.5$, and $P(E|F) = 0.1$ [E and F are dependent since $P(E) \neq P(E|F)$], then $P(E \text{ and } F) = 0.5 \cdot 0.1 = 0.05$. Since $P(E \text{ and } F)$ is greater than zero, E and F are intersecting or nonmutually exclusive events.

(2) If $P(A) = 0.3$, $P(B) = 0.4$, and $P(B|A) = 0$ [A and B are dependent since $P(B) \neq P(B|A)$], then $P(A \text{ and } B) = 0.3 \cdot 0 = 0$. Since $P(A \text{ and } B) = 0$, A and B do not intersect or are mutually exclusive events.

Exercises

4–29 If $P(A) = 0.3$ and $P(B) = 0.4$, and A and B are independent events, what is each of these probabilities?

(a) $P(A \text{ and } B)$?   (b) $P(B|A)$?   (c) $P(A|B)$?

4–30 Suppose $P(A) = 0.3$, $P(B) = 0.4$, and $P(A \text{ and } B) = 0.12$.

(a) What is $P(A|B)$?   (b) What is $P(B|A)$?
(c) Are A and B independent?

4–31 Suppose $P(A) = 0.3$, $P(B) = 0.4$, and $P(A \text{ and } B) = 0.20$.

(a) What is $P(A|B)$?   (b) What is $P(B|A)$?
(c) Are A and B independent?

4-32 Determine whether or not each of the following pairs of events is independent:

(a) rolling a pair of dice and observing a "1" on the first die and a "1" on the second die

(b) drawing a "spade" from a regular deck of playing cards and then drawing another "spade" from the same deck without replacing the first card

(c) same as (b) except the first card is returned to the deck before the second drawing

(d) owning a red automobile and having blonde hair

(e) owning a red automobile and having a flat tire today

(f) studying for an exam and passing the exam

4-33 Records show that the probability that it rains in our town on a single day is 0.35. If rain on different days constitutes independent events, find these probabilities:

(a) that it rains on neither of two given days

(b) that it rains on both of two given days

(c) that it rains on exactly one of the two given days

4-34 In a survey, which questioned high school students about their attitudes towards business, 0.85 said "Honesty is the best policy" and 0.28 stated "Success in business requires some dishonesty." Assuming that the survey results both accurately reflect the attitudes of all high school students and that the answers to the two attitudes are independent of one another, what is the probability that:

(a) a student did not say "honesty is the best policy" and agreed that "success in business requires some dishonesty?"

(b) a student said "honesty is the best policy" or agreed "success in business requires some dishonesty?"

4-35 A box contains four red and three blue poker chips. What is the probability when three are selected randomly that all three will be red if we select each chip

(a) with replacement?   (b) without replacement?

4-36 In the United States, the proportion of all births that involve Caesarean section deliveries is 15 out of 100. When Caesarean deliveries are made, 96 out of 100 babies survive. What is the probability that a randomly selected expectant female will have a Caesarean section and that the baby will survive?

4-37 Consider the set of integers 1, 2, 3, 4, and 5.

(a) One integer is selected at random. What is the probability that it is odd?

(b) Two integers are selected at random (one at a time with replacement, so that each of the five are available for second selection).

Find the probability that (1) neither are odd; (2) exactly one of them is odd; (3) both are odd.

Section 4–7 Combining the Rules of Probability

Many probability problems can be represented by tree diagrams. In these instances, the addition and multiplication rules can be applied quite readily.

To illustrate the use of tree diagrams in the solution of probability problems, let's use experiment 4–5. Two poker chips are drawn from a box containing one each of red, blue, and white chips. The tree diagram representing this experiment (figure 4–11) shows a first drawing and then a second drawing. One chip was drawn on each drawing and not replaced.

After the tree has been drawn and labeled, we need to assign probabilities to each branch of the tree. If we assume that it is equally likely that any chip would be drawn at each stage, we can assign a probability to each branch segment of the tree, as shown in figure 4–12. Notice that a set of branches that initiate from a single point have a total probability of 1. In this diagram there are four such sets of branch segments. The tree diagram shows six distinct outcomes. Reading down: branch (1) shows (R, B), branch (2) shows (R, W), and so on. (*Note*: Each outcome for the experiment is represented by a branch that begins at the common starting point and ends at the terminal points at the right.)

The probability associated with outcome (R, B), $P(R$ on 1st drawing and B on 2d drawing), is found by multiplying $P(R$ on 1st drawing) by $P(B$ on 2d drawing$|R$ on 1st drawing). These are the two probabilities $\frac{1}{3}$ and $\frac{1}{2}$ shown on the two branch segments of branch (1) in figure 4–12. The $\frac{1}{2}$ is the conditional probability asked for by the multiplication rule. Thus we will multiply along the branches.

Some events will be made up of more than one outcome from our experiment. For example, in the illustration above, suppose that we had asked

FIGURE 4–11
All Possible Combinations That Can Be Drawn

FIGURE 4–12
Probabilities of All Possible Combinations

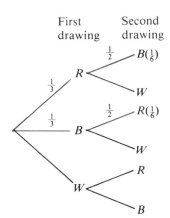

First drawing   Second drawing

FIGURE 4–13
Multiply Along the Branches

for the probability that one red chip and one blue chip are drawn. You will find two outcomes that satisfy this event, branch (1) or branch (3). With "or" we will use the addition rule (4–4b). Since the branches of a tree diagram represent mutually exclusive events, we have

$$P(\text{one } R \text{ and one } B) = \left(\frac{1}{3}\right)\left(\frac{1}{2}\right) + \left(\frac{1}{3}\right)\left(\frac{1}{2}\right) = \frac{1}{6} + \frac{1}{6} = \frac{1}{3}$$

NOTES:

1. Multiply along the branches (figure 4–13).
2. Add across the branches.

Now let's consider an example that places all the rules in perspective.

Illustration 4–10

A firm plans to test a new product in one randomly selected market area. The market areas can be categorized on the basis of location and population density. The number of markets in each category is presented in table 4–5.

TABLE 4–5
Number of Markets, by Location and by Population Density

	Population Density		
Location	Urban, U	Rural, R	Total
East, E	25	50	75
West, W	20	30	50
Total	45	80	125

What is the probability that the test market selected is in the East, $P(E)$? In the West, $P(W)$? What is the probability that the test market is in an urban area, $P(U)$? In a rural area, $P(R)$? What is the probability that the market is a western rural area, $P(W \text{ and } R)$? What is the probability it is an eastern or urban area, $P(E \text{ or } U)$? What is the probability that if it is in the East, it is an urban area, $P(U|E)$? Are "location" and "population density" independent? (What do we mean by independence or dependence in this situation?)

Solution   The first four probabilities, $P(E)$, $P(W)$, $P(U)$, and $P(R)$, represent "or" questions. For example, $P(E)$ means that the area is an eastern urban area or an eastern rural area. Since in this and the other three cases, the two components are mutually exclusive (an area can't be both urban and rural), the desired probabilities can be found by simply adding. In each case the probabilities are added across all the rows or columns of the table, therefore the totals are found in the total column or row.

$$P(E) = \frac{75}{125} \quad \text{(total for East divided by total number of markets)}$$

$$P(W) = \frac{50}{125} \quad \text{(total for West divided by total number of markets)}$$

$$P(U) = \frac{45}{125} \quad \text{(total for urban divided by total number of markets)}$$

$$P(R) = \frac{80}{125} \quad \text{(total for rural divided by total number of markets)}$$

Now we solve for $P(W$ and $R)$. There are 30 western rural markets and 125 markets total. Thus

$$P(W \text{ and } R) = \frac{30}{125}$$

Note that $P(W) \cdot P(R)$ does *not* give the right answer. [(50/125)(80/125) = 32/125.] Therefore, "location" and "population density" are dependent events.

$P(E$ or $U)$ can be solved in several different ways. The most direct way is to simply examine the table and count the number of markets that satisfy the condition that they are in the East or they are urban. We find 95 (25 + 50 + 20). Thus

$$P(E \text{ or } U) = \frac{95}{125}$$

Note that the first 25 markets were both in the East and urban; thus E and U are not mutually exclusive events.

Another way to solve for $P(E$ or $U)$ is to use the addition formula:

$$P(E \text{ or } U) = P(E) + P(U) - P(E \text{ and } U)$$

which yields

$$\frac{75}{125} + \frac{45}{125} - \frac{25}{125} = \frac{95}{125}$$

A third way to solve the problem is to recognize that the complement of $(E$ or $U)$ is $(W$ and $R)$. Thus $P(E$ or $U) = 1 - P(W$ and $R)$. Using the previous calculation, we get $1 - 30/125 = $ **95/125**.

Finally, we solve for $P(U|E)$. Looking at the table we see that there are 75 markets in the East. Of the 75 eastern markets, 25 are urban. Thus

$$P(U|E) = \frac{25}{75}$$

The conditional probability formula could also be used:

$$P(U|E) = \frac{P(U \text{ and } E)}{P(E)} = \frac{25/125}{75/125} = \frac{25}{75}$$

"Location" and "population density" are not independent events. They are dependent. This means that the probability of these events is affected by the occurrence of the other. □

While each rule for computing compound probabilities has been discussed separately, you should not think they are only used separately. In many cases they are combined to solve problems. Consider the following two illustrations.

Illustration 4–11

A production process produces light bulbs. On the average, 20% of all bulbs produced are defective. Each item is inspected before being shipped. The inspector misclasses an item 10% of the time; that is,

$$P(\text{classified good}|\text{defective item}) = P(\text{classified defective}|\text{good item})$$

$$= 0.10$$

What proportion of the items will be classified good?

Solution   What do we mean by the event "classified good"?

$$G = \text{item good}$$
$$D = \text{item defective}$$
$$CG = \text{item called good by inspector}$$
$$CD = \text{item called defective by inspector}$$

CG consists of two possibilities: "the item is good and is correctly classified good" or "the item is defective and is misclassified good." Thus

$$P(CG) = P[(CG \text{ and } G) \text{ or } (CG \text{ and } D)]$$

Since the two possibilities are mutually exclusive, we can start by using the addition rule, formula (4–4b).

$$P(CG) = P(CG \text{ and } G) + P(CG \text{ and } D)$$

The condition of a bulb and its classification by the inspector are not independent. The multiplication rule for dependent events must be used. Therefore,

$$P(CG) = [P(G) \cdot P(CG|G)] + [P(D) \cdot P(CG|D)]$$

Substituting the known probabilities, we get

$$P(CG) = [(0.8)(0.9)] + [(0.2)(0.1)] = 0.72 + 0.02 = \mathbf{0.74}$$

That is, 74% of the items are classified good. □

Illustration 4–12

Reconsider the previous example. Suppose only items that pass inspection are shipped. Items not classified good are scrapped. What is the quality of the shipped items? That is, what percentage of the items shipped are good, $P(G|CG)$?

Solution   Using the conditional probability formula (4–6),

$$P(G|CG) = \frac{P(G \text{ and } CG)}{P(CG)}$$

In illustration 4–11 we found $P(CG)$; $P(G \text{ and } CG)$ was also found in the solution of illustration 4–11. Thus

$$P(G|CG) = \frac{P(G) \cdot P(CG|G)}{P(CG)}$$

$$= \frac{(0.8)(0.9)}{0.74} = 0.9729 = \mathbf{0.973}$$

In other words, 97.3% of all items shipped will be good. Inspection increases the quality of items sold from 80% good to 97.3% good. □

Exercises

4–38   If $P(A) = 0.4$ and $P(B) = 0.5$, and if A and B are independent events, find $P(A \text{ or } B)$.

4–39 $P(G) = 0.5$, $P(H) = 0.4$, and $P(G \text{ and } H) = 0.1$ (see the diagram).

(a) Find $P(G|H)$.   (b) Find $P(H|G)$.   (c) Find $P(\bar{H})$.

(d) Find $P(G \text{ or } H)$.   (e) Find $P(G \text{ or } \bar{H})$.

(f) Are events G and H mutually exclusive? Explain.

(g) Are events G and H independent? Explain.

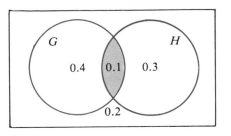

4-40 $P(R) = 0.5$, $P(S) = 0.3$, and events R and S are independent.

(a) Find $P(R \text{ and } S)$.   (b) Find $P(R \text{ or } S)$.   (c) Find $P(\bar{S})$.

(d) Find $P(R|S)$.   (e) Find $P(\bar{S}|R)$.

(f) Are events R and S mutually exclusive? Explain.

4-41 $P(M) = 0.3$, $P(N) = 0.4$, and events M and N are mutually exclusive.

(a) Find $P(M \text{ and } N)$.   (b) Find $P(M \text{ or } N)$.

(c) Find $P(M \text{ or } \bar{N})$.   (d) Find $P(M|N)$.

(e) Find $P(M|\bar{N})$.

(f) Are events M and N independent? Explain.

4-42   In a certain community, 90% of the individuals own a color television set. In this same community, 70% of the individuals own some type of stereo set. Seventy percent of the color television owners also own a stereo.

(a) What is the probability that a randomly selected individual will own both a color television and a stereo set?

(b) What is the probability that a randomly selected individual will own a color television or a stereo set?

4-43   Two flower seeds are randomly selected from a package that contains five seeds for red flowers and three seeds for white flowers.

(a) What is the probability that both seeds will result in red flowers?

(b) What is the probability that one of each color is selected?

(c) What is the probability that both seeds are for white flowers?

4-44 A, B, and C each in turn toss a balanced coin. The first one to throw a head wins.

(a) What are their respective chances of winning if each tosses only one time?

(b) What are their respective chances of winning if they continue given a maximum of two tosses each?

4-45   The probability that a certain door is locked is 0.6. The key to the door is one of five unidentified keys hanging on a key rack. Two keys are randomly selected before approaching the door. What is the probability that the door may be opened without returning for another key? (*Hint*: Draw a tree diagram.)

4-46 Suppose that when automobiles are spot-checked by a policeman for three safety features, they are found unsatisfactory with the following probabilities: tires unsafe, 15% of the time; exhaust system unsafe, 10% of the time; lights defective, 20% of the time. Assume independence (perhaps unreasonable) among these events and find the probability that a given car, when inspected,

(a) has all three items unsatisfactory

(b) has exactly two items unsatisfactory

(c) has exactly one item unsatisfactory

(d) passes the three-item inspection

4-47 Of the patients examined at a local clinic, 0.24 had high blood pressure, 0.36 had excessive weight, and 0.12 had both.

(a) If one of these patients is selected at random, what is the probability that he or she has at least one of these characteristics?

(b) Are the events "have high blood pressure" and "have excessive weight" independent? Explain.

4-48 New Motor Car Company has raised the price of its cars four out of the past five years. When New Motor Car raised its price, its chief competitor, Better Car Manufacturers, correspondingly raised its price three out of four times. Better Car has never raised its price unless New Motor Car first increased its price. Assume future events follow past trends.

(a) What is the probability that New Motor Car Company raises its price next year?

(b) What is the probability that New Motor Car and Better Car both raise their prices next year?

(c) What is the probability that only Better Car Manufacturers raises its price?

(d) What is the probability that New Motor Car or Better Car raises its price next year?

(e) What is the probability that neither one raises its price next year?

(f) Are the events "New Car raises its price" and "Better Car raises its price" independent? Mutually exclusive? Explain.

In Retrospect

You have now studied the basic concepts of probability. These fundamentals need to be understood to allow us to continue our study of statistics. Probability is the vehicle of statistics, and we have begun to see how probabilistic events occur. We have explored theoretical and experimental probabilities for the same event. Does the experimental probability turn out to have the same value as the theoretical? Not exactly, but over the long run we have seen that it does have approximately the same value.

You must, of course, know and understand the basic definition of probability as well as understanding the properties of mutual exclusiveness and independence as they apply to the concepts presented in this chapter.

From reading this chapter you are expected to be able to relate the "and" and the "or" of compound events to the multiplication and addition rules. You should also be able to calculate conditional probabilities.

In the next three chapters we will look at distributions associated with probabilistic events. This will prepare us for the statistics that will follow. We must be able to predict the variability that the sample will show with respect to the population before we will be successful at "inferential statistics," in which we describe the population based on the sample statistics available.

Chapter Exercises

4-49 Explain briefly how you would decide which is the more unusual of the following two events:

A: a 90° day in Vermont

B: a 100° day in Florida

4-50 A fair coin is tossed 100 million times. Let B be the total number of heads observed. Identify each of the following statements as true or false. Explain your answer (no computations required).

(a) It is very probable that B is very close (within a few thousand) to 50 million.

(b) It is very probable that $\frac{B}{100}$ million is very close to $\frac{1}{2}$; maybe between 0.49 and 0.51.

(c) Since Hs and Ts fall with complete irregularity, one cannot say anything about what happens in 100 million tosses. It is equally likely that B is equal to one value as to any other value; for example, $P(B = 2) - P(B = 50$ million).

(d) Hs and Ts fall about equally often. Therefore, if on the first 10 tosses we get all Hs, it is more probable that the eleventh toss will yield a T than a H.

4-51 Suppose a certain ophthalmic trait is associated with eye color. Three hundred randomly selected individuals are studied with results as follows:

Eye Color

Trait	Blue	Brown	Other	Totals
Yes	70	30	20	120
No	20	110	50	180
Totals	90	140	70	300

(a) What is the probability that a person selected at random has blue eyes?

(b) What is the probability that a person selected at random has the trait?

(c) Are events A (has blue eyes) and B (has the trait) independent? Justify your answer.

(d) How are the two events A (has blue eyes) and C (has brown eyes) related (independent, mutually exclusive, complementary, or all inclusive)? Explain why or why not each term applies.

4-52 Events R and S are defined on the same sample space. If $P(R) = 0.2$ and $P(S) = 0.5$, explain why each of the following statements are either true or false.

(a) If R and S are mutually exclusive, then $P(R \text{ or } S) = 0.10$.

(b) If R and S are independent, then $P(R \text{ or } S) = 0.6$.

(c) If R and S are mutually exclusive, then $P(R \text{ and } S) = 0.7$.

(d) If R and S are mutually exclusive, then $P(R \text{ or } S) = 0.6$.

4-53 Let's assume that there are three traffic lights between your house and a friend's house. As you arrive at each light, it may be red (R) or green (G).

(a) List the sample space showing all possible sequences of red and green lights that could occur on a trip from your house to your friend's. (RGG represents red at the first light and green at the other two.) Assuming that each element of the sample space is equally likely to occur:

(b) What is the probability that on your next trip to your friend's house you will have to stop for exactly one red light?

(c) What is the probability that you will have to stop for at least one red light?

4-54 The probability that person A will be alive in 20 years is 0.6 and the probability that person B will be alive in 20 years is 0.85.

(a) What is the probability that both will be alive 20 years from now?
(b) What is the probability that neither will be alive 20 years from now?

(c) To answer parts (a) and (b), you must make what assumption about the two events "A is alive" and "B is alive" 20 years from now?

4-55 Suppose that when a job candidate comes to interview for a job at RJB Enterprises, the probability that the candidate will want the job (A) after the interview is 0.68. Also, the probability that RJB wants the candidate (B) is 0.36. The probability $P(A \mid B)$ is 0.88.

(a) Find $P(A \text{ and } B)$.   (b) Find $P(B \mid A)$.

(c) Are events A and B independent? Explain.

(d) Are events A and B mutually exclusive? Explain.

(e) What would it mean to say A and B are mutually exclusive events in this exercise?

4-56 A traffic analysis at a busy traffic circle in Washington, D.C. showed that 0.8 of the autos using the circle entered from Connecticut Avenue. Of those entering the traffic circle from Connecticut Avenue, 0.7 continued on

Connecticut Avenue at the opposite side of the circle. What is the probability that a randomly selected auto observed in the traffic circle entered from Connecticut and will continue on Connecticut?

4–57 Three flower seeds are randomly selected from a package that contains eight seeds for yellow flowers, eight seeds for red flowers, and four seeds for white flowers.

(a) What is the probability that all three seeds will produce yellow flowers?

(b) What is the probability that the three seeds selected are all for the same flower color?

4–58 The probabilities of females being employed in certain occupational groups are listed below.

Occupational Group	Probability
Professional—technical	0.16
Managerial—administrative	0.06
Sales	0.07
Clerical	0.34
Craft	0.02
Operatives, including transport	0.12
Non-farm laborers	0.01
Service, except private households	0.18
Private households	0.03
Farm	0.01
Total	1.00

(a) What is the probability that a randomly selected female is either a professional or a managerial employee?

(b) What is the probability that two randomly selected females will both be clerical employees?

(c) What is the probability that a randomly selected female will be both a sales and a clerical employee?

4–59 At a certain college, 15% of the students received an A in their required mathematics course, 10% received an A in their required chemistry course, and 5% received an A in both. A student is selected at random.

(a) What is the probability that he received an A in either chemistry or math?

(b) What is the probability that he received an A in chemistry, if he had an A in math?

(c) What is the probability that he received an A in math, if he had an A in chemistry?

(d) Are the events "A in math" and "A in chemistry" independent? Explain.

(*Hint*: Use a Venn diagram.)

4–60 One thousand persons screened for a certain disease are given a clinical exam. As a result of the exam, the sample of 1000 persons is distributed according to height and disease status.

Height	None	Mild	Moderate	Severe	Totals
Tall	122	78	139	61	400
Medium	74	51	90	35	250
Short	104	71	121	54	350
Totals	300	200	350	150	1000

Disease Status

Use the above information to estimate the probability of being medium or short in height and of having moderate or severe disease status.

4–61 Tires salvaged from a train wreck are on sale at the Getrich Tire Company. Of the 15 tires offered in the sale five tires have suffered internal damage and the remaining ten are damage free. If you were to randomly select and purchase two of these tires:

(a) What is the probability that the tires you purchase are both damage free?

(b) What is the probability that exactly one of the tires you purchase is damage free?

(c) What is the probability that at least one of the tires you purchase is damage free?

4–62 A supermarket is planning a sale of unmarked boxes of cookies, pretzels, and crackers. The mixed unmarked lot contains 300 boxes of cookies, 300 boxes of pretzels, and 400 boxes of crackers. Consider the first box selected by the first shopper and answer the following:

(a) What is the probability of selecting a box of pretzels?

(b) What is the probability of not getting a box of cookies?

(c) Are the two events "pretzels" and "not cookies" mutually exclusive events?

(d) Are the two events "pretzels" and "cookies" mutually exclusive?

(e) Are the two events "pretzels" and "cookies" independent?

(f) Are the two events "pretzels" and "not cookies" independent?

4–63 A shipment of grapefruit arrived containing the following proportions of types: 10% pink seedless, 20% white seedless, 30% pink with seeds, 40% white with seeds. A grapefruit is selected at random from the shipment. Find these probabilities:

(a) that it is seedless   (b) that it is white

(c) that it is pink and seedless   (d) that it is pink or seedless

(e) that it is pink, given that it is seedless

(f) that it is seedless, given that it is pink

4-64 The accompanying table shows the sentiments of 2500 wage-earning employees at the Spruce Company on a proposal to emphasize fringe benefits rather than wage increases during their impending contract discussions.

| | | Opinion | | |
Employee	Favor	Neutral	Opposed	Total
Male	800	200	500	1500
Female	400	100	500	1000
Total	1200	300	1000	2500

(a) Calculate the probability that an employee selected at random from this group will be opposed.

(b) Calculate the probability that an employee selected at random from this group will be female.

(c) Calculate the probability that an employee selected at random from this group will be opposed, given that the person is male.

(d) Are the events "opposed" and "female" independent? Explain.

4-65 The following information was generated from a recent survey of females employed by a predominantly female company when asked the question "Is college more important for a male than a female?"

Is College More Important for a Male than a Female?

Age	Yes	No	Total
18–24	0.17	0.43	0.60
Over 24	0.26	0.14	0.40
Total	0.43	0.57	1.00

Using this information:

(a) What is the probability that a randomly selected female employee felt college was more important for a male than a female?

(b) What is the probability that a randomly selected female answered "Yes" or was "over 24"?

(c) What is the probability that a randomly selected female was "over 24" given that she answered "No" to the question?

4-66 Use the information reported in the news article at the beginning of this chapter. Estimate the probability that a person will favor the proposed state laws prohibiting smoking in all public places, provided the person is picked at random from the registered voters who

(a) live in New York State   (b) live in New York City

(c) are female

4-67 A person owns both his own business and his own home and purchases a variety of insurances to protect them. In any one year the probability of an insurance claim on the business is 0.4 and on the home is 0.05. Assuming that these are independent events, find the probability that this person will have an insurance claim at

(a) both the business and at home

(b) exactly one place   (c) neither place

4-68 A box contains ten ball-point pens, eight of which write properly and two of which do not write. You are going to randomly select two of these pens.

(a) Describe the sample space with a tree diagram.

(b) Assign probabilities to each branch of your tree diagram. Consider these three events: A, a good pen is drawn on the first drawing; B, a good pen is drawn on the second drawing; C, at least one defective pen is selected. Find these probabilities:

(c) $P(A)$ (d) $P(B)$ (e) $P(C)$

(f) $P(\bar{A})$ (g) $P(\bar{B})$ (h) $P(\bar{C})$

(i) $P(A \text{ and } B)$ (j) $P(A \text{ and } C)$ (k) $P(B \text{ and } C)$

(l) $P(A \text{ or } B)$ (m) $P(A \text{ or } C)$ (n) $P(B \text{ or } C)$

(o) Are events A and B mutually exclusive? Explain.

(p) Are events A and C mutually exclusive? Explain.

(q) Are events A and B independent? Explain.

(r) Are events A and C independent? Explain.

4-69 In order to fly safely, a four-engine airliner needs at least three engines operating properly. If the probability that an individual engine operates properly for an entire flight is 0.9 and the engines operate independently, what is the probability that the engines operate successfully for a safe flight?

4-70 According to automobile accident statistics, one out of every six accidents results in an insurance claim of $100 or less in property damage. Three cars insured by an insurance company are involved in different accidents. Consider the following two events:

     A: The majority of claims exceed $100.

     B: Exactly two claims are $100 or less.

(a) List the sample points for this experiment.

(b) Are the sample points equally likely?

(c) Find $P(A)$ and $P(B)$.

(d) Are A and B independent? Justify your answer.

4-71 Salesman Adams and saleswoman Jones call on 3 and 4 customers, respectively, on a given day. Adams could make 0, 1, 2, or 3 sales, while Jones could make 0, 1, 2, 3, or 4 sales. The sample space listing the number of possible sales for each person on a given day is given in the accompanying table.

| | Jones | | | | |
Adams	0	1	2	3	4
0	0, 0	1, 0	2, 0	3, 0	4, 0
1	0, 1	1, 1	2, 1	3, 1	4, 1
2	0, 2	1, 2	2, 2	3, 2	4, 2
3	0, 3	1, 3	2, 3	3, 3	4, 3

(3, 1 stands for 3 sales by Jones and 1 sale by Adams.) Assume that each sample point is equally likely.

Let's define these events:

A = at least one of the salespersons makes no sales

B = together they made exactly three sales

C = each made the same number of sales

D = Adams made exactly one sale

Find the following probabilities by *counting* sample points.

(a) $P(A)$ (b) $P(B)$ (c) $P(C)$

(d) $P(D)$ (e) $P(A \text{ and } B)$ (f) $P(B \text{ and } C)$

(g) $P(A \text{ or } B)$ (h) $P(B \text{ or } C)$ (i) $P(A|B)$

(j) $P(B|D)$ (k) $P(C|B)$ (l) $P(B|\bar{A})$

(m) $P(C|\bar{A})$ (n) $P(A \text{ or } B \text{ or } C)$

Are the following pairs of events mutually exclusive? Explain.

(o) A and B (p) B and C (q) B and D

Are the following pairs of events independent? Explain.

(r) A and B (s) B and C (t) B and D

4-72 A certain coin is biased in favor of heads with the odds being 3 to 1. This coin is tossed three times, find the probability of:

(a) obtaining exactly 1 head in the three tosses

(b) obtaining exactly 2 heads in the three tosses

(c) obtaining exactly 3 heads in the three tosses

(d) at least 1 head in the three tosses

(e) no more than 2 heads in the three tosses

4–73 A testing organization wishes to rate a particular brand of television. Six TVs are selected at random from stock. If nothing is found wrong with any of the six, the brand is judged satisfactory.

(a) What is the probability that the brand will be rated satisfactory if 10 percent of the TVs actually are defective?

(b) What is the probability that the brand will be rated satisfactory if 20 percent of the TVs actually are defective?

(c) What is the probability that the brand will be rated satisfactory if 40 percent of the TVs actually are defective?

4–74 Coin A is loaded in such a way that P(heads) is 0.6. Coin B is a balanced coin. Both coins are tossed. Find the following:

(a) the sample space that represents this experiment and assign a probability measure to each outcome

(b) P(both show heads)

(c) P(exactly one head shows)

(d) P(neither coin shows a head)

(e) P(both show heads|coin A shows a head)

(f) P(both show heads|coin B shows a head)

(g) P(heads on coin A|exactly one head shows)

4–75 On a slot machine there are three reels with digits 0, 1, 2, 3, 4, and 5 and a flower on each reel. When a coin is inserted and the lever pulled, each of the three reels spins independently and comes to rest on one of the seven positions mentioned. Find these probabilities:

(a) that a flower shows on all three reels

(b) that a flower shows on exactly one reel

(c) that a flower shows on exactly two reels

(d) that no flowers show

(e) that a three-digit sequence (in order) appears

(f) that exactly one flower shows and the other two are odd integers

4–76 Consider an experiment where the die to be used is weighted in a manner that causes each number's probability to be proportional to its number [that is, $P(2) = 2P(1)$, $P(3) = 3P(1)$, and so on].

(a) List the sample space.

(b) Assign probabilities to each outcome.

(c) Find P (even number).   (d) Find $P(4)$.

(e) Find $P(4|\text{even})$.

(f) Are events "even" and "4" independent? Explain.

(g) Are events "even" and "4" mutually exclusive? Explain.

4-77 Professor French forgets to set his alarm with a probability of 0.3. If he sets the alarm it rings with a probability of 0.8. If the alarm rings, it will wake him on time to make his first class with a probability of 0.9. If the alarm does not ring, he wakes in time for his first class with a probability of 0.2. What is the probability that Professor French wakes in time to make his first class tomorrow?

4-78 The probability that a married man regularly watches professional football on television is 0.6. The probability that a married woman regularly watches is 0.2. The probability that a married man watches regularly given that his wife does is 0.9. Find these probabilities:

(a) that they both watch regularly

(b) that the wife watches regularly knowing that the husband watches regularly

(c) that neither watch regularly

4-79 A two-page typed report contains an error on one of the pages. The proof-readers review the copy. Each has an 80% chance of catching the error. What is the probability that the error will be identified if

(a) each reads a different page?

(b) they each read both pages?

(c) the first proofreader randomly selects a page to read, then the second proofreader randomly selects a page unaware of which page the first selected?

Vocabulary List

Be able to define each term. In addition, describe, in your own words, and give an example of each term. Your examples should not be ones given in class or in the textbook.

The terms followed by bracketed numbers were defined in earlier chapters (the bracketed numbers are those of the earlier chapters), but we ask you to define them again to show your increased understanding of their meaning.

addition rule
all-inclusive events
complementary event
compound event
conditional probability
dependent events
empirical probability
equally likely events
event
experiment [1]
experimental probability
independent events
intersection
law of large numbers
listing

long-term average
multiplication rule
mutually exclusive events
odds
outcome
probability of an event
relative frequency [2]
sample point
sample space
subjective probability
theoretical probability
tree diagram
Venn diagram

Quiz

Answer " True" if the statement is always true. If the statement is not always true, replace the words shown in color with words that make the statement always true.

4–1 The probability of an event is a **whole number**.

4–2 The concepts of probability and relative frequency as related to an event are very **similar**.

4–3 The **sample space** is the theoretical population for probability problems.

4–4 The sample points of a sample space are **equally likely** events.

4–5 The value found for experimental probability will **always** be exactly equal to the theoretical probability assigned to the same event.

4–6 The probabilities of complementary events always **are equal**.

4–7 If two events are mutually exclusive, they are also **independent**.

4–8 If events A and B are **mutually exclusive**, the sum of their probabilities must be exactly 1.

4–9 If the sets of sample points belonging to two different events do not intersect, the events are **independent**.

4–10 A compound event formed by use of the word "and" requires the use of the **addition rule**.

5 Probability Distributions (Discrete Variables)

Alphanumeric Terminal Market Overview

PERIPHERALS DIGEST

The alphanumeric terminal market had a relatively calm year. Competition was tougher than 1981, but technical innovations were fewer and less spectacular than in previous years. The definitions that have characterized terminal types are becoming increasing blurred. With distribution channels becoming more numerous, thanks to retailers, mail-order suppliers and industrial distributors, market analysts have a tougher time, but buyers have more options.

The terms "dumb terminal" and "low-end terminal" used to mean the same thing, but because of rampant price cutting, the less-than-$1000 low end now encompasses a number of smart or editing terminals. At the high-end, distinctions between intelligent terminals and desk-top microcomputers blur as more micros gain flexible communications capabilities and as micro sales induce terminal vendors to re-label their high-end terminals as microcomputer systems.

Alphanumeric terminal price distribution

Data: Mini-Micro Systems

Terminal prices are falling and becoming more homogeneous. *Taken from the product tables that follow, these data mirror the intense competition in the dumb and low-end editing markets in which features are standardized, and price differentials of only a few percentage points affect large- and small-volume buying decisions.*

Chapter Objectives

Chapter 2 dealt with frequency distributions of data sets and chapter 4 with the fundamentals of probability. Now we are ready to combine these ideas to form probability distributions, which are much like relative frequency distributions. The basic difference between probability and relative frequency distributions is the use of the random variable. The random variable of a probability distribution corresponds to the response variable of a frequency distribution.

In this chapter we will investigate discrete probability distributions and study measures of central tendency and dispersion for such distributions. Special emphasis will be given to the binomial random variable and its probability distribution since it is the most important discrete random variable encountered in most fields of application.

Section 5-1 Random Variables

The events in a probability experiment must be defined in such a manner that each event can be assigned a numerical value. This numerical value will be the value of the *random variable* under study.

random variable

Random Variable

A variable that assumes a unique numerical value for each of the outcomes in the sample space of a probability experiment.

In other words, a random variable is used to denote the outcome of a probability experiment. It can take on any numerical value that belongs to the set of all possible outcomes of the experiment. (It is called "random" because the value it assumes is the result of a chance or random event.)

Each event in a probability experiment must also be defined in such a way that only one value of the random variable is assigned to it, and each event must have a value assigned to it. Typically, the **discrete random variable** is a count of something.

discrete random variable

The following illustrations demonstrate what we mean by random variable.

Illustration 5-1

We toss five coins and observe the number of heads visible. The random variable x is the number of heads observed and may take on integer values from 0 to 5. □

Illustration 5-2

In experiment 4-7 we rolled two dice and observed the sum of the dots on both dice. If a random variable had been assigned, it would have been the total number of dots showing. In this case the random variable x could take on integer values from 2 to 12. □

Illustration 5–3

Let the number of phone calls received per day by a company be a random variable. It could take on integer values ranging from zero to some very large number. □

Illustration 5–4

The age of automobiles owned by the faculty of our college could be used as a random variable. An observed probability distribution would result if the data were presented as a relative frequency distribution.

For example, suppose we find that 0.20 of the faculty-owned cars are new this year, 0.25 are last year's models, 0.28 are two years old, 0.15 are three years old, and so on; see figure 5–1. The random variable "age of the automobile" is actually a continuous random variable since it measures time. It only appears to be discrete because of the context in which it is presented. Continuous random variables and their distributions are the topics of chapter 6.

FIGURE 5–1
*Age of Automobiles Owned
by Our College Faculty*

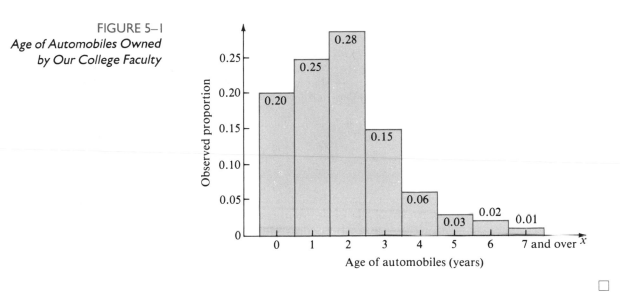

Exercises

5–1 Consider the experiment of tossing a coin four times and constructing its sample space. Suppose that we define the random variable x to be the number of heads seen.

(a) Make a list of the possible values for x.

(b) Assign a value of the random variable to each of the sixteen outcomes making up your sample space.

(c) Is x a discrete or a continuous variable?

5–2 Two dice are rolled and the random variable of interest is the total number of dots seen on the two dice.

(a) Make a list of the various values that the random variable can attain.

(b) Use the sample space shown in figure 4–4 to assign a value of the random variable to each sample point.

(c) Is this random variable discrete or continuous?

5–3 An archer shoots eight arrows at a bull's-eye of one target. Identify the random variable most likely to be of interest and list its possible values.

5–4 Radios are packed in cartons of 12. A carton of radios is inspected and the number of defective radios found is recorded. Identify the random variable used and list its possible values.

Section 5–2 Probability Distributions of a Discrete Random Variable

Recall the experiment of tossing two coins that was used at the beginning of section 4–1. Two coins were tossed and no heads, one head, or two heads were observed. If we define the random variable x to be the number of heads that are observed when two coins are tossed, x can take on the values 0, 1, or 2. The probability of each of these three events is the same as we calculated in chapter 4.

$$P(x = 0) = P(0H) = \tfrac{1}{4}$$

$$P(x = 1) = P(1H) = \tfrac{1}{2}$$

$$P(x = 2) = P(2H) = \tfrac{1}{4}$$

These probabilities can be listed in any number of ways, but they are best displayed in the form illustrated by table 5–1. Can you see why the name "probability distribution" is used?

TABLE 5–1
Probability Distribution;
Tossing a Coin

x	$P(x)$
0	$\tfrac{1}{4}$
1	$\tfrac{1}{2}$
2	$\tfrac{1}{4}$

probability distribution

Probability Distribution

A distribution of the probabilities associated with each of the values of a random variable. The probability distribution represents a theoretical population.

In the experiment in which a single die is rolled and the number of dots on the top surface is observed, the random variable is the number observed. The probability distribution for this random variable is shown in table 5–2.

TABLE 5–2
Probability Distribution; Rolling a Die

x	P(x)
1	$\frac{1}{6}$
2	$\frac{1}{6}$
3	$\frac{1}{6}$
4	$\frac{1}{6}$
5	$\frac{1}{6}$
6	$\frac{1}{6}$

Sometimes it is convenient to write a rule that expresses the probability of an event in terms of the value of the random variable. This expression is typically written in formula form and is called a "probability function."

probability function

Probability Function

A rule that assigns probabilities to the values of the random variables.

A probability function can be as simple as a list pairing the values of a random variable with their probabilities. Tables 5–1 and 5–2 show two such listings. However, a probability function is most often expressed in formula form.

Consider a die that has been modified so it has one face with one dot, two faces with two dots, and three faces with three dots. Let x be the number of dots observed when this die is rolled. The probability distribution for this experiment is presented in table 5–3. Each of the probabilities can be represented by the value of x divided by 6. That is, each $P(x)$ is equal to the value of x divided by 6, where $x = 1, 2,$ or 3. Thus $P(x) = x/6$, for $x = 1, 2,$ and 3, is the formula expression of the probability function of this experiment.

TABLE 5–3
Probability Distribution; Rolling the Modified Die

x	P(x)
1	$\frac{1}{6}$
2	$\frac{2}{6}$
3	$\frac{3}{6}$

The probability function of the experiment of rolling one ordinary die is $P(x) = \frac{1}{6}$ for $x = 1, 2, 3, 4, 5,$ and 6. This particular function is called a **constant function** because the value of $P(x)$ does not change as x changes.

Every probability function must display the two basic properties of probability. These two properties are (1) the probability assigned to each value of the random variable must be between 0 and 1 inclusive, that is,

$$0 \leq \text{each } P(x) \leq 1$$

and (2) the sum of the probabilities assigned to all the values of the random variable must equal 1, that is,

$$\sum_{\text{all } x} P(x) = 1$$

Is $P(x) = x/10$, for $x = 1, 2, 3$, and 4, a probability function? To answer this question we need only test the function in terms of the two basic properties. The probability distribution is shown in table 5–4. Property 1 is

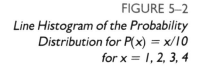

x	P(x)
1	$\frac{1}{10}$
2	$\frac{2}{10}$
3	$\frac{3}{10}$
4	$\frac{4}{10}$
Total	$\frac{10}{10} = 1$

satisfied, since $\frac{1}{10}$, $\frac{2}{10}$, $\frac{3}{10}$, and $\frac{4}{10}$ are all numerical values between 0 and 1. Property 2 is also satisfied, since the sum of all four probabilities is exactly 1. Since both properties are satisfied, we can conclude that $P(x) = x/10$ for $x = 1, 2, 3, 4$ *is* a probability function.

What about $x = 5$ (or any value other than 1, 2, 3, or 4) in the function $P(x) = x/10$ for $x = 1, 2, 3$, and 4? $P(x = 5)$ is considered to be zero. That is, the probability function provides a probability of zero for all values of x other than the values specified.

Figure 5–2 presents a probability distribution graphically. Regardless of the specific representation, the values of the random variable are plotted

FIGURE 5–2

Line Histogram of the Probability Distribution for P(x) = x/10 for x = 1, 2, 3, 4

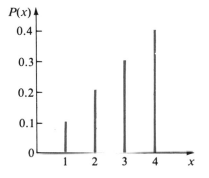

on the horizontal scale, and the probability associated with each value of the random variable is plotted on the vertical scale. A discrete random variable should really be presented by a **line histogram**, since the variable can assume only discrete values. Figure 5–2 shows the probability distribution of $P(x) = x/10$ for $x = 1, 2, 3$, and 4. This line representation makes sense because the variable x can assume *only* the values 1, 2, 3, and 4. The length of the vertical line represents the value of the probability.

line histogram

probability histogram

However, a regular **histogram** is more frequently used to present probability distributions. Figure 5–3 shows the same probability distribution of figure 5–2 but in histogram form. This representation suggests that x can assume all numerical values (fractions, decimals, and so on). The histogram probability distribution uses the physical area of each "bar" to represent its assigned probability. The bar for $x = 2$ is 1 unit wide (from 1.5 to 2.5) and is 0.2 unit high. Therefore, its area is $(1)(0.2) = 0.2$, the probability assigned to $x = 2$. The areas of the other bars can be determined in similar fashion. This area representation will be an important concept in chapter 6 when we begin to work with continuous random variables.

FIGURE 5–3
Histogram of the Probability Distribution of P(x) = x/10 for x = 1, 2, 3, 4

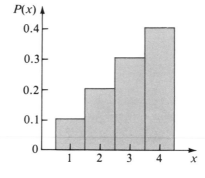

Exercises

5–5   Test the following function to determine whether it is a probability function. If it is not, try to make it into a probability function. List the distribution of probabilities and sketch a histogram.

$$P(x) = \frac{5 - x}{10} \quad \text{for} \quad x = 1, 2, 3, 4$$

5–6   Test the following function to determine whether it is a probability function. If it is not, try to make it into a probability function. List the distribution of probabilities and sketch a histogram.

$$Q(x) = \frac{x^2 - 1}{40} \quad \text{for} \quad x = 2, 3, 4, 5$$

5-7 Test the following function to determine whether it is a probability function. If it is not, try to make it into a probability function. $R(x) = 0.2$ for $x = 0, 1, 2, 3, 4$.

(a) List the distribution of probabilities.

(b) Sketch a histogram.

5-8 Test the following function to determine whether it is a probability function. If it is not, try to make it into a probability function.

$$S(x) = \frac{6 - |x - 7|}{36} \quad \text{for} \quad x = 2, 3, 4, 5, 6, 7, \ldots, 11, 12$$

(a) List the distribution of probabilities and sketch a histogram.

(b) Do you recognize $S(x)$? If so, identify it.

Section 5-3 Mean and Variance of a Discrete Probability Distribution

Recall that the mean of a frequency distribution is

$$\bar{x} = \frac{\sum xf}{n}$$

which also can be written as

$$\bar{x} = \sum \left(x \cdot \frac{f}{n} \right)$$

We learned in our study of probability that the probability of an event is the expected relative frequency of its occurrence. Thus if we replace f/n with $P(x)$, we can find the mean of a theoretical probability distribution:

$$\text{mean value of } x = \sum [x \cdot P(x)] \tag{5-1}$$

NOTES:

1. $\bar{x}$ is the mean of a sample.
2. s is the standard deviation of the individual elements of the sample.
3. $\bar{x}$ and s are called **sample statistics**.
4. μ (Greek letter mu) is the mean of the population under consideration.
5. σ (Greek letter sigma) is the standard deviation of the individual elements of the population under consideration.
6. μ and σ are called **population parameters**. (A parameter is a constant. μ and σ are typically unknown values.)

Recall that the "expected relative frequency" represents what will occur in the long run. The mean of x is therefore the mean of the entire population of experimental outcomes. The symbol for the **mean** value of x (in a probability distribution) is μ and formula (5–1) can be written as

mean of probability distribution

$$\mu = \sum [x \cdot P(x)] \tag{5–2}$$

Illustration 5–5

Let's return to the previous probability function: $P(x) = x/10$ for $x = 1, 2, 3,$ and 4. We can find its mean by use of formula (5–2) after we compile a probability distribution table (see table 5–5). The mean μ, or the sum of the products of x times $P(x)$, is $\frac{30}{10}$, or **3.0**. [The sum of the probabilities, $\sum P(x)$, must be 1.0. Use this as a check.]

TABLE 5–5
Probability Distribution;
$P(x) = x/10$, $x = 1, 2, 3, 4$

	$P(x)$	$x \cdot P(x)$
1	$\frac{1}{10}$	$\frac{1}{10}$
2	$\frac{2}{10}$	$\frac{4}{10}$
3	$\frac{3}{10}$	$\frac{9}{10}$
4	$\frac{4}{10}$	$\frac{16}{10}$
Total	$\frac{10}{10} = 1.0$ Ⓒⓚ	$\mu = \frac{30}{10} = 3.0$

variance of probability distribution

The **variance** of discrete probability distributions is defined in much the same way as the variance of sample data.

$$\sigma^2 = \sum [(x - \mu)^2 \cdot P(x)] \tag{5–3}$$

The variance of x for the probability distribution discussed in illustration 5–5 is found as shown in table 5–6. The variance σ^2 is $\frac{10}{10}$, or **1.0**. ($\mu = 3$ was found in table 5–5.)

Formula (5–3) is often not convenient to use; fortunately it can be reworked to appear in the form

$$\sigma^2 = \sum [x^2 \cdot P(x)] - \{\sum [x \cdot P(x)]\}^2 \tag{5–4a}$$

or

$$\sigma^2 = \sum [x^2 P(x)] - \mu^2 \tag{5–4b}$$

TABLE 5–6
Finding the Variance for
$P(x) = x/10$, $x = 1, 2, 3, 4$

x	$x - \mu$	$(x - \mu)^2$	$P(x)$	$(x - \mu)^2 P(x)$
1	-2	4	$\frac{1}{10}$	$\frac{4}{10}$
2	-1	1	$\frac{2}{10}$	$\frac{2}{10}$
3	0	0	$\frac{3}{10}$	$\frac{0}{10}$
4	1	1	$\frac{4}{10}$	$\frac{4}{10}$
Total				$\sigma^2 = \frac{10}{10}$

It is left for you (exercise 5–14) to verify formula (5–4a) and (5–4b) are equivalent to formula (5–3). □

Illustration 5–6

To find the variance of the probability distribution in illustration 5–5 by using formula (5–4a), we will need to add two additional columns to table 5–5. These columns are shown in table 5–7.

TABLE 5–7
Calculations Needed to Find the
Variance for $P(x) = x/10$

x	$P(x)$	$x \cdot P(x)$	x^2	$x^2 \cdot P(x)$
1	$\frac{1}{10}$	$\frac{1}{10}$	1	$\frac{1}{10}$
2	$\frac{2}{10}$	$\frac{4}{10}$	4	$\frac{8}{10}$
3	$\frac{3}{10}$	$\frac{9}{10}$	9	$\frac{27}{10}$
4	$\frac{4}{10}$	$\frac{16}{10}$	16	$\frac{64}{10}$
Total	$\frac{10}{10}$ Ⓒⓚ	$\frac{30}{10}$		$\frac{100}{10}$

The variance is

$$\sigma^2 = \sum [x^2 \cdot P(x)] - \{\sum [x \cdot P(x)]\}^2$$

$$= \frac{100}{10} - \left(\frac{30}{10}\right)^2 = 10 - (3)^2 = 10 - 9 = 1.0 \qquad \square$$

**standard deviation
of probability
distribution**

The **standard deviation** is the square root of variance; therefore, $\sigma = 1.0$.

Illustration 5–7

A coin is tossed three times. Let the number of heads occurring in those three tosses be the random variable x, which can take on the values 0, 1, 2, or 3. There are eight possible outcomes to this experiment: one results in $x = 0$, three in $x = 1$, three in $x = 2$, and one in $x = 3$. Therefore, the probabilities for this random variable are $\frac{1}{8}$, $\frac{3}{8}$, $\frac{3}{8}$, and $\frac{1}{8}$. The probability distribution associated with this experiment is shown in figure 5–4 and in

FIGURE 5–4
Probability Distribution
for Illustration 5–7

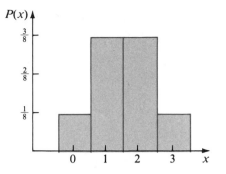

table 5–8. The necessary extensions and summations for the calculation of its mean and standard deviation are also shown in table 5–8.

TABLE 5–8
Probability Distribution and
Extensions, Illustration 5–7

x	$P(x)$	$x \cdot P(x)$	x^2	$x^2 \cdot P(x)$
0	$\frac{1}{8}$	$\frac{0}{8}$	0	$\frac{0}{8}$
1	$\frac{3}{8}$	$\frac{3}{8}$	1	$\frac{3}{8}$
2	$\frac{3}{8}$	$\frac{6}{8}$	4	$\frac{12}{8}$
3	$\frac{1}{8}$	$\frac{3}{8}$	9	$\frac{9}{8}$
Total	$\frac{8}{8}$ Ⓒ𝕂	$\mu = \frac{12}{8} = 1.5$		$\frac{24}{8} = 3.0$

The mean is found with the aid of formula (5–2):

$$\mu = \sum [x \cdot P(x)] = 1.5$$

This result, 1.5, is the mean number of heads expected per experiment. The variance is found with the aid of formula (5–4a):

$$\sigma^2 = \sum [x^2 \cdot P(x)] - \{\sum [x \cdot P(x)]\}^2$$
$$= 3.0 - (1.5)^2 = 3.0 - 2.25 = 0.75$$

The standard deviation is the positive square root of the variance:

$$\sigma = \sqrt{0.75} = 0.866 = 0.87$$

That is, 0.87 is the standard deviation expected among the number of heads observed per experiment. ☐

Exercises

5-9 Given the probability function

$$P(x) = \frac{5 - x}{10} \quad \text{for} \quad x = 1, 2, 3, 4,$$

find the mean and standard deviation.

5-10 Given the probability function

$$R(x) = 0.2 \quad \text{for} \quad x = 0, 1, 2, 3, 4,$$

find the mean and standard deviation.

5-11 Given the probability function

$$S(x) = \frac{6 - |x - 7|}{36} \quad \text{for} \quad x = 2, 3, 4, 5, 6, 7, \ldots, 11, 12,$$

find the mean and standard deviation.

5-12 Based on past history, the distribution of sales of 1-pound containers of cottage cheese at a convenience store are the following:

Number of Cartons Sold per Day, x	Probability
10	0.2
11	0.4
12	0.2
13	0.1
14	0.1
Total	1.0

Determine the mean and the standard deviation of x, the number of cartons sold per day.

5-13 (a) Draw a histogram of the probability distribution for the single-digit random numbers $(0, 1, 2, \ldots, 9)$.

(b) Calculate the mean and standard deviation associated with the population of single-digit random numbers.

5-14 Verify that formulas (5-4a) and (5-4b) are equivalent to formula (5-3).

Section 5-4 The Binomial Probability Distribution

Consider the following probability experiment. You are given a six-question multiple-choice quiz. You have not studied the material to be quizzed and therefore decide to answer the six questions by randomly guessing the answers without reading the question or the answers.

Answer Page to Quiz

Directions: Circle the best answer to each question.

1. A   B   C
2. A   B   C
3. A   B   C
4. A   B   C
5. A   B   C
6. A   B   C

Circle your answers before continuing.

Before we look at the correct answers to the quiz, let's think about some of the things we might consider when a quiz is answered in this way.

1. How many of the six questions do you think you answered correctly?

2. If an entire class answered the quiz by guessing, what do you think the "average" number of correct answers would be?

3. What is the probability that you selected the correct answers to all six questions?

4. What is the probability that you selected wrong answers for all six questions?

Let's construct the probability distribution associated with this experiment. Let x be the number of correct answers on your paper; x may then take on the values $0, 1, 2, \ldots, 6$. Since each individual question has three possible answers and since there is only one correct answer, the probability of selecting the correct answer to an individual question is $\frac{1}{3}$. The probability that a wrong answer is selected is $\frac{2}{3}$. $P(x = 0)$ is the probability that all questions were answered incorrectly.

$$P(x = 0) = \left(\frac{2}{3}\right)\left(\frac{2}{3}\right)\left(\frac{2}{3}\right)\left(\frac{2}{3}\right)\left(\frac{2}{3}\right)\left(\frac{2}{3}\right) = \frac{64}{729} \approx 0.09$$

These events are independent because each question is a separate event. We can therefore multiply the probabilities according to formula (4–8b). Also,

$$P(6) = \left(\frac{1}{3}\right)^6 = \frac{1}{729} \approx 0.001$$

Before we find the other five probabilities, let's look at the events on a tree diagram, figure 5–5. Notice that the event $x = 0$, "zero correct answers," is shown by the bottom branch, and that event $x = 6$, "six correct answers," is shown by the top branch. The other events, "one correct answer," "two correct answers," and so on, are represented by several branches of the tree.

Although not all branches are shown in figure 5–5, if we did show all branches, we would find that the event $x = 1$ occurs on 6 different branches, event $x = 2$ occurs on 15 branches, $x = 3$ occurs on 20 branches, $x = 4$ on 15 branches, and $x = 5$ on 6 branches. Thus the remaining probabilities can be calculated by finding the probability of a single branch representing the event and multiplying it by the number of branches that represent that event.

$$P(x = 1) = \left(\frac{1}{3}\right)\left(\frac{2}{3}\right)\left(\frac{2}{3}\right)\left(\frac{2}{3}\right)\left(\frac{2}{3}\right)\left(\frac{2}{3}\right)(6) = \left(\frac{1}{3}\right)^1\left(\frac{2}{3}\right)^5(6) = \frac{192}{729} \approx 0.26$$

$$P(x = 2) = \left(\frac{1}{3}\right)^2\left(\frac{2}{3}\right)^4(15) = \frac{240}{729} \approx 0.33$$

$$P(x = 3) = \left(\frac{1}{3}\right)^3\left(\frac{2}{3}\right)^3(20) = \frac{160}{729} \approx 0.22$$

$$P(x = 4) = \left(\frac{1}{3}\right)^4\left(\frac{2}{3}\right)^2(15) = \frac{60}{729} \approx 0.08$$

$$P(x = 5) = \left(\frac{1}{3}\right)^5\left(\frac{2}{3}\right)^1(6) = \frac{12}{729} \approx 0.02$$

The answer to question 1 on page 181 is suggested by the probability distribution in table 5–9. The most likely occurrence would be to get two correct answers. One, two, or three correct answers are expected to result approximately 80% of the time ($0.26 + 0.33 + 0.22 = 0.81$). We might also argue that $\frac{1}{3}$ (that is, $\frac{2}{6}$) of the questions would be expected to be answered correctly. This information provides a reason for answering both questions 1 and 2 page 181 with the value of 2. Questions 3 and 4 were asked with the expectation that you would feel the chance of "all correct" was very slight, as it is, 0.001. "All wrong" is not too likely, but it is not as rare an event as "all correct." It has a probability of 0.09.

The correct answers to the quiz are B, C, B, A, C, and C. How many correct answers did you have? You might ask several people to answer the quiz for you by guessing at the answers. Then construct an observed relative frequency distribution and compare it to the distribution shown in table 5–9.

FIGURE 5–5

Tree Diagram for the
Multiple-Choice Quiz

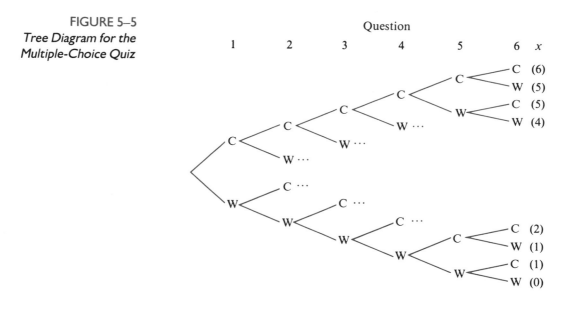

Many experiments result in outcomes that can be classified in one of two categories, **success or failure**. Examples of such experiments include the previous experiments of tossing coins or thumbtacks and other more practical experiments, such as determining if a flashbulb flashes or does not flash, if a product did its prescribed job or did not do it, if quiz question answers are correct or wrong. There are other experiments that have many outcomes which, under the right conditions, may fit this general description of being classified in one of two categories. For example, when we roll a single die, we usually consider six possible outcomes. However, if we are only interested in knowing whether an ace (a 1) shows or not, there are really only two outcomes: the ace shows or something else shows.

success or failure

TABLE 5–9

Probability Distribution
for the Quiz Example

x	P(x)
0	0.09
1	0.26
2	0.33
3	0.22
4	0.08
5	0.02
6	0.001
Total	1.001 (ck) (because of round-off error, the sum is not exactly 1)

The experiments described above are usually called "binomial probability experiments."

---

binomial experiment

Binomial Probability Experiment

An experiment that is made up of repeated trials of a basic experimental event. The binomial experiment must possess the following properties:

trial

1. Each **trial** has two possible outcomes (success, failure).
2. There are n repeated independent trials.
3. $P(\text{success}) = p$, $P(\text{failure}) = q$, and $p + q = 1$.

binomial random variable

4. The **binomial random variable** x is the count of the number of successful trials that occur; x may take on any integer value from 0 to n.

NOTE: $p = P(\text{success})$ on *each* individual trial and $q = P(\text{failure})$ on *each* trial.

The binomial experiment *must* demonstrate the two basic properties 1 and 2. Properties 3 and 4 concern notation and the identification of the variables involved.

NOTE: It is of utmost importance that a probability p be assigned to the particular outcome that is considered to be the "success" since the binomial random variable indicates the number of "successes" that occur.

Illustration 5–8

Consider the roll of a die, where "ace" or "something else" occurs. In this experiment we would count the number of "aces." The random variable x would be assigned a value that is the number of times that an ace is observed in the n trials. Since "ace" is considered "success," $P(\text{ace}) = p$ and $P(\text{not ace}) = q$. $\square$

Illustration 5–9

If you were an inspector on a production line in a plant where television sets were manufactured, you would be concerned with identifying the number of defective television sets. You would probably define "success" as the occurrence of a defective television. This is not what we normally think of as success, but if we count defective sets in a binomial experiment, we must define "success" as a "defective." The random variable x indicates the number of defective sets found per lot of n sets. p represents $P(\text{television is defective})$ and q is $P(\text{television is good})$. $\square$

NOTE: "Independent" trials mean that the result of one trial does not affect the probability of success of any other trial in the experiment. In other words, the probability of "success" remains constant throughout the entire experiment.

Each binomial probability experiment has its own specific probability function. However, all such functions have a common format and all such distributions display some similarities. Let's look at a relatively simple binomial experiment and discuss its probability function.

Let's reconsider illustration 5–7; a coin is tossed three times and we observe the number of heads that occur in the three tosses. This is a binomial experiment because it displays all the properties of a binomial experiment:

1. Each trial (one toss of the coin) has two outcomes: success (heads, what we are counting) and failure (tails).

2. There are n (three) repeated trials that are independent (each is a separate toss and the outcome of any one trial has no effect on the outcome of another).

3. The probability of success (heads) is p $(\frac{1}{2})$ and the probability of failure (tails) is q $(\frac{1}{2})$, and $p + q = 1$.

4. The random variable x is the number of successes (heads) that occur in n (three) trials. x can assume the values 0, 1, 2, or 3.

Let's consider first the probability of $x = 1$.

$$P(x = 1) = P(\text{exactly one head is observed in three tosses})$$

We want to find this probability. What does it mean to say "exactly one head in three tosses"? The probability that one head appears on one toss is $\frac{1}{2}$. The probability that heads do not occur on the other two tosses is $\frac{1}{2} \times \frac{1}{2} = \frac{1}{4}$. The probability that exactly one head occurs in three tosses, then, must be related to $\frac{1}{2} \times \frac{1}{4} = \frac{1}{8}$ [from the multiplication rule, formula (4–8a)]. But there is one other consideration: In how many different ways can exactly one of the three tosses result in a head?

1. The first toss could be a head and the others tails.

2. The second toss could be a head and the first and last tails.

3. The first two tosses could be tails and the last toss a head.

These are the three ways in which three trials can result in exactly one success. This means that the $\frac{1}{8}$ found above can occur three different ways, resulting in $P(x = 1) = 3(\frac{1}{8}) = \frac{3}{8}$.

If we look at the case $x = 2$, we find a similar situation: there are three ways to obtain exactly two heads. Thus $P(x = 2) = \frac{3}{8}$, also. We can find only one way for $x = 0$ and $x = 3$ to occur—all tails or all heads. In both cases the probability is equal to $(\frac{1}{2})^3 = \frac{1}{8}$. The rest of the probability distribution is exactly as shown in table 5–8.

If we take a careful look at the probability of each case shown in the tree diagram in figure 5–5, we can argue that the probability that a random variable x takes on a particular value in a binomial experiment is always the product of three basic factors. These three factors are as follows:

1. the probability of exactly x successes, p^x

2. the probability that failure will occur on the remaining $(n - x)$ trials, q^{n-x}

3. the number of ways that exactly x successes can occur in n trials

The number of ways that exactly x successes can occur in a set of n trials is represented by the symbol

$$\binom{n}{x}$$

binomial coefficient   which must always be a positive integer. This is termed the **binomial coefficient**. The binomial coefficient is found by using the formula

$$\binom{n}{x} = \frac{n!}{x!(n-x)!} \tag{5-5}$$

If the factorial notation, $n!$, is unfamiliar, see the study guide. Also, see the study guide for general information on the binomial coefficient. The values for $n!$ when n is equal to or smaller than 20 are found in table 2 of appendix D. The values for $\binom{n}{x}$ when n is equal to or smaller than 20 are found in table 3 of appendix D.

This information allows us to form a general binomial probability function.

---

binomial probability function   **Binomial Probability Function**

If, for a binomial experiment, p is the probability of a success and q is the probability of a failure on a single trial, then the probability $P(x)$ that there will be exactly x successes in n trials is

$$P(x) = \binom{n}{x} \cdot p^x \cdot q^{n-x} \qquad \text{for} \qquad x = 0, 1, 2, \ldots, n \tag{5-6}$$

When this general binomial probability function is applied to the illustration with three coins, we find

$$P(x) = \binom{3}{x} \cdot \left(\frac{1}{2}\right)^x \cdot \left(\frac{1}{2}\right)^{3-x} \qquad \text{for} \quad x = 0, 1, 2, 3$$

Let's calculate each of the probabilities and see if they form a probability distribution:

$$P(x = 0) = \binom{3}{0} \cdot \left(\frac{1}{2}\right)^0 \cdot \left(\frac{1}{2}\right)^3 = 1 \cdot 1 \cdot \frac{1}{8} = \mathbf{\frac{1}{8}}$$

$P(x = 0) = \frac{1}{8}$ is the probability of no heads occurring in three tosses of a coin. (*Note*: $\left(\frac{1}{2}\right)^0 = 1$.)

$$P(x = 1) = \binom{3}{1} \cdot \left(\frac{1}{2}\right)^1 \cdot \left(\frac{1}{2}\right)^2 = 3 \cdot \frac{1}{2} \cdot \frac{1}{4} = \mathbf{\frac{3}{8}}$$

$P(x = 1) = \frac{3}{8}$ is the probability of one head occurring in three tosses of a coin.

$$P(x = 2) = \binom{3}{2} \cdot \left(\frac{1}{2}\right)^2 \cdot \left(\frac{1}{2}\right)^1 = 3 \cdot \frac{1}{4} \cdot \frac{1}{2} = \mathbf{\frac{3}{8}}$$

$P(x = 2) = \frac{3}{8}$ is the probability of two heads occurring in three tosses of a coin.

$$P(x = 3) = \binom{3}{3} \cdot \left(\frac{1}{2}\right)^3 \cdot \left(\frac{1}{2}\right)^0 = 1 \cdot \frac{1}{8} \cdot 1 = \mathbf{\frac{1}{8}}$$

$P(x = 3) = \frac{1}{8}$ is the probability of three heads occurring in three tosses of a coin. Since each of these probabilities is between 0 and 1, and the sum of all the probabilities is exactly 1, this is a probability distribution. □

Let's look at another example. Consider an experiment that calls for drawing five cards, one at a time with replacement, from a well-shuffled deck of playing cards. The drawn card is identified as a spade or not a spade, returned to the deck, the deck reshuffled, and so on. The random variable x is the number of spades observed in five drawings. Is this a binomial experiment? Let's identify the various properties.

1. Each drawing is a trial and each drawing has two outcomes, "spade" or "not spade."
2. There are five repeated drawings, so $n = 5$. These individual trials are independent since the drawn card is returned to the deck and the deck reshuffled before the next drawing.
3. $p = \frac{1}{4}$ and $q = \frac{3}{4}$.
4. x is the number of spades recorded in the five trials. The binomial probability function is

$$P(x) = \binom{5}{x} \cdot \left(\frac{1}{4}\right)^x \cdot \left(\frac{3}{4}\right)^{5-x}$$

for $x = 0, 1, 2, 3, 4,$ and 5. The probabilities are

$$P(0) = \binom{5}{0} \cdot \left(\frac{1}{4}\right)^0 \cdot \left(\frac{3}{4}\right)^5 = 1 \cdot 1 \cdot \left(\frac{3}{4}\right)^5 = 0.2373$$

$$P(1) = \binom{5}{1} \cdot \left(\frac{1}{4}\right)^1 \cdot \left(\frac{3}{4}\right)^4 = 5 \cdot \left(\frac{1}{4}\right) \cdot \left(\frac{3}{4}\right)^4 = 0.3955$$

$$P(2) = \binom{5}{2} \cdot \left(\frac{1}{4}\right)^2 \cdot \left(\frac{3}{4}\right)^3 = 10 \cdot \left(\frac{1}{4}\right)^2 \cdot \left(\frac{3}{4}\right)^3 = 0.2637$$

$$P(3) = \binom{5}{3} \cdot \left(\frac{1}{4}\right)^3 \cdot \left(\frac{3}{4}\right)^2 = 10 \cdot \left(\frac{1}{4}\right)^3 \cdot \left(\frac{3}{4}\right)^2 = 0.0879$$

$$P(4) = \binom{5}{4} \cdot \left(\frac{1}{4}\right)^4 \cdot \left(\frac{3}{4}\right)^1 = 5 \cdot \left(\frac{1}{4}\right)^4 \cdot \left(\frac{3}{4}\right)^1 = 0.0146$$

$$P(5) = \binom{5}{5} \cdot \left(\frac{1}{4}\right)^5 \cdot \left(\frac{3}{4}\right)^0 = 1 \cdot \left(\frac{1}{4}\right)^5 \cdot \left(\frac{3}{4}\right)^0 = 0.0010$$

$$\overline{}$$

$$\sum P(x) = 1.0000$$

The preceding distribution of probabilities indicates that the single most likely value of x is 1, the event of observing exactly one spade in a hand of five cards. What is the least likely number of spades that would be observed? Let's consider one more binomial probability problem.

Illustration 5–10

The manager of Steve's Food Market guarantees that none of his cartons of one dozen eggs will contain more than one bad egg. If a carton contains more than one bad egg, he will replace the whole dozen and allow the customer to keep the original eggs. The probability that an individual egg is bad is 0.05. What is the probability that Steve will have to replace a given carton of eggs?

Solution   Is this a binomial experiment? To find out, let x be the number of bad eggs found in a carton of a dozen eggs. $p = 0.05$, and let the inspection of each egg be a trial resulting in finding a "bad" or "not bad" egg. To find the probability that the manager will have to make good on his guarantee, we need the probability function associated with this experiment:

$$P(x) = \binom{12}{x} \cdot (0.05)^x \cdot (0.95)^{12-x} \quad \text{for} \quad x = 0, 1, 2, \ldots, 12$$

The probability that Steve will replace a dozen eggs is the probability that $x = 2, 3, 4, 5, \ldots,$ or 12. Recall that $\sum P(x) = 1$, that is,

$$P(x = 0) + P(x - 1) + P(x = 2) + \cdots + P(x = 12) = 1$$

Therefore,

$$P(x = 2) + P(x = 3) + \cdots + P(x = 12) = 1 - [P(x = 0) + P(x = 1)]$$

Finding $P(x = 0)$ and $P(x = 1)$ and subtracting them from 1 is easier than finding each of the other probabilities.

$$P(x) = \binom{12}{x} \cdot (0.05)^x \cdot (0.95)^{12-x}$$

$$P(0) = \binom{12}{0} \cdot (0.05)^0 (0.95)^{12} = \mathbf{0.540}$$

$$P(1) = \binom{12}{1} \cdot (0.05)^1 (0.95)^{11} = \mathbf{0.341}$$

NOTE: The value of many binomial probabilities, for small values of n and common values of p, are found in table 4 of appendix D. In this example we have $n = 12$ and $p = 0.05$, and we want the probabilities for $x = 0$ and 1. We need to locate the section of table 4 where $n = 12$, find the column marked $p = 0.05$, and read the numbers opposite $x = 0$ and 1. We find 540 and 341, as shown in table 5–10. (Look these values up in table 4.) The decimal point was left out of the table to save space, and we must replace it. It belongs at the front of each table entry. Therefore, our values are 0.540 and 0.341.

Now let's return to our illustration:

$$P(\text{replacement}) = 1 - (0.540 + 0.341) = \mathbf{0.119}$$

If $p = 0.05$ is correct, Steve will be busy replacing cartons of eggs. If he replaces 11.9% of all the cartons of eggs he sells, he certainly will be giving away a substantial proportion of his eggs. This suggests that he should

TABLE 5–10

Abbreviated Portion of Table 4 in Appendix D, Binomial Probabilities

n	x	0.01	0.05	0.10	0.20	0.30	0.40	p 0.50	0.60	0.70	0.80	0.90	0.95	0.99	x
12	0	886	540	282	069	014	002	0+	0+	0+	0+	0+	0+	0+	0
	1	107	341	377	206	071	017	003	0+	0+	0+	0+	0+	0+	1
	2	006	099	230	283	168	064	016	002	0+	0+	0+	0+	0+	2
	3	0+	017	085	236	240	142	054	012	001	0+	0+	0+	0+	3
	4	0+	002	021	133	231	213	121	042	008	001	0+	0+	0+	4

adjust his guarantee. For example, if he were to replace a carton of eggs only when four or more were found bad, he would expect to replace only 0.003 cartons [1.0 − (0.540 + 0.341 + 0.099 + 0.017)], or 0.3% of the cartons sold. Notice that he will be able to control his "risk" (probability of replacement) if he adjusts the value of the random variable stated in his guarantee. ☐

Exercises

5–15 Evaluate each of the following:

(a) 4!   (b) 7!   (c) 0!

(d) $\dfrac{6!}{2!}$ (e) $\dfrac{5!}{3!2!}$ (f) $\dfrac{6!}{4!(6-4)!}$

(g) $(0.3)^4$ (h) $\dbinom{7}{3}$ (i) $\dbinom{5}{2}$

(j) $\dbinom{3}{0}$ (k) $\dbinom{4}{1}(0.2)^1(0.8)^3$ (l) $\dbinom{5}{0}(0.3)^0(0.7)^5$

5–16 (a) Find the sample space of an experiment where a coin is tossed four times and heads or tails is observed on each toss.

(b) Find $P(x)$ for $x = 0, 1, 2, 3,$ and 4, where x is the number of heads observed in the four tosses of the coin.

5–17 Consider a binomial experiment made up of three trials with outcomes of success S and failure F, where $P(S) = p$ and $P(F) = q$.

(a) Complete the accompanying tree diagram. Label all branches completely.

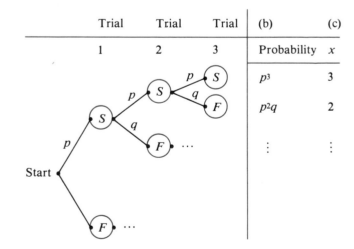

(b) In column (b) of the tree diagram, express the probability of each outcome represented by the branches as a product of powers of p and q.

(c) Let x be the random variable, the number of successes observed. In column (c) identify the value of x for each branch of the tree diagram.

(d) Notice that all the products in column (b) are made up of three factors and that the value of the random variable is the same as the exponent for the number p.

(e) Write the equation for the binomial probability function for this situation.

5-18 In reference to the six-question multiple-choice "quiz" on page 181 show that the guessing of the answers is actually a binomial experiment.

(a) Specify exactly how this experiment satisfies the four properties of a binomial experiment.

(b) Complete the tree diagram (figure 5-5) showing all possible outcomes for the quiz. There should be a total of 64 outcomes shown. How many outcomes are there for each of the events "0 correct answers" "1 correct answer," "2 correct answers," ..., "5 correct answers," "6 correct answers"?

(c) Write the equation of the binomial probability function for this experiment.

5-19 If x is a binomial random variable, calculate the probability of x for each case.

(a) $n = 4$, $x = 1$, $p = 0.3$ (b) $n = 3$, $x = 2$, $p = 0.8$

(c) $n = 2$, $x = 0$, $p = \frac{1}{4}$ (d) $n = 5$, $x = 2$, $p = \frac{1}{3}$

(e) $n = 4$, $x = 2$, $p = 0.5$ (f) $n = 3$, $x = 3$, $p = \frac{1}{6}$

5-20 If x is a binomial random variable, use table 4 to determine the probability of x for each case.

(a) $n = 10$, $x = 8$, $p = 0.3$ (b) $n = 8$, $x = 7$, $p = 0.95$

(c) $n = 15$, $x = 3$, $p = 0.05$ (d) $n = 12$, $x = 12$, $p = 0.99$

(e) $n = 9$, $x = 0$, $p = 0.5$ (f) $n = 6$, $x = 1$, $p = 0.01$

(g) Explain the meaning of the symbol $0+$ that appears in table 4.

5-21 Test the following function to determine whether or not it is a binomial probability function. List the distribution of probabilities and sketch a histogram.

$$T(x) = \binom{5}{x} \cdot \left(\frac{1}{2}\right)^x \cdot \left(\frac{1}{2}\right)^{5-x} \quad \text{for} \quad x = 0, 1, 2, 3, 4, 5$$

5-22 Consider the experiment of tossing a coin five times and counting the number of heads that appear as the random variable (x).

(a) Is this a binomial experiment? Describe all the properties and assign the variable.

(b) Write the equation of the binomial probability function.

(c) Find the various binomial probabilities to form the probability distribution.

5–23 Let x be a random variable with the following probability distribution:

x	0	1	2	3
$P(x)$	0.4	0.3	0.2	0.1

Does x have a binomial distribution? Justify your answer.

5–24 State a very practical reason why the defective item in an industrial situation would be defined to be the "success" in a binomial experiment.

5–25 The probability of a snowstorm on any given day during the ten-day period from Christmas Eve to the day after New Year's Day is equal to p.

(a) What is the probability of at least one snowstorm during that period? (Express the answer in terms of p since you have no specific value for p.)

(b) If $p = 0.1$, what is the probability of at least one snowstorm during that period?

(c) If $p = 0.4$, what is the probability of at least one snowstorm during that period?

(d) If $p = 0.7$, what is the probability of at least one snowstorm during that period?

5–26 A true-false quiz has ten questions. What is the probability that from seven to ten questions are answered correctly by random guessing? (Use table 4 of appendix D.)

5–27 In the biathlon event of the Olympic Games, a participant skis cross-country and on four intermittent occasions stops at a rifle range and shoots a set of five shots. If the center of the target is hit, no penalty points are assessed. If a particular man has a history of hitting the center of the target with 90% of his shots, what is the probability that he will hit the center of the target with

(a) all five of his next set of five shots?

(b) at least four of his next set of five shots?
(Assume independence.)

5–28 A basketball player has a history of making 80% of the foul shots taken during games. What is the probability that three of the next five foul shots he takes are missed?

5–29 The survival rate of a risky operation used for patients with no other hope of survival is 80%. What is the probability that exactly four of the next five patients survive this operation?

5–30 According to studies in 1982, one out of every 15 individuals entering a department store will attempt to shoplift something before leaving the store. Assuming that the binomial distribution is appropriate, what is the probability that 1 out of 3 randomly selected individuals, who have entered the store, will attempt to shoplift something before leaving the store?

Section 5–5 Mean and Standard Deviation of the Binomial Distribution

mean and standard deviation of binomial distribution

The **mean** and **standard deviation** of a theoretical binomial probability distribution can be found by using these two formulas:

$$\mu = np \tag{5-7}$$

$$\sigma = \sqrt{npq} \tag{5-8}$$

The formula for the mean seems appropriate as the number of trials multiplied by the probability that a "success" will occur. (We previously learned that the mean number of correct answers on the binomial quiz was expected to be $\frac{1}{3}$ of 6, $6 \cdot \frac{1}{3}$, or np) The formula for the standard deviation is not as easily understood. Thus at this point it is appropriate to look at an example that demonstrates that formulas (5–7) and (5–8) yield the same results as formulas (5–2) and (5–4).

Referring to the case of tossing three coins in illustration 5–7, $n = 3$ and $p = \frac{1}{2}$. Using formulas (5–7) and (5–8), we find

$$\mu = np = (3)(\tfrac{1}{2}) = \textbf{1.5}$$

$\mu = 1.5$ is the mean value of the random variable x.

$$\sigma = \sqrt{npq} = \sqrt{(3)(\tfrac{1}{2})(\tfrac{1}{2})} = \sqrt{\tfrac{3}{4}} = \sqrt{0.75} = 0.866 = \textbf{0.87}$$

$\sigma = 0.87$ is the standard deviation of the random variable x. Look back at the solution for illustration 5–7 and compare the use of formulas (5–2) and (5–4a) with formulas (5–7) and (5–8). Note that the results are the same, regardless of the formula you use. However, formulas (5–7) and (5–8) are much easier to use when x is a binomial random variable.

Illustration 5–11

Find the mean and standard deviation of the binomial distribution where $n = 20$ and $p = \frac{1}{5}$.

$$\mu = np = (20)(\tfrac{1}{5}) = \textbf{4.0}$$

$$\sigma = \sqrt{npq} = \sqrt{(20)(\tfrac{1}{5})(\tfrac{4}{5})} = \sqrt{\tfrac{80}{25}}$$

$$= \frac{(4\sqrt{5})}{5} = \textbf{1.79} \qquad \square$$

Exercises

5–31   Given the binomial probability function

$$T(x) = \binom{5}{x} \cdot \left(\frac{1}{2}\right)^x \cdot \left(\frac{1}{2}\right)^{5-x} \quad \text{for} \quad x = 0, 1, 2, 3, 4, 5$$

(a) Calculate the mean and standard deviation of the random variable by using formulas (5–2) and (5–4a).

(b) Calculate the mean and standard deviation using formulas (5–7) and (5–8).

(c) Compare the results of (a) and (b).

5–32   Seventy-five percent of the foreign-made autos sold in the United States in 1978 are now falling apart.

(a) Determine the probability distribution of x, the number of these autos that are falling apart in a random sample of size 5.

(b) Draw a histogram of the distribution.

(c) Calculate the mean and the standard deviation of this distribution.

5–33   Find the mean and standard deviation of x for each of the following binomial random variables:

(a) The number of aces seen in 100 draws from a well-shuffled bridge deck (with replacement).

(b) The number of cars found to have unsafe tires among the 400 cars stopped at a road block for inspection. Assume that 6% of all cars have one or more unsafe tires.

(c) The number of melon seeds that germinate when a package of 50 seeds is planted. The package states that the probability of germination is 0.88.

(d) The number of defective televisions in a shipment of 125. The manufacturer claimed that 98% of the sets were operative.

In Retrospect

In this chapter we combined concepts of probability with some of the ideas presented in chapter 2. We now are able to deal with distributions of probability values and find means, standard deviations, and so on.

In chapter 4 we explored the concepts of mutually exclusive events and independent events. The addition and multiplication rules were used on several occasions in this chapter but very little was said about mutual exclusiveness or independence. Recall that every time we add probabilities together, as we did in each of the probability distributions, we need to know that the associated events are mutually exclusive. If you look back over the chapter, you will notice that the random variable actually requires events to be mutually exclusive; therefore, no real emphasis was placed on this concept. *The same basic comment can be made in reference to the multiplication of probabilities and the concept of independent events. Throughout the chapter probabilities were multiplied together and occasionally independence was mentioned. Independence, of course, is necessary in order to be able to multiply probabilities together.*

If we were to take a close look at some of the sets of data in chapter 2, we would see that several of these problems could be reorganized to form probability distributions. For example: (1) Let x be the number of credit hours that a student is registered for in this semester, with the percentage of the student body being reported for each value of x. (2) Let x be the number of correct passageways through which an experimental laboratory animal passes before taking a wrong one. (3) Let x be the price of an alphanumeric terminal (see the news article at the beginning of this chapter). (4) Let x be the annual income of a family as discussed in the article at the beginning of chapter 2. (This is an example of a continuous random variable.) The list of examples is endless.

We are now ready to extend these concepts to continuous random variables, which we will do in chapter 6.

Chapter Exercises

5-34   What are the two basic properties of every probability distribution?

5-35   Verify whether or not the following is a probability function. State your conclusion and explain.

$$f(x) = \frac{\frac{3}{4}}{x!(3-x)!} \quad \text{for} \quad x = 0, 1, 2, 3$$

5-36   Calculate the mean and standard deviation, to the nearest tenth, for this probability distribution of x.

x	P(x)
0	0.1
1	0.1
4	0.4
6	0.2
8	0.2

5-37   One alphanumeric terminal will be selected at random from the product tables discussed in the news article at the beginning of the chapter. What is the probability that the terminal selected has a price

(a) between $500 and $1000?

(b) between $1000 and $2500?

5-38   A "wheel of fortune" in a professor's office produces uniformly distributed random integer values between zero and 100. Suppose that each student in this professor's class is assigned his grade by giving the wheel a whirl and that grades are determined by the following distribution.

0 to	20	is F
21 to	40	is D
41 to	62	is C
63 to	87	is B
88 to	100	is A

(a) Find the probability associated with each letter grade.

(b) Construct a histogram depicting the random variable x, the random integer generated by the wheel.

5-39   An experiment consists of having two persons rank a movie from 1 to 4, as shown in the sample space in the accompanying table. Define x to be the random variable total of the two rankings. (Assume that each rank is equally likely.)

Second Person	First Person			
	1	2	3	4
1	1, 1	2, 1	3, 1	4, 1
2	1, 2	2, 2	3, 2	4, 2
3	1, 3	2, 3	3, 3	4, 3
4	1, 4	2, 4	3, 4	4, 4

(a) Construct a probability distribution for x.

(b) Calculate the mean value of x.

(c) Calculate the standard deviation of x.

5-40   A local elementary school holds four open houses annually. Records show that the probability that a child's parents (one or both) attend from 0 to 4 of the open houses is as shown in the accompanying table.

Number of Open Houses Attended, x	0	1	2	3	4
Probability	0.12	0.38	0.30	0.12	0.08

(a) Is this a probability distibution? Explain.

(b) What is the probability that an individual child's parents attend at least one of these functions?

(c) Find the mean and the standard deviation for this distribution.

5-41 The ratio of men to women at a college is 2:1. Give the binomial probability function and distribution for the random variable "number of men" in a randomly selected group of four taken from the student body.

5-42 If boys and girls are equally likely to be born, what is the probability that in a randomly selected family of six children, three will be boys? (Find the answer by use of a formula.)

5-43 One-fourth of a certain breed of rabbits are born with long hair. What is the probability that in a litter of six rabbits, exactly three will have long hair? (Find by use of a formula.)

5-44 Suppose that you take a five-question multiple-choice quiz by guessing. Each question has exactly one correct answer of the four alternatives given. What is the probability that you guess more than one-half of the answers correctly? (Find by use of a formula.)

5-45 As a quality control inspector for toy trucks, you have observed that wooden wheels that are bored off-center occur about three percent of the time. If six wooden wheels are used on each toy truck produced, what is the probability that a randomly selected set of wheels has no off-center wheels?

5-46 The town council has 9 members. A proposal must have at least $\frac{2}{3}$ of the votes to be accepted. A proposal to establish a new industry in this town has been tabled. If we know that two members of the town council are opposed, and the others randomly vote "in favor" and "against," what is the probability that the proposal will be accepted?

5-47 Ninety percent of the trees planted by a landscaping firm survive. What is the probability that eight or more of the ten trees they just planted will survive? (Find by use of a table.)

5-48 In California, 30% of the people have a certain blood type. What is the probability that exactly 5 out of a randomly selected group of 14 Californians will have that blood type? (Find by use of a table.)

5-49 On the average, 1 out of every 10 boards purchased by a cabinet manufacturer is unusable for building cabinets. What is the probability that 8, 9, or 10 of a set of 11 such boards are usable? (Find by use of a table.)

5-50 A local polling organization maintains that 90% of the eligible voters have never heard of John Anderson, who was a presidential candidate in 1980. If this is so, what is the probability that in a randomly selected sample of 12 eligible voters two or fewer have heard of John Anderson?

5-51 In 1981, a prediction model for the Miss America Pageant listed the 10 top-ranked contestants. Since there were 50 contestants in the pageant, the probability of a randomly selected individual ranking in the top 10 is 0.20. The prediction model had 6 of its selection ranked in the top 10. Use the binomial probability function to estimate the probability of randomly selecting 6 of 10. Could the prediction model be considered successful?

5-52   A doctor knows from experience that 10% of the patients to whom he gives a certain drug will have undesirable side effects. Find the probabilities that among the 10 patients to whom he gives the drug;

(a) at most 2 will have undesirable side effects

(b) at least 2 will have undesirable side effects

5-53   In a recent survey of females, 90% admitted that they had never looked at a copy of *Vogue* magazine. Assuming that this is accurate information, what is the probability that a random sample of 3 females will show that fewer than 2 have looked at the magazine?

5-54   Of the Heisman Trophy winners in collegiate football, it is considered that only 0.35 have had successful professional football careers.

(a) What is the probability that none of the next 5 winners will have successful pro-careers?

(b) What is the probability that more than 3 of the next 5 winners will have successful pro-careers?

5-55   Seventy percent of those seeking a driver's license admitted that they would not report the person next to them if he or she copied some of their answers during the written exam. You have just entered the room and see 10 people waiting to take the written exam. What is the probability that, if the incident happened, 5 of the 10 would not report copying answers?

5-56   Prior to the elimination of its credit cards in 1982, ARCO's research indicated that 63% of gasoline purchasers said that price was more important than brand name.

(a) What is the probability that 3 or more of 5 randomly selected individuals make their purchases based on price?

(b) What is the probability that 6 or more of 10 randomly selected individuals make their purchases based on price?

(c) What is the probability that 9 or more of 15 randomly selected individuals make their purchases based on price?

(d) Why aren't these answers all the same?

5-57   The engines on an airliner operate independently. The probability that an individual engine operates for a given trip is 0.95. A plane will be able to complete a trip successfully if at least one-half of its engines operate for the entire trip. Determine whether a four-engine or a two-engine plane has the higher probability of a successful trip.

5-58   A large shipment of radios is accepted upon delivery if an inspection of ten randomly selected radios yields no more than one defective radio.

(a) Find the probability that this shipment is accepted if 5% of the total shipment is defective.

(b) Find the probability that this shipment is not accepted if 20% of this shipment is defective.

5-59　A business firm is considering two investments of which it will choose the one that promises the greater payoff. Which of the investments should it accept? (Let the mean profit measure the payoff.)

Invest in Tool Shop		Invest in Book Store	
Profit	Probability	Profit	Probability
$100,000	0.10	$400,000	0.20
50,000	0.30	90,000	0.10
20,000	0.30	− 20,000	0.40
− 80,000	0.30	− 250,000	0.30
	—		—
Total	1.00	Total	1.00

5-60　Experiment has shown that a pair of beetles produces black-eyed offspring 30% of the time. The eye color of one offspring is not related to that of another, and we are concerned with the "number of black-eyed beetles" in a litter of 12.

(a) Describe how this experiment exhibits the properties of a binomial experiment.

(b) Find the probability that at least one-half of this litter is black-eyed.

5-61　An experiment consists of placing each of ten numbered objects into one of two containers, A or B. After the objects have been placed in the containers, the results are recorded by listing the set of objects that are in container A.

(a) How many different sets could be found in container A at the conclusion of the experiment?

(b) How many of these contain exactly no elements? One element? Two elements? Three elements? . . . Nine elements? Ten elements?

(c) If the objects were placed into the containers randomly, what is the probability of each of the possible sets in part (a)?

(d) Define a random variable x to be the number of objects in container A and construct the probability distribution for x.

(e) Find the mean and standard deviation of x.

5-62　A box contains eight identical balls numbered from 1 to 8. An experiment consists of selecting two of these balls, without replacement, and identifying their numbers.

(a) How many different sets of two balls could be selected?

(b) List the sample space for the experiment.

(c) If the balls are selected randomly, what is the probability of each of the possible sets in the sample space?

(d) Define the random variable x to be the total of the numbers on the two balls. Construct the probability distribution for x.

(e) Find the mean and standard deviation of x.

5–63   An advertising agency has drawn six new ads for one of its customers. The agency has ranked each of these new ads from 1 to 6 according to its opinion of the ad's effectiveness in selling the product. The customer, the manufacturer of the product, is to select two of these ads for use in promoting the product.

(a) Construct the sample space showing the possible selections that can be made. (Use the agency's rank number to identify each ad.)

(b) Assume that the customer is selecting in a random manner. Find the probability distribution for the random variable x, the sum of the two rank numbers for the ads selected.

(c) Find the mean and the standard deviation of x.

5–64   Bill has completed a ten-question multiple-choice test on which he answered seven questions correctly. Each question had one correct answer to be chosen from five alternatives. Bill says that he answered the test by randomly guessing the answers without reading the questions or answers.

(a) Define the random variable x to be the number of correct answers on this test and construct the probability distribution if the answers were obtained by random guessing.

(b) What is the probability that Bill guessed seven of the ten answers correctly?

(c) What is the probability that anybody can guess six or more answers correctly?

(d) Do you believe that Bill actually randomly guessed as he claims? Explain. (Retain this solution for use in chapter 8.)

Vocabulary List

Be able to define each term. In addition, describe, in your own words, and give an example of each term. Your examples should not be ones given in class or in the textbook

The terms shown followed by bracketed numbers are those of earlier chapters (the bracketed numbers are those of the earlier chapters), but we ask you to define them again to show increased understanding of their meaning.

addition of probabilities [4]
binomial coefficient
binomial experiment
binomial probability function
binomial random variable
constant function
discrete random variable
experiment [1, 4]
failure
independent events [4]
mean of probability distribution
multiplication of probabilities [4]
mutually exclusive events [4]
population parameter [1]

probability distribution
probability function
probability histogram
probability line histogram
random variable
relative frequency [2, 4]
sample statistic [1]
standard deviation of probability
   distribution
success
trial
variance of probability
   distribution

Quiz

Answer "True" if the statement is always true. If the statement is not always true, replace the words shown in color with words that make the statement always true.

5–1 The number of hours that you waited in line to register this semester is an example of a **discrete** random variable.

5–2 The number of automobile accidents that you were involved in as a driver last year is an example of a **discrete** random variable.

5–3 The sum of all the probabilities in any probability distribution is always exactly **two**.

5–4 An observed probability distribution is actually a **relative frequency** distribution of data in which there is a random variable.

5–5 A binomial experiment always has **three or more** possible outcomes to each trial.

5–6 The formula $\mu = np$ may be used to compute the mean of a **discrete** population.

5–7 A binomial probability distribution is **always** symmetrical about its mean.

5–8 A parameter is a statistical measure of some aspect of a **sample**.

5–9 **Sample statistics** are represented by letters from the Greek alphabet.

5–10 The probability of events A or B is equal to the sum of the probability of event A and the probability of event B when A and B are **mutually exclusive** events.

6 The Normal Probability Distribution

Color Multiplexing on an Optical Fiber

Michael Flanagan
MITRE Corp.
Bedford, MA 01730

Fiber optic communications systems in use today have the common characteristic that a single color source transmits the data. A consequence of this single color mode of operation is that most of the optical fiber's bandwidth is unused.

So how should an experimenter use this "unused" bandwidth? Conceptually, the answer is simple: He should use many colors on a single fiber optic cable. Each color will correspond to a new channel on an existing fiber optic cable. Using multiple colors rather than a single color offers many advantages; the advantages of multiple color operation on a fiber optic cable outweigh the new problem areas. For instance, a typical, installed, single color, fiber optic communication system for digital data, operating at a data rate of 10^7 bits per second, could have its data rate increased to 10^8 bits per second by adding nine more colors onto the existing cable. Or, nine more users could use the same cable simultaneously with no interaction, with each user having his own dedicated color. A fully loaded voice channel could similarly have its capacity increased by a factor of ten. These increases could be accomplished with no cable installation or construction problems, merely by adding circuitry at the sending and receiving ends of the system. Contrast this with what would be required to extend the capacity of an existing coaxial cable system by a similar amount.

Background

An investigation into the subject areas unique to color multiplexing has been made (as part of a graduate thesis). The investigation was divided into six different areas, as noted below.

- *Sources*—At present, the number of different color sources available is limited. A wide variety of spectrally narrow sources is required for this scheme.

Sources

LEDs were used for sources because they are available in multiple colors, they are simple to drive, and they are inexpensive. Only three standard LED colors, green, red and yellow, were used for these tests. However, in practice, any present-day LED source is capable of supplying many subcolors, because of its wide spectral width. A typical spectral distribution curve for an infrared LED is shown in **Fig. 3**. At the 50% level, visible LED spectral widths are typically 30 to 50 nm, while infrared diodes have spectral widths of about 80 nm. By using narrowband optical filters, suitably spaced to provide guard bands, multiple colors can be obtained from each source type. (Such filters offer typical in-band transmission of about 50%.) For instance, a 50 nm wide LED type might supply ten colors (with guard bands) using 1 nm filters, while a wider IR source type might supply 15 to 20 different colors.

Lasers would be better sources, since they are spectrally narrow to begin with and filtering might not be necessary. In the future many different color lasers with small spectral widths should be available, but at present, lasers are available in few colors. They are expensive and are not as easy to use as LEDs.

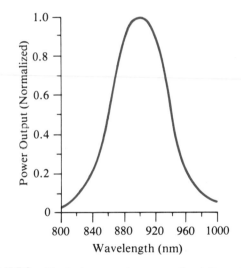

FIGURE 3   *Typical spectral output of an infrared LED.*

Chapter Objectives

Until now we have considered distributions of discrete variables only. In this chapter we will examine one particular probability distribution of major importance whose domain is the set of all real numbers. This distribution is called the normal, *the* bell-shaped, *or the* Gaussian *distribution. "Normal" is simply the traditional title of this particular distribution and is not a descriptive name meaning "typical." Although there are many other types of continuous distributions (rectangular, triangular, skewed, and so on), many variables have an approximately normal distribution. For example, several of the histograms drawn in chapter 2 suggested a normal distribution. A mounded histogram that is approximately symmetric is an indication of such a distribution.*

In addition to learning what a normal distribution is, we will consider (1) how probabilities are found, (2) how they are represented, and (3) how the normal distribution is used. Although there are other distributions of continuous variables, the normal distribution is the most important.

Section 6–1 The Normal Probability Distribution

normal distribution

As in all other probability distributions, there are formulas that give us information about the **normal distribution**. Previously the probability function was the only function of interest. However, for the normal distribution there are two functions that define and describe it. Each formula uses the random variable as an independent variable. Following this paragraph you will find two formulas. Formula (6–1) gives an ordinate (y value) for each point on the graph of the normal distribution for each given abscissa (x value). Formula (6–2) yields the probability associated with x when x is in the interval between the values a and b. Note that the random variable x is a continuous variable. For continuous variables we will discuss the probability that x has a value between the two extreme values of the interval, and we will depict the probability as an area under the graph of the probability distribution.

$$y = f(x) = \frac{1}{\sigma\sqrt{2\pi}} \cdot \exp\left[-\frac{1}{2}\left(\frac{x-\mu}{\sigma}\right)^2\right] \qquad \text{for all real } x \qquad (6\text{–}1)$$

$$P(a \leq x \leq b) = \int_a^b f(x)\,dx \qquad (6\text{–}2)$$

If you don't understand these formulas, don't worry about it. We will not be using them in this book. However, a few comments about their meaning will be helpful if you ever use a probability table in any of the standard reference books. The directions for the use of these tables are often described with the aid of mathematical formulas such as (6–1) and (6–2).

Formula (6–1) and its associated table can be used to graph a normal distribution with a given mean and standard deviation. [The table associated with formula (6–1) is not included in this text, but it is available in many other textbooks and in all standard books of statistical tables.] In a more practical sense, we can draw a normal or **bell-shaped curve** and then label it to approximate scale, as indicated in figure 6–1.

To talk intelligently about formula (6–2), we need to know what the symbols used in the formula represent. The $\int_a^b f(x)\, dx$ is called a "definite integral" and it comes from calculus. The definite integral is a number that is the measure of the area under the curve of $f(x)$. This area is bounded by a on the left, b on the right, $f(x)$ at the top, and the x-axis at the bottom. The equation of the normal curve in formula (6–2) gives the measure of the shaded area in figure 6–2.

Since $P(a \leq x \leq b)$ (read "the probability that x is between a and b") is given by $\int_a^b f(x)\, dx$, the measure of this probability is also the measure of the probability that a continuous random variable will assume a value within a specified interval. As a student of elementary statistics, you are not expected to have studied calculus, so we will say nothing more about the definite integral in this text.

The values calculated by using these formulas [(6–1) and (6–2)] are also found in a table, with the probabilities expressed to four decimal places. Like all other tables, this table must be read in the manner in which it was constructed.

However, before we show you how to read the table, we must point out that the table is expressed in a "standardized" form. It is standardized so that it is not necessary to have a different table for every different pair of values of the mean and standard deviation. For example, the area under the normal curve of a distribution with a mean of 15 and a standard deviation of 3 must be somehow related to the area under the curve of a normal distribution with a mean of 113 and a standard deviation of 38.5. If you will recall, the empirical rule concerns the percentage of a distribution within a certain number of standard deviations of the mean (see page 64).

NOTE: Percentage and probability are the same relative frequency concept. **Percentage** is usually used when talking about a **proportion** of a population. **Probability** is usually used when talking about the **chance** that the next individual item will display a particular characteristic.

The empirical rule is a fairly crude measuring device. With it we are able to find probabilities associated only with whole-number multiples of the standard deviation (within one standard deviation of the mean, and so on). We will often be interested in the probabilities associated with fractional parts of the standard deviation. For example, we might want to know the probability that x is within 1.85 standard deviations of the mean. Therefore, we must refine the empirical rule so that we can deal with more precise measurements. This refinement is discussed in the next section.

bell-shaped curve

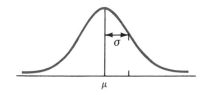

FIGURE 6–1
The Normal Distribution

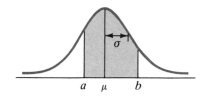

FIGURE 6–2
Shaded Area Under the Curve Is
$P(a \leq x \leq b)$

percentage—proportion

Section 6–2 The Standard Normal Distribution

The key to working with the normal distribution is the **standard score** z. The measures of probabilities associated with the random variable x are determined by the relative position of x with respect to the mean and the standard deviation of the distribution. The empirical rule tells us that approximately 68 % of the data are within one standard deviation of the mean. The value of the mean and the size of the standard deviation do not change this fact.

Recall that z was defined in chapter 2 to be

$$\frac{x - (\text{mean of } x)}{(\text{standard deviation of } x)}$$

Therefore, we can write the formula for z as

$$z = \frac{x - \mu}{\sigma} \tag{6-3}$$

(Note that if $x = \mu$, then $z = 0$.)

standard normal distribution

The z score is known as a "standardized" variable because its units are standard deviations. The normal probability distribution associated with this standard score z is called the **standard normal distribution**. Table 5 in appendix D lists the probabilities associated with the intervals from the mean for specific positive values of z. Other probabilities must be found by addition, subtraction, and so on, based on the concept of symmetry that exists in the normal distribution.

Let's look at an illustration to see how we read table 5. The z scores are in the margins: the left margin has the units digit and the tenths digit; the top has the hundredths digit. Let's look up a z score of 1.52. We find 0.4357, as shown in table 6–1. Now, exactly what is this 0.4357? It is the measure of the area under the standard normal curve between $z = 0$ (which locates the mean) and $z = 1.52$ (a number 1.52 standard deviations larger than the mean). See figure 6–3. **This area is also the measure of the probability associated with the same interval,** that is,

area representation for probability

$$P(0 < z < 1.52) = 0.4357$$

TABLE 6–1 A Portion of Table 5, Appendix D				
z	0.00	0.01	0.02	...
⋮				
1.5			0.4357	
⋮				

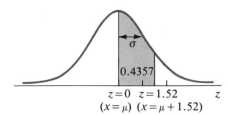

FIGURE 6–3

Area Under Normal Curve from z = 0 to z = 1.52 Is 0.4357

Read this result as "the probability that a value picked at random falls between the mean and 1.52 standard deviations above the mean is 0.4357" or, equivalently, "the probability that a z score picked at random will fall between 0 and 1.52 is 0.4357."

Recall that one of the basic properties of probability is that the sum of all probabilities is exactly 1.0. Since the area under the normal curve represents the measure of probability, the total area under the bell-shaped curve is exactly 1 unit. This distribution is symmetric with respect to the vertical line drawn through $z = 0$, which cuts the area exactly in half at the mean. Can you verify this fact by inspecting formula (6–1)? That is, the area under the curve to the right of the mean is exactly one-half unit (0.5), and the area to the left is also one-half unit (0.5). Areas (probabilities) not given directly in the table can be found by relying on these facts.

Now let's look at some illustrations.

Illustration 6–1

Find the area under the normal curve to the right of $z = 1.52$.

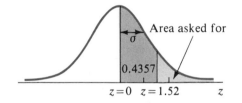

Solution   The area to the right of the mean (all the shading in the figure) is exactly 0.5000. The question asks for the shaded area that is not included in the 0.4357. Therefore, subtract 0.4357 from 0.5000.

$$0.5000 - 0.4357 = \mathbf{0.0643}$$ □

SUGGESTION: As we have done here, always draw and label a sketch. It is most helpful.

Illustration 6–2

Find the area to the left of $z = 1.52$.

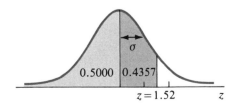

Solution   The total shaded area is made up of the 0.4357 found in the table and the 0.5000 that is to the left of the mean. Therefore, add 0.4357 to 0.5000.

$$0.4357 + 0.5000 = \textbf{0.9357}$$ □

NOTE: The addition and subtraction done in illustrations 6–1 and 6–2 are correct because the "areas" represent mutually exclusive events (discussed in section 4–5).

The symmetry of the normal distribution is a key factor in determining probabilities associated with values below (to the left of) the mean. The area between the mean and $z = -1.52$ is exactly the same as the area between the mean and $z = +1.52$. This fact allows us to find values related to the left side of the distribution.

Illustration 6–3

The area between the mean ($z = 0$) and $z = -2.1$ is the same as the area between $z = 0$ and $z = +2.1$.

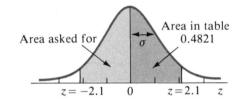

Thus we have

$$P(-2.1 < z < 0) = \textbf{0.4821}$$ □

Illustration 6–4

The area to the left of $z = -1.35$ is found by subtracting 0.4115 from 0.5000.

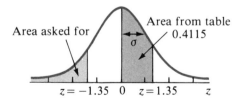

Therefore, we obtain

$$P(z < -1.35) = \textbf{0.0885}$$ □

Illustration 6–5

The area between $z = -1.5$ and $z = 2.1$ is found by adding the two areas together.

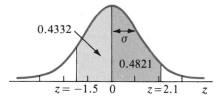

Therefore we obtain

$$P(-1.5 < z < 0) + P(0 < z < 2.1) = 0.4332 + 0.4821 = \mathbf{0.9153} \quad \square$$

Illustration 6–6

The area between $z = 0.7$ and $z = 2.1$ is found by subtracting. The area between $z = 0$ and $z = 2.1$ contains all the area between $z = 0$ and $z = 0.7$. The area between $z = 0$ and $z = 0.7$ is therefore subtracted from the area between $z = 0$ and $z = 2.1$.

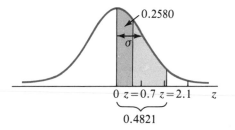

Thus we have

$$P(0.7 < z < 2.1) = 0.4821 - 0.2580 = \mathbf{0.2241} \quad \square$$

The normal distribution table can also be used to determine a z score if we are given an area. The next illustration considers this idea.

Illustration 6–7

What is the z score associated with the 75th percentile? (Assume the distribution is normal.) See figure 6–4.

FIGURE 6–4
P_{75} and Its Associated z Score

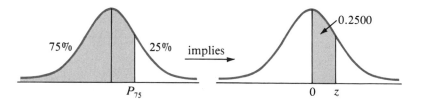

Solution   To find this z score, look in table 5, appendix D, and find the "area" entry that is closest to 0.2500; this area entry is 0.2486. Now read the z score that corresponds to this area. From the table the z score is found to

z	$\cdots$ 0.07	0.08 $\cdots$
⋮		
0.6	0.2486	0.2517
⋮		

be $z = \mathbf{0.67}$. This says that the 75th percentile in a normal distribution is 0.67 (approximately $\frac{2}{3}$) standard deviation above the mean.   □

Illustration 6-8

What z scores bound the middle 95% of a normal distribution? See figure 6-5.

FIGURE 6-5
Middle 95% of Distribution and Its Associated z Score

z	$\cdots$ 0.06 $\cdots$
⋮	
1.9	0.4750
⋮	

Solution   The 95% is split into two equal parts by the mean; 0.4750 is the area (percentage) between $z = 0$, the mean, and the z score at the right boundary. Since we have the area, we look for the entry in table 5 closest to 0.4750 (it happens to be exactly 0.4750) and read the z score in the margin. We obtain $z = 1.96$. Therefore, $z = -1.96$ and $z = 1.96$ bound the middle 95% of a normal distribution.   □

Exercises

6-1   Describe the distribution of the standard normal score z.

6-2   Find the area under the normal curve that lies between the following pairs of z values:

(a) $z = 0$ to $z = 1.30$ (b) $z = 0$ to $z = 1.28$
(c) $z = 0$ to $z = -3.20$ (d) $z = 0$ to $z = -1.98$

6-3   Find the probability that a piece of data picked at random from a normal population will have a standard score (z) that lies between the following pairs of z values:

(a) $z = 0$ to $z = 2.10$ (b) $z = 0$ to $z = 2.57$
(c) $z = 0$ to $z = -1.20$ (d) $z = 0$ to $z = -1.57$

6-4 Find the area under the normal curve that lies between the following pairs of z values:

(a) $z = -1.20$ to $z = 1.22$ (b) $z = -1.75$ to $z = 1.54$

(c) $z = -1.30$ to $z = 2.58$ (d) $z = -3.5$ to $z = -0.35$

6-5 Find the probability that a piece of data picked at random from a normal population will have a standard score (z) that lies between the following pairs of z values:

(a) $z = -2.75$ to $z = 1.38$ (b) $z = 0.67$ to $z = 2.95$

(c) $z = -2.95$ to $z = -1.18$

6-6 Find the following areas under the normal curve.

(a) to the right of $z = 0.00$

(b) to the right of $z = 1.05$

(c) to the right of $z = -2.30$

(d) to the left of $z = 1.60$

(e) to the left of $z = -1.60$

6-7 Find the probability that a piece of data picked at random from a normally distributed population will have a standard score that is as follows:

(a) less than 3.00   (b) greater than -1.55

(c) less than -0.75 (d) less than 1.25

(e) greater than -1.25

6-8 Find these probabilities:

(a) $P(0.00 < z < 2.35)$ (b) $P(-2.10 < z < 2.34)$

(c) $P(z > 0.13)$ (d) $P(z < 1.48)$

6-9 Find the z score for the standard normal distribution shown on each of the diagrams.

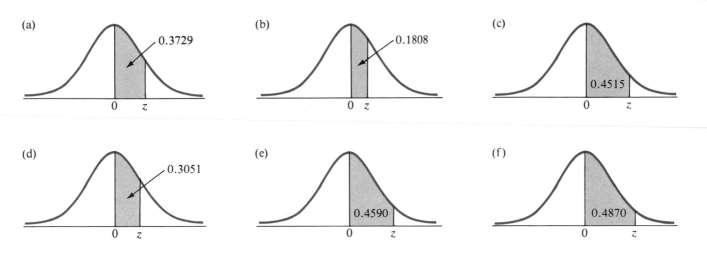

6-10 Find the z score for the standard normal distribution shown in each of the diagrams.

(a)

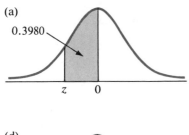

0.3980

z 0

(b)

0.2422

z 0

(c)

0.1844

z 0

(d)

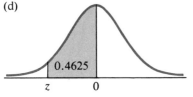

0.4625

z 0

(e)

0.4410

z 0

(f)

0.0915

z 0

6-11 Find a value of z such that 40% of the distribution lies between it and the mean. (There are two possible answers.)

6-12 Find the standard score z such that:

(a) eighty percent of the distribution is below (to the left of) this value

(b) the area to the right of this value is 0.15

6-13 Find the two standard scores z such that:

(a) the middle 90% of a normal distribution is bounded by them

(b) the middle 98% of a normal distribution is bounded by them

Section 6–3 Applications of the Normal Distribution

The probabilities associated with any normal distribution can be found by applying the techniques discussed in section 6–2. First, however, we must "standardize" the given information. When dealing with a normal distribution, we need to know its mean μ and its standard deviation σ. Once these values are known, any value of the random variable x can be easily converted to the standard score z by use of formula (6–3):

$$z = \frac{x - \mu}{\sigma}$$

Illustration 6–9

Consider the intelligence quotient (IQ) scores for people. IQs are normally distributed with a mean of 100 and a standard deviation of 10. If a person is picked at random, what is the probability that his IQ is between 100 and 115, that is, what is $P(100 < x < 115)$?

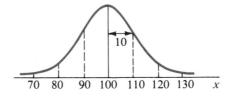

Solution $P(100 < x < 115)$ is represented by the shaded area in the figure.

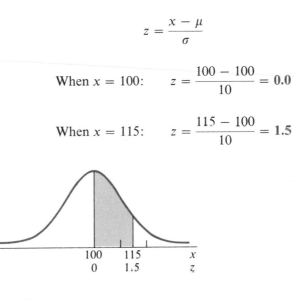

The variable x must be standardized by using formula (6–3). The z values are shown on the next diagram.

$$z = \frac{x - \mu}{\sigma}$$

When $x = 100$: $z = \dfrac{100 - 100}{10} = \mathbf{0.0}$

When $x = 115$: $z = \dfrac{115 - 100}{10} = \mathbf{1.5}$

Therefore

$$P(100 < x < 115) = P(0.0 < z < 1.5) = \mathbf{0.4332}$$

(The value 0.4332 is found by using table 5.) Thus the probability is 0.4332 that a person picked at random has an IQ between 100 and 115.   □

Illustration 6–10

Find the probability that a person selected at random will have an **IQ** greater than 95.

Solution

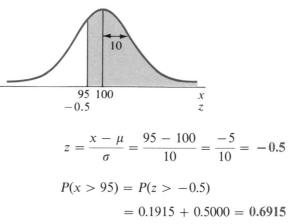

$$z = \frac{x - \mu}{\sigma} = \frac{95 - 100}{10} = \frac{-5}{10} = -0.5$$

$$P(x > 95) = P(z > -0.5)$$
$$= 0.1915 + 0.5000 = \mathbf{0.6915}$$

Thus the probability is 0.6915 that a person selected at random will have an **IQ** greater than 95. □

The normal table can be used to answer many kinds of questions that involve a normal distribution. Many times a problem will call for the location of a "cutoff point," that is, a particular value of x such that there is exactly a certain percentage in a specified area. The following illustrations concern some of these problems.

Illustration 6–11

In a large class, suppose your instructor told you that you needed to obtain a grade in the top 10% of your class to get an A on a particular exam. From past experience she is able to estimate that the mean and standard deviation on this exam will be 72 and 13, respectively. What will be the minimum grade needed to obtain an A? (Assume that the grades will be approximately normally distributed.)

Solution

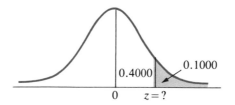

$$10\% = 0.1000 = 0.5000 - 0.4000$$

Look in the table to find the value of z that is closest to 0.4000; it is $z = 1.28$. Thus

$$P(z > 1.28) = 0.10 \qquad \text{(information given)}$$

Now find the x value that corresponds to $z = 1.28$ by using formula (6–3):

$$z = \frac{x - \mu}{\sigma} \quad \text{or} \quad 1.28 = \frac{x - 72}{13}$$

$$x - 72 = (13)(1.28)$$

$$x = 72 + (13)(1.28) = 72 + 16.64 = 88.64, \text{ or } \mathbf{89}$$

Thus if you receive an 89 or higher, you can expect to be in the top 10% (which means an A). □

Illustration 6–12

Find the 33d percentile for IQ scores ($\mu = 100$ and $\sigma = 10$ from illustration 6–9).

Solution

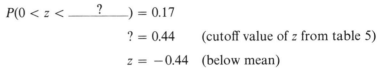

$$P(0 < z < \underline{\quad ? \quad}) = 0.17$$

$$? = 0.44 \qquad \text{(cutoff value of } z \text{ from table 5)}$$

$$z = -0.44 \quad \text{(below mean)}$$

Now we convert the z score of -0.44 to an x score:

$$z = \frac{x - \mu}{\sigma} \quad \text{[formula (6–3)]}$$

$$-0.44 = \frac{x - 100}{10}$$

$$-4.4 = x - 100$$

$$100 - 4.4 = x$$

$$x = \mathbf{95.6}$$

Thus 95.6 is the 33d percentile for IQ scores. □

Illustration 6–13 concerns a situation where we are asked to find the mean μ when given the related information.

Illustration 6–13

The incomes of junior executives in a large corporation are normally distributed with a standard deviation of $1,200. A cutback is pending, at which time those who earn less than $18,000 will be discharged. If such a cut represents 10% of the junior executives, what is the current mean salary of the group of junior executives?

Solution   If 10% of the salaries are below $18,000, then 40% (or 0.4000) are between $18,000 and the mean μ. Table 5 indicates that $z = -1.28$ is the standard score that occurs at $x = \$18,000$. Using formula (6–3) we can find the value of μ:

$$-1.28 = \frac{18,000 - \mu}{1,200}$$

$$-1,536 = 18,000 - \mu$$

$$\mu = \$19,536$$

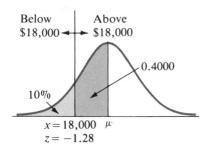

Below $18,000 ⟷ Above $18,000

0.4000

10%

$x = 18,000$ μ
$z = -1.28$

That is, the current mean salary of junior executives is $19,536.   ☐

Referring again to the IQ scores, what is the probability that a person picked at random has an IQ of 125, $P(x = 125)$? (IQ scores are normally distributed with a mean of 100 and a standard deviation of 10.) This situation has two interpretations: (1) theoretical and (2) practical. Let's look at the theoretical interpretation first. Recall that the probability associated with a continuous random variable is represented by the area under the curve. That is, $P(a < x < b)$ is equal to the area between a and b under the curve. $P(x = 125)$—that is, x is exactly 125—is then $P(125 < x < 125)$, or the area of the vertical line segment at $x = 125$. This area is zero. However, this is not the *practical* meaning of $x = 125$. It generally means 125 to the nearest integer value. Thus $P(x = 125)$ would most likely be interpreted as

$$P(124.5 < x < 125.5)$$

The interval from 124.5 to 125.5 under the curve has a measurable area and is then nonzero. In situations of this nature, you must be sure of the meaning being used.

Exercises

6–14   Given that x is a normally distributed random variable with a mean of 60 and a standard deviation of 10, find the following probabilities:

(a) $P(x > 60)$ (b) $P(60 < x < 72)$

(c) $P(57 < x < 83)$ (d) $P(65 < x < 82)$

(e) $P(38 < x < 78)$ (f) $P(x < 38)$

6–15   Let y be a normally distributed random variable with a mean of 8 and a standard deviation of 3. If a single value of y is taken from this population at random, what is the probability that it will be between 10 and 13?

6–16   Let h be a normally distributed random variable with a mean of 26.7 and a standard deviation of 3.4. Find the probability that an individual value of h, selected at random, will fall in the following intervals:

(a) between 26.7 and 31.7   (b) between 28.0 and 30.0

(c) between 20.0 and 24.0   (d) between 20.0 and 30.0

6–17   A survey was taken among the freshmen at our college to find out how many hours a week they spent studying for courses in which they were currently enrolled. The data were normally distributed with a mean of 28 hours and a standard deviation of 11 hours. What percentage of these freshmen spend the following times studying per week?

(a) more than 40 hours

(b) less than 15 hours

(c) more than 25 hours

(d) more than 20 but less than 36 hours

6–18   Final averages are typically approximately normally distributed with a mean of 72 and a standard deviation of 12.5. Your professor says that the top 8% of the class will receive A, the next 20% B, the next 42% C, the next 18% D, and the bottom 12% F.

(a) What average must you exceed to obtain an A?

(b) What average must you exceed to receive a grade better than a C?

(c) What average must you obtain to pass the course? (You'll need a D or better.)

6–19   The length of useful life of a fluorescent tube used for indoor gardening is normally distributed. The useful life has a mean of 600 hours and a standard deviation of 40 hours. Determine:

(a) the probability that a tube chosen at random will last between 620 and 680 hours

(b) the probability that such a tube will last more than 740 hours

6–20   At Pacific Freight Lines, bonuses are given to billing clerks when they complete 300 or more freight bills during an eight-hour day. The number of bills completed per clerk per eight-hour day is approximately normally distributed with a mean of 270 and a standard deviation of 16. What proportion of the time should a randomly selected billing clerk expect to receive a bonus?

6–21   The waiting time x at a certain bank is approximately normally distributed with a mean of 3.7 minutes and a standard deviation of 1.4 minutes.

(a) Find the probability that a randomly selected customer has to wait less than 2.0 minutes.

(b) Find the probability that a randomly selected customer has to wait more than 6 minutes.

(c) Find the value of the 75th percentile for x.

6-22 The height of a certain species of ornamental shrubs has an average of 2.3 feet and a standard deviation of 0.3 feet. Assume that the mature height of these shrubs is normally distributed.

(a) Find the probability that a given shrub will grow to be less than 1.5 feet tall.

(b) Find the percentage of these shrubs that grow to be more than 2 feet tall.

6-23 A brewery filling machine is adjusted to fill quart bottles with a mean of 32 ounces of ale, and a variance of 0.003. Periodically a bottle is checked and the amount of ale contained is noted. Assuming that the amount of fill is normally distributed, what is the probability that the next randomly checked bottle contains more than 32.02 ounces?

6-24 The weights of ripe watermelons grown at Mr Smith's farm are normally distributed with a standard deviation of 2.8 pounds. Find the mean weight of Mr. Smith's ripe watermelons if only 3% weigh less than 15 pounds.

6-25 A radar unit is used to measure the speed of automobiles on an expressway during rush hour traffic. The speeds of individual automobiles are normally distributed with a mean of 62 miles per hour.

(a) Find the standard deviation of all speeds if 3% of the automobiles travel faster than 72 miles per hour.

(b) Using the standard deviation found in (a), find the percentage of these cars that are traveling less than 55 miles per hour.

(c) Using the standard deviation found in (a), find the 95th percentile for the variable "speed."

Section 6–4 Notation

When working with the standard score z, it is often helpful and necessary to identify z with an area under the normal curve. One standard procedure is to use the area under the curve and to the right of the z. We will write this area within parentheses following the z.

Illustration 6–14

$z(0.05)$ is the value of z such that exactly 0.05 of the area under the curve lies to its right, as shown in figure 6–6.

FIGURE 6-6
Area Associated with z (0.05)

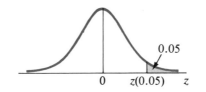

Now let's find the value of $z(0.05)$. We must convert this information into a value that can be read from table 5; see the areas shown in figure 6–7.

FIGURE 6–7
Finding the Value of z (0.05)

Now we look in table 5 in appendix D and find an area as close as possible to 0.4500.

z	···	0.04	0.05	···
⋮				
1.6	···	0.4495	0.4505	
⋮				

Therefore, $z(0.05) = 1.65$ (*Note*: We will use the z that corresponds to the area closest in value. If the value happens to be exactly halfway between the table entries, always round up to the larger value of z.) □

$z(0.60)$ is that value of z such that 0.60 of the area that lies to its right, as shown in figure 6–8.

FIGURE 6–8
Area Associated with z (0.60)

Illustration 6–15

Find the value of $z(0.60)$.

Solution   The value 0.60 is related to table 5 by use of the area 0.1000, as shown in the diagram.

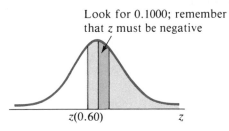

The closest values in table 5 are 0.0987 and 0.1026.

z	···	0.05	0.06	···
⋮				
0.2		0.0987	0.1026	
⋮				

Therefore, $z(0.60)$ is related to 0.25. Since $z(0.60)$ is below the mean, we conclude that $z(0.60) = -0.25$ ☐

In later chapters we will use this notation on a regular basis. The two values of z that will be used regularly come from one of the following situations: (1) the z score such that there is a specified area in one tail of the normal distribution, or (2) the z scores that bound a specified middle proportion of the normal distribution.

Illustration 6–14 showed a commonly used one-tail situation; $z(0.05) = 1.65$ is located so that 0.05 of the area under the normal distribution curve is in the tail to the right.

Illustration 6–16

Find $z(0.95)$.

Solution $z(0.95)$ is located on the left-hand side of the normal distribution since the area to the right is 0.95. The area in the tail to the left then contains the other 0.05, as shown in figure 6–9. Because of the symmetrical nature of the normal distribution, $z(0.95)$ is $-z(0.05)$; that is, $z(0.05)$ with its sign changed. Thus $z(0.95) = -z(0.05) = -1.65$. ☐

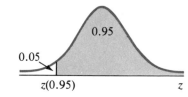

FIGURE 6–9
Area Associated with z (0.95)

When the middle proportion of a normal distribution is specified, we can still use the "area to the right" notation to identify the specific z score involved.

Illustration 6–17

Find the z scores that bound the middle 0.95 of the normal distribution.

Area Associated with Middle 0.95

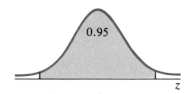

Solution   Given 0.95 as the area in the middle, the two tails must contain a total of 0.05. Therefore each tail contains 0.025, as shown in figure 6–10. In

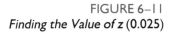

FIGURE 6–10
Finding z scores for Middle 0.95

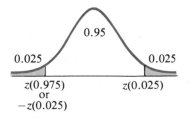

order to find $z(0.025)$ in table 5, we must determine the area between the mean and $z(0.025)$. It is 0.4750 as shown in figure 6–11.

FIGURE 6–11
Finding the Value of z (0.025)

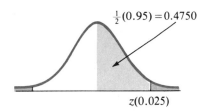

From table 5,

	...	**0.06**	...
$\vdots$			
1.9		0.4750	

Therefore, $z(0.025) = 1.96$ and $z(0.975) = -z(0.025) = -1.96$. The middle 0.95 of the normal distribution is bounded by **−1.96 and 1.96**. ☐

Exercises

6–26 Use the notation defined in this section to identify the *z* score shown in each of the diagrams.

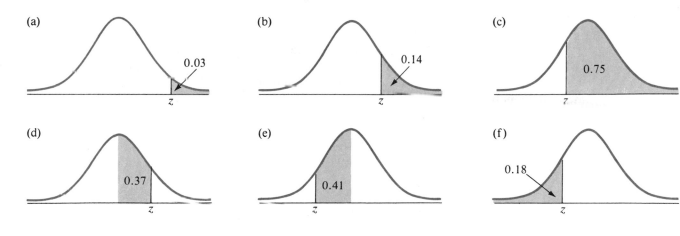

(a)

0.03

z

(b)

0.14

z

(c)

0.75

z

(d)

0.37

z

(e)

0.41

z

(f)

0.18

z

6–27 Use table 5 to find the following values of z.
  (a) z(0.05)   (b) z(0.01)   (c) z(0.025)
  (d) z(0.975)   (e) z(0.98)

6–28 Complete the following charts of z scores. The area A given in the tables is the area to the right under the normal distribution in the figures.
  (a) z scores associated with the right-hand tail: Given the area A, find z(A).

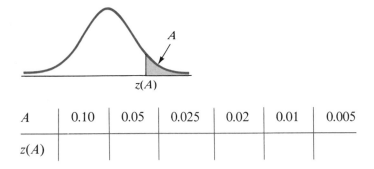

A	0.10	0.05	0.025	0.02	0.01	0.005
z(A)						

  (b) z scores associated with the left-hand tail: Given the area A, find z(A).

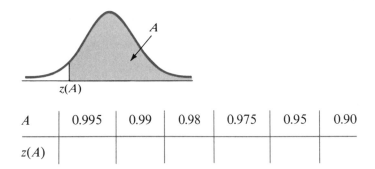

A	0.995	0.99	0.98	0.975	0.95	0.90
z(A)						

Section 6–5 Normal Approximation of the Binomial

In chapter 5 we introduced the binomial distribution. Recall that the binomial distribution is a probability distribution of the discrete random variable x, the number of successes observed in n repeated independent trials. We will now **binomial probability**   see how **binomial probabilities**; that is, probabilities associated with a binomial distribution, can be reasonably estimated by use of the normal probability distribution.

Let's look first at a few specific binomial distributions. Figures 6–12(a), 6–12(b), and 6–12(c) show the probabilities of x for 0 to n for three situations: n = 4, n = 8, and n = 24. For each of these distributions, the probability of success for one trial is 0.5. Notice that as n becomes larger, the distribution appears more and more like the normal distribution.

FIGURE 6–12
Binomial Distributions

(a) Distribution for $n=4$, $p=0.5$

(b) Distribution for $n=8$, $p=0.5$

(c) Distribution for $n=24$, $p=0.5$

To make the desired approximation, we need to take into account one major difference between the binomial and the normal probability distributions. The binomial random variable is discrete, whereas the normal random variable is continuous. Recall that in chapter 5 it was demonstrated that the probability assigned to a particular value of x should be shown on a diagram by means of a straight-line segment whose length represents the probability (as in figure 6–12). It was suggested, however, that we also can use a histogram in which the area of each bar is equal to the probability of x.

Let's look at the distribution of the binomial variable x, where $n = 14$ and $p = 0.5$. The probabilities for each x value can be obtained from table 4 in appendix D. This distribution of x is shown in figure 6–13. In histogram form we see the very same distribution in figure 6–14.

FIGURE 6–13
Line Histogram for the Distribution of x when $n = 14$, $p = 0.5$

FIGURE 6–14
Histogram for the Distribution of x when $n = 14$, $p = 0.5$

Let's examine $P(x = 4)$ for $n = 14$ and $p = 0.5$ to study the approximation technique. $P(x = 4)$ is equal to 0.061 (see table 4 of appendix D), the area of the bar above $x = 4$ in figure 6–15. Area is the product of width and height. In this case the height is 0.061 and the width is 1.0; thus the area is 0.061. Let's take a closer look at the width. For $x = 4$, the bar starts at 3.5 and ends at 4.5, so we are looking at an area bounded by $x = 3.5$ and $x = 4.5$.

The addition and subtraction of 0.5 to the x value is commonly called the **continuity correction factor**. It is our method of converting a discrete into a continuous variable.

Now let's look at the normal distribution related to this situation. We will first need a normal distribution with a mean and a standard deviation equal to those of the binomial distribution we are discussing. Formulas (5–7) and (5–8) give us these values.

$$\mu = np = (14)(0.5) = \mathbf{7.0}$$

$$\sigma = \sqrt{npq} = \sqrt{(14)(0.5)(0.5)} = \sqrt{3.5} = \mathbf{1.87}$$

The probability that $x = 4$ is approximated by the area under the normal curve between $x = 3.5$ and $x = 4.5$, as shown in figure 6–16. Figure 6–17 shows the entire distribution of the binomial variable x with a normal distribution of the same mean and standard deviation superimposed. Notice that the "bars" and the "interval areas" under the curve cover nearly the same area.

The probability that x is between 3.5 and 4.5 under this normal curve is found by using table 5 and the methods outlined in section 6–3.

$$P(3.5 < x < 4.5) = P\left(\frac{3.5 - 7.0}{1.87} < z < \frac{4.5 - 7.0}{1.87}\right)$$

$$= P(-1.87 < z < -1.34)$$

$$= 0.4693 - 0.4099 = \mathbf{0.0594}$$

continuity correction factor

FIGURE 6–15
Area of Bar Above $x = 4$ Is 0.061 when $n = 14$, $p = 0.5$

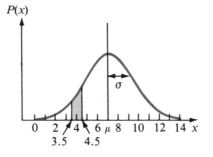

FIGURE 6–16
Probability That $x = 4$ Is Approximated by Shaded Area

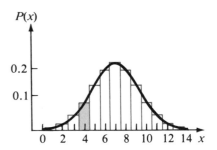

FIGURE 6–17
Normal Distribution Superimposed over Distribution for Binomial Variable x

Since the binomial probability of 0.061 and the normal probability of 0.0594 are reasonably close in value, the normal probability distribution seems to be a reasonable approximation of the binomial distribution.

By now you may be thinking, "So what? I will just use the binomial table and find the probabilities directly and avoid all the extra work." But consider for a moment a situation such as presented in illustration 6–18.

Illustration 6–18

An unnoticed mechanical failure has caused a machine shop's production of 5000 rifle firing pins to be $\frac{1}{3}$ defective. What is the probability that an inspector will find no more than 3 defective firing pins in a random sample of 25?

Solution   In this illustration of a binomial experiment, x is the number of defectives found in the sample, $n = 25$, and $p = P(\text{defective}) = \frac{1}{3}$. To answer the question by using the binomial distribution, we will need to use the binomial probability function [formula (5–6)]:

$$P(x) = \binom{25}{x} \cdot \left(\frac{1}{3}\right)^x \cdot \left(\frac{2}{3}\right)^{25-x} \quad \text{for} \quad x = 0, 1, 2, \ldots, 25$$

We must calculate the values for $P(0)$, $P(1)$, $P(2)$, and $P(3)$ since they do not appear in table 4. This is a very tedious job because of the size of the exponent. In situations such as this, we can use the normal approximation method.

Now let's find $P(x \leq 3)$ by using the normal approximation method. We first need to find the mean and standard deviation of x [formulas (5–7) and (5–8)]:

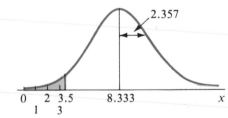

$$\mu = np = (25)(\tfrac{1}{3}) = \mathbf{8.333}$$

$$\sigma = \sqrt{npq} = \sqrt{(25(\tfrac{1}{3})(\tfrac{2}{3})} = \mathbf{2.357}$$

These values are shown in the diagram. The measure of the shaded area $(x < 3.5)$ in the diagram represents the probability of $x = 0, 1, 2,$ or 3. Remember that $x = 3$, the discrete binomial variable, covers the continuous interval from 2.5 to 3.5.

$$P(x \text{ is no more than } 3) = P(x \leq 3) \quad [\text{for discrete variable } x]$$
$$= P(x < 3.5) \quad [\text{using a continuous variable } x]$$

$$P(x < 3.5) = P\left(z < \frac{3.5 - 8.333}{2.357}\right) = P(z < -2.05)$$

$$= 0.5000 - 0.4798 = \mathbf{0.0202}$$

Thus $P(\text{no more than 3 defectives})$ is approximately 0.02. □

The normal approximation of the binomial distribution is also useful for values of p that are not close to 0.5. The binomial probability distributions shown in figures 6–18 and 6–19 suggest that binomial probabilities can be approximated by use of the normal distribution. Notice that as n increases

FIGURE 6–18
Binomial Distributions

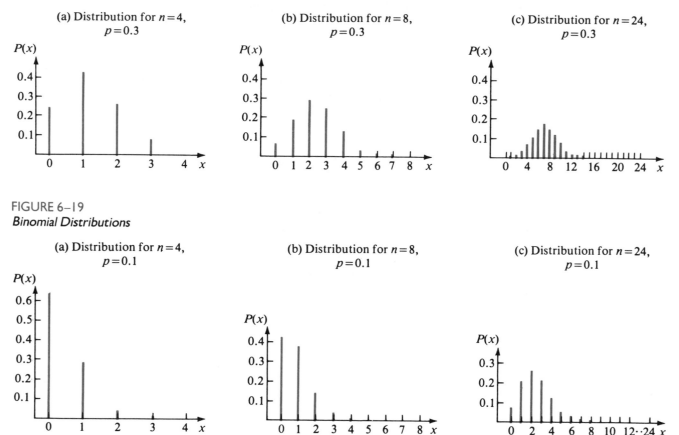

in size, the binomial distribution begins to look like the normal distribution. As the value of p moves away from 0.5, a larger n will be needed in order for the normal approximation to be reasonable. The following "rule of thumb" is generally used as a guideline.

Rule

The normal distribution provides a reasonable approximation to a binomial probability distribution whenever the values of np and $n(1 - p)$ both equal or exceed 5.

Exercises

6-29 Find the normal approximation for the binomial probability $P(x = 6)$, where $n = 12$ and $p = 0.6$. Compare this to the value of $P(x = 6)$ obtained from table 4.

6-30 Find the normal approximation for the binomial probability $P(x = 4, 5)$, where $n = 14$ and $p = 0.5$. Compare this to the value of $P(x = 4, 5)$ obtained from table 4.

6-31 Find the normal approximation for the binomial probability $P(x \leq 8)$, where $n = 14$ and $p = 0.4$. Compare this to the value of $P(x \leq 8)$ obtained from table 4.

6-32 A production run of tee shirts is expected to produce seconds 5% of the time. If you randomly select 100 tee shirts from a large run,

(a) What is the mean and the standard deviation of the binomial distribution when samples of 100 are taken?

(b) What is the probability that your sample will contain no tee shirt seconds?

(c) What is the probability that your sample will contain no more than two tee shirt seconds?

6-33 A company asserts that 80% of the customers who purchase its special lawn mower will have no repairs during the first two years of ownership. Your personal study has shown that only 70 of the 100 in your sample lasted the two years without repair expenses. What is the probability of your sample outcome or less, if the actual expenses-free percentage is 80%?

6-34 Surveys have shown that 40% of the United States population have used Crest toothpaste at one time or another. A researcher, seeking to substantiate that 40% is correct, took a properly conducted random sample and found that 220 of the 600 individuals contacted had used Crest. What is the probability that a sample of 600 would have fewer than 221 individuals, who had used Crest, if the 40% is really correct?

6-35 If 30% of all students entering a certain university drop out during or at the end of their first year, what is the probability that more than 600 of this year's entering class of 1800 will drop out during or at the end of their first year?

6-36 In our state 30% of the registered voters are Republicans and 70% are Democrats. In a random sample of 100 voters, what is the probability that half or more are Democrats?

6-37 Consider a production process that results in a 10% defective product. A sample of size 60 is taken at random.

(a) What is the probability that no more than 3 defective items are found?

(b) What is the probability that more than 8 defective items are found?

6-38 It is believed that the student body is equally split on a "new pub program" proposal. Assuming that to be the case, what is the probability that the Student Senate's straw poll of 100 student opinions shows at least 60% favoring the new proposal?

In Retrospect

We now know what a normal distribution is, how to use it, and how it can help us. Let's again consider the news article at the beginning of the chapter, which illustrated a normally distributed continuous random variable. The random variable is the wavelength of a typical infrared LED. The idea of color multiplexing takes advantage of the fact that the vast majority of the values of a normally distributed variable are clustered about the mean value. We have seen many illustrations of normal distributions.

In the next chapter we will examine sampling distributions and learn how to apply the normal distribution to additional applications.

Chapter Exercises

6-39 The article at the beginning of this chapter includes a figure of a normalized distribution. What continuous random variable is shown in figure 3 in the article?

6-40 The middle 60% of a normally distributed population lies between what two standard scores?

6-41 Find the standard score z such that the area above the mean and below z under the normal curve is
(a) 0.3962 (b) 0.4846 (c) 0.3712

6-42 Find the standard score z such that the area below the mean and above z under the normal curve is
(a) 0.3212 (b) 0.4788 (c) 0.2700

6-43 Find the standard score of a normally distributed variable such that 42% of the distribution falls between the mean and this particular value.

6-44 Given that z is the standard normal variable, find the value of k such that:
(a) $P(|z| > 1.68) = k$ (b) $P(|z| < 2.15) = k$

6-45 Given that z is the standard normal variable, find the value of c such that:
(a) $P(|z| > c) = 0.0384$ (b) $P(|z| < c) = 0.8740$

6-46 Find the following values of z:
(a) $z(0.12)$ (b) $z(0.28)$ (c) $z(0.85)$ (d) $z(0.99)$

6-47 During daylight hours on weekdays, airplanes arriving at a certain airport average 36 arrivals per hour. Given that the variance of arrivals per hour is 36 and assuming that the number of arrivals per hour is approximately normally distributed; what is the probability that during a randomly selected 1-hour period,

(a) fewer than 30 airplanes will arrive?

(b) more than 45 will arrive?

6-48 Find the boundaries of the range of the middle 0.9 of population IQ scores; $\mu = 100$ and $\sigma = 10$.

6-49 Let x be the length of the life of the battery in your automobile. The manufacturer guaranteed that it would last 3 years. The battery was manufactured so that the mean life was to be 3.25 years, with a standard deviation of 3 months (0.25 year). Find the probability with which the manufacturer expects the product to fail, that is, $P(x \leq 3$ years$)$.

6-50 A soft drink vending machine can be regulated so as to dispense an average of μ ounces of soft drink per glass. If the ounces dispensed per glass are normally distributed with a standard deviation of 0.2 ounce, find the setting for μ that will allow a 6-ounce glass to hold (without overflowing) the amount dispensed 99% of the time.

6-51 The length of life of a certain type of refrigerator is approximately normally distributed with a mean of 4.8 years and a standard deviation of 1.3 years.

(a) If this machine is guaranteed for 2 years, what is the probability that the machine you purchased will require replacement under the guarantee?

(b) What period of time should the manufacturer give as a guarantee if he is willing to replace only 0.5% of the machines?

6-52 The average length of time required for completing a certain academic achievement test is believed to be 150 minutes, and the standard deviation is 20 minutes. If we wish to allow sufficient time for only 80% to complete the test, when should the test be terminated? (Assume that the lengths of time required to complete this test are normally distributed.)

6-53 In a large industrial complex the maintenance department has been instructed to replace light bulbs before they burn out. It is known that the life of light bulbs is normally distributed with a mean life of 900 hours of use and a standard deviation of 75 hours. When should the light bulbs be replaced so that no more than 10% of them will burn out while in use?

6-54 For a certain brand of TV-dinners, the mean weight of the meat in the chopped beef dinners is 3.6 ounces and the standard deviation is 0.2 ounce. The beef portions are measured mechanically and any deviation from 3.6 ounces is a mechanical error. The result of this is that the weights of the chopped beef are approximately normally distributed.

(a) In what proportion of the dinners will the chopped beef weigh less than 3.5 ounces?

(b) In what proportion of the dinners will the chopped beef weigh over 3.8 ounces?

(c) What proportion will have weights between 3.5 and 3.8 ounces?

(d) Eighty-five percent of the dinners will contain more than what weight of chopped beef?

6-55 Assume that the number of cases sold per week in December by a large group of beverage salesmen is quite close to a normal distribution. The mean is 684 cases and the standard deviation is 60.

(a) Given that the top 16% of the salesmen receive bonuses, how many cases must a salesman sell in order to receive a bonus?

(b) If the break-even number of cases is 600 per salesman, what proportion of the salesmen are above the break-even value?

6-56 A test-scoring machine is known to record an incorrect grade on 5% of the exams it grades. Find, by the appropriate method, the probability that the machine records

(a) 3 wrong grades in a set of 5 exams

(b) no more than 3 wrong grades in a set of 5 exams

(c) no more than 3 wrong grades in a set of 15 exams

(d) no more than 3 wrong grades in a set of 150 exams

6-57 A small airline has a 0.10 probability of arriving at a flight destination more than 20 minutes late. This percentage is based on historical information gathered by the airline's research division. Assuming that the estimate is valid, what is the probability that fewer than 5 of the next 60 scheduled flights will arrive at their destinations more than 20 minutes late?

6-58 It is believed that 58% of the married couples with children agree on methods of disciplining their children. Assuming that to be the case, what is the probability that in a random survey of 200 married couples, we would find the following?

(a) exactly 110 couples who agree

(b) less than 110 couples who agree

(c) more than 100 couples who agree

6-59 A new drug is supposed to be 85% effective in treating a particular illness. (That is, 85% of the patients with this illness respond favorably to the drug.) Let x be the number of patients out of every group of 50 who respond favorably. Use the normal approximation method to find these probabilities:

(a) $P(x > 45)$ (b) $P(40 < x < 50)$ (c) $P(x < 35)$

6-60 The grades on an examination whose mean score is 525 and whose standard deviation is 80 are normally distributed.

(a) Anyone who scores below 350 will be retested. What percentage does this represent?

(b) The top 12% are to receive a special commendation. What score must be surpassed to receive this special commendation?

(c) The interquartile range of a distribution is the difference between Q_1 and Q_3, $Q_3 - Q_1$. Find the interquartile range for the grades on this examination.

(d) Find the grade such that only 1 out of 500 will score above it.

6-61 The designers for Custom Sport Coupe (illustration 1–3) want to make the roof in their new model high enough so that there is at least 1 inch of clearance between the headliner (ceiling) and the top of the driver's head for 90% of the drivers. If the heights of sports car enthusiasts are normally distributed with a mean of 68 inches and a standard deviation of 2.5 inches, what height must the designers work with?

Vocabulary List

Be able to define each term. In addition, describe, in your own words, and give an example of each term. Your examples should not be ones given in class or in the textbook.

The terms followed by bracketed numbers were defined in earlier chapters (the bracketed numbers are those of the earlier chapters), but we ask you to define them again to show your increased understanding of their meaning.

area representation
  for probability
bell-shaped curve
binomial distribution [5]
binomial probability
continuity correction factor
discrete random variable [1, 5]
normal approximation
  of binomial

normal curve
normal distribution
percentage
probability [4]
proportion
random variable [5]
standard normal distribution
standard score [2]
z score [2]

Quiz

Answer "True" if the statement is always true. If the statement is not always true replace the words shown in color with words that make the statement always true.

6-1 The normal probability distribution is symmetric about 0.

6-2 The total area under the curve of any normal distribution is 1.

6-3 The theoretical probability that a particular value of a **continuous** random variable will occur is exactly 0.

6-4 The unit of measure of the standard score is **the same as** the unit of measure of the data.

6-5 All **normal distributions** have the same general probability function and distribution.

6-6 When using the notation $z(0.05)$, the number in the parentheses is the measure of the area to the **left** of the z score.

6–7   The standard normal score has a mean of **1** and a standard deviation of **0**.

6–8   Probability distributions of **all** continuous random variables are normal.

6–9   We are able to add and subtract the areas under the curve because these areas represent the probabilities of **independent events**.

6–10 The most common distribution of a continuous random variable is the **binomial probability**.

7 Sample Variability

Galluping Attitudes

Since its founding in 1935, the American Institute of Public Opinion, better known as the Gallup poll, has put approximately 20,000 questions to more than two million people.

One of the most interesting social changes to follow through the years in these pages is that of Americans' attitudes toward women at work and in politics. Herewith, an abbreviated reading of Gallup's progress report.

A Woman President

1937

Would you vote for a woman for President, if she qualified in every other respect?

Yes.................................... 34%
No 66

By Sex	Yes	No
Men	27%	73%
Women	41	59

1955

If the party whose candidate you most often support nominated a woman for President of the United States, would you vote for her if she seemed best qualified for the job?

Yes.................................... 52%
No 44

By Sex	Yes	No
Men	47%	48%
Women	57	40

1971

If your party nominated a woman for President, would you vote for her if she were qualified for the job?

Yes.................................... 66%
No 29

By Sex	Yes	No
Men	65%	35%
Women	67	33

1978

If your party nominated a woman for president, would you vote for her if she were qualified for the job?

Yes.................................... 76%
No 19

By Sex	Yes	No
Men	76%	19%
Women	77	18

Chapter Objectives

In chapters 2 and 3 we discussed how to obtain and describe a sample. The description of the sample data is accomplished by using three basic concepts: (1) measures of central tendency (the mean is the most popularly used sample statistic), (2) measures of dispersion (the standard deviation is most commonly used), and (3) kind of distribution (normal, skewed, rectangular, and so on). The question that seems to follow is this: What can be deduced about the statistical population from which a sample is taken?

To put this query at a more practical level, suppose that you have just taken a sample of 25 rivets made for the construction of airplanes. The rivets were tested for shearing strength, and the force required to break each rivet was the response variable. The various descriptive measures—mean, standard deviation, type of distribution—can be found for this sample. However, it is not the sample itself that we are interested in. The rivets that were tested were destroyed during the test, so they can no longer be used in the construction of airplanes. What we are trying to find out is information about the total population, and we certainly cannot test every rivet that is produced (there would be none left for construction). Therefore, somehow we must deduce information, or make inferences, about the population based on the results observed in the sample.

Suppose that we take another sample of 25 rivets and test them by the same procedure. Do you think that we would obtain the same sample mean from the second sample that we obtained from the first? The same standard deviation?

After considering these questions we might suspect that we would need to investigate the variability in the sample statistics obtained from **repeated sampling**. *Thus we need to find (1) measures of central tendency for the sample statistics of importance, (2) measures of dispersion for the sample statistics, and (3) the pattern of variability (distribution) of the sample statistics. Once we have this information, we will be better able to predict the population parameters.*

The objective of this chapter is to study the measures and the patterns of variability for the distribution formed by repeatedly observed values of a sample mean.

Section 7–1 Sampling Distributions

To make inferences about a population, we need to discuss sample results a little more. A sample mean, $\bar{x}$, is obtained from a sample. Do you expect that this value, $\bar{x}$, is exactly equal to the value of the population mean μ? Your answer should be "no." We do not expect that to happen, but we will be satisfied with our sample results if the sample mean is "close" to the value of the population mean. Let's consider a second question: If a second sample is taken, will the second sample have a mean equal to the population mean? Equal to the first sample mean? Again, no, we do not expect it to be equal to the population mean, nor do we expect the sample mean to repeat.

We do, however, again expect the values to be "close." (This argument should hold for any other sample statistic and its corresponding population value.)

The next questions should already have come to mind: What is "close"? How do we determine (and measure) this closeness? Just how would repeated sample statistics be distributed? To answer these questions we must take a look at a "sampling distribution."

sampling distribution

Sampling Distribution of a Sample Statistic

The distribution of values for that sample statistic obtained from all possible samples of a population. The samples must all be the same size, and the sample statistic could be any descriptive sample statistic.

Illustration 7–1

To illustrate the concept of a sampling distribution, let's consider the mean of each sample of size 2 that can be drawn with replacement from the set of even single-digit integers, (0, 2, 4, 6, 8). There are 25 possible samples of size 2:

(0, 0)	(2, 0)	(4, 0)	(6, 0)	(8, 0)
(0, 2)	(2, 2)	(4, 2)	(6, 2)	(8, 2)
(0, 4)	(2, 4)	(4, 4)	(6, 4)	(8, 4)
(0, 6)	(2, 6)	(4, 6)	(6, 6)	(8, 6)
(0, 8)	(2, 8)	(4, 8)	(6, 8)	(8, 8)

Each of these samples has a mean $\bar{x}$. These means are, respectively,

0	1	2	3	4
1	2	3	4	5
2	3	4	5	6
3	4	5	6	7
4	5	6	7	8

TABLE 7–1
Sampling Distribution of Sample Means

$\bar{x}$	$P(\bar{x})$
0	0.04
1	0.08
2	0.12
3	0.16
4	0.20
5	0.16
6	0.12
7	0.08
8	0.04

Each of these samples is equally likely, and thus each of the 25 sample means can be assigned a probability of $\frac{1}{25} = 0.04$. [Why? See exercise 7–1(c).] The sampling distribution for the sample mean then becomes as shown in table 7–1. This is a probability distribution of $\bar{x}$. (See figure 7–1.)

For the same set of all possible samples of size 2, let's find the sampling distribution for sample ranges. Each sample has a range R. These ranges are, respectively,

0	2	4	6	8
2	0	2	4	6
4	2	0	2	4
6	4	2	0	2
8	6	4	2	0

FIGURE 7–1
Histogram: Sampling Distribution of
Sample Means

FIGURE 7–2
Histogram: Sampling Distribution of
Sample Ranges

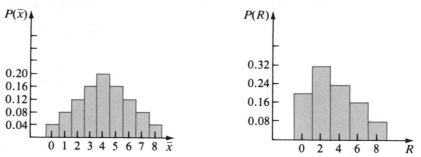

Again, each of these 25 sample ranges has a probability of 0.04, and table 7–2 shows the sampling distribution of sample ranges. This is a probability distribution of R. (See figure 7–2.)

TABLE 7–2
Sampling Distribution of Sample
Ranges

R	P(R)
0	0.20
2	0.32
4	0.24
6	0.16
8	0.08

Most populations that are sampled are much larger than the one used in illustration 7–1, and it would be a very tedious job to list all the possible samples. With this in mind, let's investigate a sampling distribution empirically (that is, by experimentation).

Illustration 7–2

Let's consider the sampling distribution of sample means for samples of size 5 obtained from the rolling of a single die. One sample will consist of 5 rolls of the die, and we will obtain a sample mean $\bar{x}$ from this sample. We repeat the experiment until 30 sample means have been obtained. Table 7–3 shows 30 such samples and their means. The resulting frequency distribution is shown in figure 7–3. This distribution seems to display characteristics of a normal distribution; it's mounded and nearly symmetric about its mean (approximately 3.5).

FIGURE 7–3

Sampling Distribution of Sample Means
for Rolling a Die

TABLE 7–3
Sample Means for Rolling
a Single Die 5 Times

Trial	Sample	$\bar{x}$	Trial	Sample	$\bar{x}$
1	1, 2, 3, 2, 2	2.0	16	5, 2, 1, 3, 5	3.2
2	4, 5, 5, 4, 5	4.6	17	6, 1, 3, 3, 5	3.6
3	3, 1, 5, 2, 4	3.0	18	6, 5, 5, 2, 6	4.8
4	5, 6, 6, 4, 2	4.6	19	1, 3, 5, 5, 6	4.0
5	5, 4, 1, 6, 4	4.0	20	3, 1, 5, 3, 1	2.6
6	3, 5, 6, 1, 5	4.0	21	5, 1, 1, 4, 3	2.8
7	2, 3, 6, 3, 2	3.2	22	4, 6, 3, 1, 2	3.2
8	5, 3, 4, 6, 2	4.0	23	1, 5, 3, 4, 5	3.6
9	1, 5, 5, 3, 4	3.6	24	3, 4, 1, 3, 3	2.8
10	4, 1, 5, 2, 6	3.6	25	1, 2, 4, 1, 4	2.4
11	5, 1, 3, 3, 2	2.8	26	5, 2, 1, 6, 3	3.4
12	1, 5, 2, 3, 1	2.4	27	4, 2, 5, 6, 3	4.0
13	2, 1, 1, 5, 3	2.4	28	4, 3, 1, 3, 4	3.0
14	5, 1, 4, 4, 6	4.0	29	2, 6, 5, 3, 3	3.8
15	5, 5, 6, 3, 3	4.4	30	6, 3, 5, 1, 1	3.2

The theory involved with sampling distributions that will be described in the remainder of this chapter requires "random sampling."

random sample

Random Sample

A sample obtained in such a way that each of the possible samples of fixed size has an equal chance to be selected.

Exercises

7–1 (a) What is the sampling distribution of sample means?

(b) A sample of size 3 is taken from a population and the sample mean found. Describe how this sample mean is related to the sampling distribution of sample means.

(c) Why is the probability of 0.04 assigned to each of the sample mean values in illustration 7–1?

7–2 Consider the set of odd single-digit integers {1, 3, 5, 7, 9}.

(a) Make a list of all samples of size 2 that can be drawn from this set of integers. (Sample with replacement; that is, the first number is drawn, observed, then replaced before the next drawing.)

(b) Construct the sampling distribution of sample means for samples of size 2 selected from this set.

(c) Construct the sampling distributions of sample ranges for samples of size 2.

7-3 Consider the set of even single-digit integers {0, 2, 4, 6, 8}.

(a) Make a list of all the possible samples of size 3 that can be drawn from this set of integers. (Sample with replacement; that is, the first number is drawn, observed, then replaced before the next drawing.)

(b) Construct the sampling distribution of the sample medians for samples of size 3.

(c) Construct the sampling distribution of the sample means.

7-4 Suppose that a box contains three identical blocks numbered 2, 4, and 6. A sample of two numbers is drawn with replacement (that is, the first number is drawn, observed, and returned; then the second number is drawn). The mean of this sample is determined.

(a) Make a list that shows all the possible samples that could result from this sampling. (*Hint*: There should be 9 samples.)

(b) Determine the mean of each of these samples and form a sampling distribution of these sample means. (Express as a probability distribution.)

(c) Find the mean of this sampling distribution, $\mu_{\bar{x}}$.

(d) Find the standard deviation, $\sigma_{\bar{x}}$ for the distribution.

7-5 Using the telephone numbers listed in your local directory as your population, obtain randomly 20 samples of size 3. From each number identified as a source, take the fourth, fifth, and sixth digits. (For example, for 245-(826)8, you would take the 8, the 2, and the 6 as your sample of size 3.)

(a) Calculate the mean of the 20 samples.

(b) Draw a histogram showing the 20 sample means. (Use classes −0.5 to 0.5, 0.5 to 1. 5, 1.5 to 2.5, and so on.)

Section 7-2 The Central Limit Theorem

On the preceding pages we discussed two sampling distributions. There are many others that could be discussed; in fact, these two could themselves be discussed further. However, the only sampling distribution of concern to us here is the **sampling distribution of sample means**. The mean is the most commonly used sample statistic and thus is the most important.

The central limit theorem (CLT) tells us about the sampling distribution of sample means of random samples of size n. Recall that there are basically three kinds of information that we want about a distribution: (1) where the center is, (2) how widely it is dispersed, and (3) how it is distributed. The central limit theorem tells us all three.

Central Limit Theorem

If all possible random samples, each of size n, are taken from any population with a mean μ and standard deviation σ, the sampling distribution of sample means will

1. have a mean $\mu_{\bar{x}}$ equal to μ
2. have a standard deviation $\sigma_{\bar{x}}$ equal to $\sigma/\sqrt{n}$
3. be normally distributed when the parent population is normally distributed or will be approximately normally distributed for samples of size 30 or more when the parent population is not normally distributed. The approximation to the normal distribution improves with samples of larger size.

In short, the central limit theorem states the following:

1. $\mu_{\bar{x}} = \mu$; the mean of the $\bar{x}$'s equals the mean of the x's.
2. $\sigma_{\bar{x}} = \sigma/\sqrt{n}$; the standard error of the mean (see the definition that follows) equals the standard deviation of the population divided by the square root of the sample size.
3. The sample means are approximately normally distributed (regardless of the shape of the parent population).

NOTE: The n referred to in the central limit theorem is the **size of each sample** in the sampling distribution.

Standard Error of the Mean

The standard deviation of the sampling distribution of sample means

We are unable to prove the central limit theorem without using advanced mathematics. However, it is possible to check its validity by examining two illustrations. Let's consider a population for which we can construct the theoretical sampling distribution of all possible samples. For this example let's consider all possible samples of size 2 that could be drawn from a population that contains the three numbers 2, 4, and 6.

First let's look at the population itself. To calculate the mean μ and the standard deviation σ, we must use the formulas from chapter 5 for discrete probability distributions:

$$\mu = \sum [x \cdot P(x)] \quad \text{and} \quad \sigma = \sqrt{\sum [x^2 \cdot P(x)] - \{\sum [x \cdot P(x)]\}^2}$$

These formulas are necessary because we are not drawing the samples but discussing the theoretical possibilities. See table 7–4.

TABLE 7–4

Probability Distribution and Extensions for x = 2, 4, 6

x	P(x)	x · P(x)	x² · P(x)
2	$\frac{1}{3}$	$\frac{2}{3}$	$\frac{4}{3}$
4	$\frac{1}{3}$	$\frac{4}{3}$	$\frac{16}{3}$
6	$\frac{1}{3}$	$\frac{6}{3}$	$\frac{36}{3}$
Total	$\frac{3}{3}$ ⓒ	$\frac{12}{3}$	$\frac{56}{3}$

$$\mu = \tfrac{12}{3} = \mathbf{4.0}$$

$$\sigma = \sqrt{56/3 - (12/3)^2} = \sqrt{18.66 - 16.0} = \sqrt{2.66} = \mathbf{1.63}$$

Table 7–5 gives a list of all the possible samples that could be drawn if samples of size 2 were to be drawn from this population. (One number is

TABLE 7–5

All Possible Samples of Size 2 and Their Means

Possible Samples	$\bar{x}$
2, 2	2
2, 4	3
2, 6	4
4, 2	3
4, 4	4
4, 6	5
6, 2	4
6, 4	5
6, 6	6

drawn, observed, and then returned to the population before the second number is drawn.) Table 7–5 also lists the means of these samples. The probability distribution for these means and the extensions are given in table 7–6. Thus we have

$$\mu_{\bar{x}} = \frac{36}{9} = \mathbf{4.0}$$

$$\sigma_{\bar{x}} = \sqrt{\frac{156}{9} - \left(\frac{36}{9}\right)^2} = \sqrt{17.33 - 16} = \sqrt{1.33} = \mathbf{1.15}$$

The histogram for the distribution of possible $\bar{x}$'s is shown in figure 7–4.

The CLT says that three things will occur in this sampling distribution:

1. It will be approximately normally distributed. The histogram (figure 7–4) suggests this very strongly.

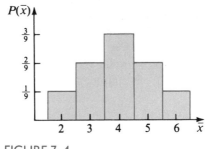

FIGURE 7–4

Histogram for Distribution of Table 7–6

242   Chapter 7   Sample Variability

TABLE 7–6

*Probability Distribution for Means of
All Possible Samples of Size 2 and the
Extensions*

$\bar{x}$	$P(\bar{x})$	$\bar{x}\cdot P(\bar{x})$	$\bar{x}^2\cdot P(\bar{x})$
2	1/9	2/9	4/9
3	2/9	6/9	18/9
4	3/9	12/9	48/9
5	2/9	10/9	50/9
6	1/9	6/9	36/9
Total	9/9 Ⓒⓚ	36/9	156/9

2. The mean $\mu_{\bar{x}}$ of the sampling distribution will equal the mean of the population. They both have the value 4.0.

3. The standard deviation $\sigma_{\bar{x}}$ of the sampling distribution will equal the standard deviation of the population divided by the square root of the sample size ($\sigma/\sqrt{n}$):

$$\sigma_{\bar{x}} = 1.15 \quad \text{and} \quad \frac{\sigma}{\sqrt{n}} = \frac{1.63}{\sqrt{2}} = \frac{1.63}{1.41} = 1.15$$

This illustration shows that the CLT is true for a **theoretical probability distribution**.

Now let's look at the empirical distribution that occurred in illustration 7–2 and see if it supports the three claims of the central limit theorem.

First, let's look at the theoretical probability distribution from which these samples were taken (table 7–7). A histogram showing the probability distribution of the tossing of a die is shown in figure 7–5. The population mean μ equals **3.5** (see table 7–7). The population standard deviation σ equals $\sqrt{15.17 - (3.5)^2}$, which is $\sqrt{2.92} = \textbf{1.71}$. (Note that this population has a uniform distribution.)

Now let's look at the empirical sampling distribution of the 30 sample means found earlier. Using the 30 values of $\bar{x}$ in table 7–3, the observed

TABLE 7–7

*Probability Distribution and
Extensions for Rolling a Die*

x	$P(x)$	$x\cdot P(x)$	$x^2\cdot P(x)$
1	$\frac{1}{6}$	$\frac{1}{6}$	$\frac{1}{6}$
2	$\frac{1}{6}$	$\frac{2}{6}$	$\frac{4}{6}$
3	$\frac{1}{6}$	$\frac{3}{6}$	$\frac{9}{6}$
4	$\frac{1}{6}$	$\frac{4}{6}$	$\frac{16}{6}$
5	$\frac{1}{6}$	$\frac{5}{6}$	$\frac{25}{6}$
6	$\frac{1}{6}$	$\frac{6}{6}$	$\frac{36}{6}$
Total	$\frac{6}{6}=1$ Ⓒⓚ	$\frac{21}{6}=3.5$	$\frac{91}{6}=15.17$

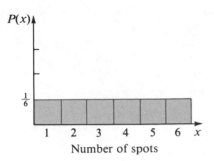

FIGURE 7–5
Probability Distribution for Rolling a Die

mean of the $\bar{x}$'s turns out to be 3.43 and the observed standard deviation $s_{\bar{x}}$ turns out to be 0.73. The histogram appears in figure 7–3.

The central limit theorem states that the $\bar{x}$'s should be approximately normally distributed, and the histogram certainly suggests this to be the case. The CLT also says that the mean $\mu_{\bar{x}}$ of the sampling distribution and the mean μ of the population are the same. The mean of the $\bar{x}$'s is 3.43 and $\mu = 3.5$; they seem to be reasonably close. Remember that we have taken only 30 samples, not all possible samples, of size 5.

The theorem says that $\sigma_{\bar{x}}$ should equal $\sigma/\sqrt{n}$. The observed standard deviation of $\bar{x}$'s is $s_{\bar{x}} = 0.73$ and the standard error of the mean is

$$\frac{\sigma}{\sqrt{n}} = \frac{1.71}{\sqrt{5}} = 0.76$$

These two values are very close.

The evidence seen in these two illustrations seems to suggest that the CLT is true, although this does not constitute a proof of the theorem, of course.

Having taken a look at these two specific illustrations that support the CLT, let's now look at four graphic illustrations that present the same information in slightly different form. In each of these graphic illustrations there are four distributions. The first is a distribution of the parent population, the distribution of the individual x values. Each of the other three graphs show a sampling distribution of sample means, using three different sample sizes. In figure 7–6 we have a uniform distribution, much like figure 7–5 for the die illustration, and the resulting distributions of sample means for

FIGURE 7–6
Uniform Distribution

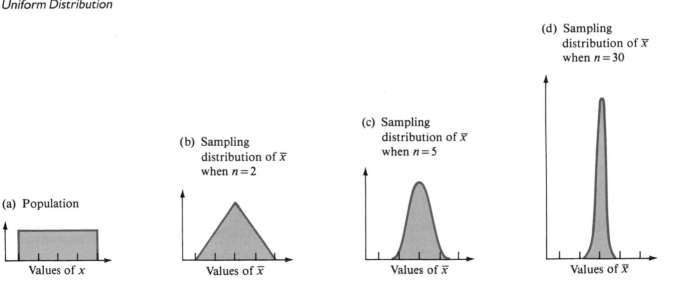

samples of size 2, 5, and 30. Figure 7–7 shows a U-shaped population and the corresponding sampling distributions. Figure 7–8 shows a J-shaped population and the three corresponding distributions. Figure 7–9 shows a normal distribution population and the three sampling distributions.

All four illustrations seem to verify the CLT. Note that the sampling distribution of the three nonnormal distributions produced sample means with an approximately normal distribution for samples of size 30. In the normal population (figure 7–9) the sampling distributions for all sample sizes appear to be normal. Thus you have seen an amazing phenomenon: no matter what the shape of a population, the sampling distribution of the

FIGURE 7–7
U-shaped Distribution

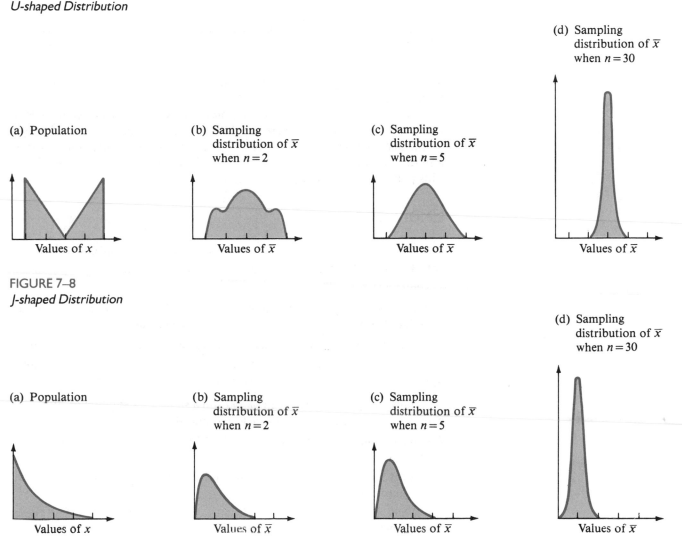

(a) Population   (b) Sampling distribution of $\bar{x}$ when $n=2$ (c) Sampling distribution of $\bar{x}$ when $n=5$ (d) Sampling distribution of $\bar{x}$ when $n=30$

Values of x Values of $\bar{x}$ Values of $\bar{x}$ Values of $\bar{x}$

FIGURE 7–8
J-shaped Distribution

(a) Population   (b) Sampling distribution of $\bar{x}$ when $n=2$ (c) Sampling distribution of $\bar{x}$ when $n=5$ (d) Sampling distribution of $\bar{x}$ when $n=30$

Values of x Values of $\bar{x}$ Values of $\bar{x}$ Values of $\bar{x}$

FIGURE 7–9
Normal Distribution

(a) Population

(b) Sampling distribution of $\bar{x}$ when $n=2$

(c) Sampling distribution of $\bar{x}$ when $n=5$

(d) Sampling distribution of $\bar{x}$ when $n=30$

mean becomes approximately normally distributed when n becomes sufficiently large.

You should notice one other point; the sample mean becomes less variable as the sample size increases. Notice that as n increases from 2 to 30, all the distributions become narrower and taller. Can you explain how this implies less variability? How does the CLT state this? (See exercise 7–6.) This point is discussed further in section 7–3.

Exercises

7–6 (a) What is the measure of the total area for any probability distribution?

(b) How does the CLT state that as n becomes larger, the sample mean becomes less variable?

7–7 A normal population has a mean of 50 and a standard deviation of 10. Several samples of size 16 are taken and their means $\bar{x}$ recorded.

(a) What value would you expect to find for the mean of the sample means?

(b) If you were to calculate the standard deviation of those sample means, approximately what value would you expect to find? Explain.

7–8 A certain population has a mean of 500 and a standard deviation of 30. Many samples of size 36 are randomly selected and their mean calculated.

(a) What value would you expect to find for the mean of all these sample means?

(b) What value would you expect to find for the standard deviation of all these sample means?

(c) What shape distribution would you expect the distribution of all these sample means to have?

7-9   If a normal population has a standard deviation σ of 25 units, what is the standard error of the mean $(\sigma_{\bar{x}})$ if samples of size 16 are used? Size 25? Size 50? Size 100?

7-10   Consider the experiment of taking a standardized mathematics test. The variable x is the raw score received. This exam has a mean of 720 and a standard deviation of 60. A group (sample) of 40 students takes the exam and the sample mean $\bar{x}$ is 725.6. A sampling distribution of means is formed from the means of all such groups of 40 students.

(a) Determine the mean of this sampling distribution.

(b) Determine the standard deviation for this sampling distribution.

Section 7-3 Application of the Central Limit Theorem

The central limit theorem tells us about the sampling distribution of sample means by describing the shape of the distribution of all possible sample means. It also specifies the relationship between the mean μ of the population and the mean $\mu_{\bar{x}}$ of the sampling distribution, and the relationship between the standard deviation σ of the population and the standard deviation $\sigma_{\bar{x}}$ of the sampling distribution. Since sample means are approximately normally distributed, we will be able to answer probability questions by using table 5 of appendix D.

Illustration 7-3

Consider a normal population with $\mu = 100$ and $\sigma = 20$. If a sample of size 16 is selected at random, what is the probability that this sample will have a mean value between 90 and 110? That is, what is $P(90 < \bar{x} < 110)$?

Solution   The CLT states that the distribution of $\bar{x}$'s is approximately normally distributed. To determine probabilities associated with a normal distribution, we will need to convert the statement $P(90 < \bar{x} < 110)$ to a probability statement concerning z in order to use table 5, the standard normal distribution table. The sampling distribution is shown in the accompanying figure, with $P(90 < \bar{x} < 110)$ represented by the shaded area.

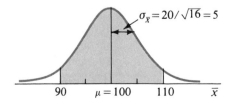

The formula for finding z, knowing a value of $\bar{x}$, is

$$z = \frac{\bar{x} - \mu_{\bar{x}}}{\sigma_{\bar{x}}} \qquad (7\text{--}1)$$

However, the CLT tells us that $\mu_{\bar{x}} = \mu$ and $\sigma_{\bar{x}} = \sigma/\sqrt{n}$. Therefore, we will rewrite formula (7–1) in terms of μ and σ:

$$z = \frac{\bar{x} - \mu}{\sigma/\sqrt{n}} \qquad (7\text{--}2)$$

Using formula (7–2) we find that $\bar{x} = 90$ has a standard score of

$$z = \frac{90 - 100}{20/\sqrt{16}} = \frac{-10}{5} = -2.0$$

$\bar{x} = 110$ has a standard score of

$$z = \frac{110 - 100}{20/\sqrt{16}} = \frac{10}{5} = 2.0$$

Therefore,

$$P(90 < \bar{x} < 110) = P(-2.0 < z < 2.0) = 2(0.4772) = 0.9544 \qquad \square$$

Before we look at more illustrations, let's consider for a moment what is implied by saying that $\sigma_{\bar{x}} = \sigma/\sqrt{n}$. For demonstration purposes let's suppose that $\sigma = 20$ and let's use a sampling distribution of samples of size 4. Now $\sigma_{\bar{x}}$ would be $20/\sqrt{4}$, or 10, and approximately 95 % (0.9544) of all such sample means should be within the interval from 20 below to 20 above the population mean (within two standard deviations of the population mean). However, if the sample size were increased to 16, $\sigma_{\bar{x}}$ would become

$$20/\sqrt{16} = 5$$

and approximately 95 % of the sampling distribution would be within 10 units of the mean; and so on. As the sample size increases, the size of $\sigma_{\bar{x}}$ becomes smaller, so that the distribution of sample means becomes much narrower. Figure 7–10 illustrates what happens to the distribution of $\bar{x}$'s as the size of the individual samples increases. Recall that the area under the normal curve is always exactly one unit of area. So as the width of the curve narrows, the height will have to increase in order to maintain this area.

FIGURE 7–10
Distributions of Sample Means

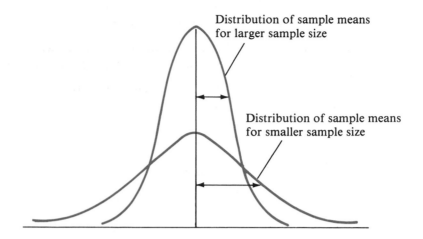

Distribution of sample means for larger sample size

Distribution of sample means for smaller sample size

Illustration 7–4

Kindergarten children have heights that are approximately normally distributed about a mean of 39 inches and a standard deviation of 2 inches. A random sample of size 25 is taken and the mean $\bar{x}$ is calculated. What is the probability that this mean value will be between 38.5 and 40 inches?

Solution We want to find $P(38.5 < \bar{x} < 40.0) = P(? < z < ?)$, where the z scores are as follows:

$$\text{When } \bar{x} = 38.5: z = \frac{38.5 - 39.0}{2/\sqrt{25}} = \frac{-0.5}{0.4} = -1.25$$

$$\text{When } \bar{x} = 40.0: z = \frac{40.0 - 39.0}{2/\sqrt{25}} = \frac{10}{0.4} = 2.5$$

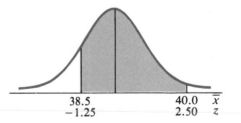

| 38.5 | 40.0 $\bar{x}$ |
| −1.25 | 2.50 z |

(See the accompanying figure.) Therefore,

$$P(38.5 < \bar{x} < 40.0) = P(-1.25 < z < 2.50)$$

$$= 0.3944 + 0.4938 = 0.8882 \qquad \square$$

Illustration 7–5

For illustration 7–4 within what limits would the middle 90 % of the sampling distribution of sample means of sample size 100 fall?

Solution   The basic formula is

$$z = \frac{\bar{x} - \mu}{\sigma/\sqrt{n}}$$

and

$$\sigma_{\bar{x}} = \frac{\sigma}{\sqrt{n}} = \frac{2}{\sqrt{100}} = \frac{2}{10} = 0.2$$

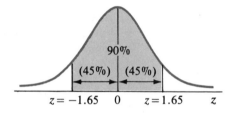

[Recall that the area to the right of the mean, 45 % (0.4500), is related to the z score of 1.65, according to table 5.] We want to find the two values for $\bar{x}$.
    If $z = -1.65$;

$$-1.65 = \frac{\bar{x} - 39}{0.2}$$

$$(-1.65)(0.2) = \bar{x} - 39$$

$$\bar{x} = 39 - 0.33 = 38.67$$

If $z = 1.65$:

$$1.65 = \frac{\bar{x} - 39}{0.2}$$

$$(1.65)(0.2) = \bar{x} - 39$$

$$\bar{x} = 39 + 0.33 = 39.33$$

Thus

$$P(38.67 < \bar{x} < 39.33) = 0.90$$

Exercises

7-11   A random sample of size 36 will be selected from a population that has a mean μ of 50 and a standard deviation σ of 10.

(a) What is the probability that this sample mean will be between 45 and 55?

(b) What is the probability that the sample mean will have a value greater than 48?

(c) What is the probability that the sample mean will be within 3 units of the mean?

7-12   Consider the approximately normal population of heights of male students at a college. Assume that the individual heights have a mean of 69 inches and a standard deviation of 4 inches. A random sample of 16 heights is obtained.

(a) Find the mean of this sampling distribution.

(b) Find the standard error of the mean.

(c) Find the shape of this sampling distribution.

(d) Find $P(\bar{x} > 70)$.

(e) Find $P(\bar{x} < 67)$.

7-13   The heights of the kindergarten children mentioned in illustration 7–4 are approximately normally distributed with $\mu = 39$ and $\sigma = 2$.

(a) If an individual kindergarten child is selected at random, what is the probability that he or she has a height between 38 and 40 inches?

(b) A classroom of 30 of these children is used as a sample. What is the probability that the class mean $\bar{x}$ is between 38 and 40 inches?

(c) If an individual kindergarten child is selected at random, what is the probability that he or she is taller than 40 inches?

(d) A classroom of 30 of these kindergarten children is used as a sample. What is the probability that the class mean $\bar{x}$ is greater than 40 inches?

7-14   At a large bank the amounts of money in personal savings accounts have a mean of $289.56 and a standard deviation of $124.00.

(a) Assuming that these amounts are approximately normally distributed, find the proportion of the accounts that have a balance in excess of $250.00.

(b) If 120 of these accounts are selected randomly, what is the probability that the sample mean will be in excess of $250.00?

7-15   A shipment of steel bars will be accepted if the mean breaking strength of a random sample of 10 steel bars is greater than 250 pounds per square inch. In the past, the breaking strengh of such bars has had a mean of 235 and a variance of 400.

(a) What is the probability, assuming that the breaking strengths are normally distributed, that one randomly selected steel bar will have a breaking strength in the range from 245 to 255?

(b) What is the probability that the shipment will be accepted?

7-16 From a sample of 50, taken in a random manner from all of the employees of a large firm, the mean weekly earnings of employed males was $259.30. Given the current wage structure, it has been estimated in labor negotiations that the standard deviation is $34.10. What is the probability of the sample outcome or less if the population mean is really $275.00, as the management negotiator insists?

7-17 In a certain ghetto area of a large U.S. city, the residents have completed an average of 9.6 years of schooling; and the standard deviation is 3.0 years.

(a) What is the probability that a random sample of 100 individuals from this ghetto area will average less than 8 years of schooling?

(b) Twenty percent of the samples of 100 will average less than what value?

7-18 A trucking firm delivers appliances for a large retail operation. The packages (or crates) have a mean weight of 300 pounds and a variance of 2,500 pounds.

(a) If a truck can carry 4,000 pounds and 25 appliances need to be picked up, what is the probability that the 25 appliances will have an aggregate weight greater than the truck's capacity? (Assume that the 25 appliances represent a random sample.)

(b) If the truck has a capacity of 8,000 pounds, what is the probability that it will be able to carry the entire lot of 25 appliances?

In Retrospect

In both chapters 6 and 7 we used the standard normal probability distribution. There are now two formulas for calculating a z score:

$$z = \frac{x - \mu}{\sigma} \quad and \quad z = \frac{\bar{x} - \mu}{\sigma/\sqrt{n}}$$

You must distinguish between these two formulas. The first gives the standard score when dealing with individual values from a normal population (x values). The second uses information provided by the central limit theorem. Sampling distributions of sample means are approximately normally distributed. Therefore, the standard scores and the probabilities in table 5 may be used in connection with sample means ($\bar{x}$ values). The key to distinguishing between the formulas is to decide whether the problem deals with individual values of x from the population or sample means from the sampling distribution. If it deals with the individual values of x, we use the first formula, as presented in chapter 6. If, on the other hand, the problem deals with the sample means, we use the second formula and proceed as illustrated in this chapter.

The basic purpose for repeated sampling, as discussed in this chapter, is to form sampling distributions. The sampling distribution is then used to describe the variability that occurs from one sample to the next. Once this pattern of variability is known and understood for a specific sample statistic, we will be able to make accurate predictions about the corresponding population parameters. The central limit theorem describes the distribution for sample means. We will begin to make inferences about populaton means in chapter 8.

There are other reasons for repeated sampling. Repeated samples are commonly used in the field of production control, in which samples are taken to determine whether a product is of the proper size or quantity. When the sample is defective, a mechanical adjustment of the machinery is necessary. The adjustment is then followed by another sampling.

Another use for repeated sampling is demonstrated by the news article at the beginning of the chapter. Public opinion is quite likely to shift from time to time. For example, a recognizable shift has taken place in the American attitude toward a woman's place in life. On the question of a female president, our abstract of the news article shows a change of opinion from 66% no to 76% yes. The Gallup poll frequently takes samples to estimate the current opinion of the population as a whole. The news article does not demonstrate a sampling distribution as such, but it does emphasize the difference between repeated sampling for control or development and repeated sampling for the analysis of sampling distributions.

The "standard error of the _____" is the name used for the standard deviation of the sampling distribution of whatever statistic is named in the blank. In this chapter we have been concerned with the standard error of the mean. However, we could also work with the standard error of the range, median, or whatever.

You should now be familiar with the concept of a sampling distribution and, in particular, with the sampling distribution of sample means. In the next chapter we will begin to make predictions about the values of the various population parameters.

Chapter Exercises

7-19   Suppose that a box contains three identical blocks numbered 2, 4, and 6. A sample of three numbers is drawn with replacement (that is, the first number is drawn, observed, and returned; then the second number is drawn; then the third). The mean of the sample is determined.

(a) Make a list that shows all the possible samples that could result from the sampling. (*Hint*: There should be 27 samples.)

(b) Determine the mean of each of these samples and form a sampling distribution of these sample means. (Express as a probability of distribution.)

(c) Construct a probability histogram of this probability distribution.

(d) Find the mean of this sampling distribution, $\mu_{\bar{x}}$.

(c) Find the standard error of the mean $\sigma_{\bar{x}}$ for this sampling distribution.

7-20   Consider the experiment of taking a standardized mathematics test. The variable x is the raw score received. The exam has a mean of 720 and a standard deviation of 60. Assume that the variable x is normally distributed.

(a) Consider the experiment of an individual student taking this exam. Describe the distribution (shape of distribution, mean, and standard deviation) for the experiment.

(b) A group of 100 students takes the exam; the mean is reported as a result. Describe the sampling distribution (shape of distribution, mean, and standard deviation) for the experiment.

(c) What is the probability that the student in (a) scored less than 725.6?

(d) What is the probability that the mean of the group in (b) is less than 725.6?

7-21 The diameters of Red Delicious apples in a certain orchard are normally distributed with a mean of 2.63 inches and a standard deviation of 0.25 inch.

(a) What percentage of the apples in this orchard have diameters less than 2.25 inches?

(b) What percentage of the apples in the orchard are larger than the 2.56 inches?

A random sample of 100 apples is gathered and the mean diameter obtained is $\bar{x} = 2.56$.

(c) If another sample of size 100 is taken, what is the probability that its sample mean will be greater than 2.56 inches?

(d) Why is the z score used in answering parts (a), (b), and (c)?

(e) Why is the formula for z used in (c) different from that used in parts (a) and (b)?

7-22 Find a value for d such that 98% of the apples in exercise 7-21 are larger than $(2.63 - d)$. That is, find d so that $P(x > 2.63 - d) = 0.98$.

7-23 Find a value for e such that 95% of the apples in exercise 7-21 are within e units of the mean 2.63. That is, find e such that $P(2.63 - e < x < 2.63 + e) = 0.95$.

7-24 Find a value for E such that 95% of the samples of 100 apples taken from the orchard in exercise 7-21 will have mean values within E units of the mean 2.63. That is, find E such that $P(2.63 - E < \bar{x} < 2.63 + E) = 0.95$.

7-25 The population of "times" measured by 3-minute egg timers is normally distributed with $\mu = 3.0$ and $\sigma = 0.4$. A 3-minute egg is considered overcooked if cooked for more than 3.50 minutes. What is the probability that a timer selected at random will allow the egg to be overcooked?

7-26 A random sample of 40 part-time employees of fast-food chain-outlets in a large metropolitan area showed a mean income of $125.60. Given that the average weekly income for all such employees is $130.60 and the standard deviation is $15.00,

(a) What is the probability of a sample mean of $125.60 or less?

(b) If the sample size is set at 40, what is the probability that a sample will have a mean between $128.00 and $135.00?

7-27　As a result of several surveys during the last three years, it was concluded that people want pollution controlled and are willing to pay pollution taxes in order to do something about it. The amount that people are willing to pay has a mean of $22.80 per year and a standard deviation of $3.00. Given these values, what is the probability that a random sample of 200 individuals will show a mean differing from $22.80 by more than $0.75?

7-28　Every year, around Halloween, many street signs in a small city are defaced. The average repair cost per sign is $68.00 and the standard deviation is $12.40.

(a) If 300 signs are damaged this year, there is a 5% chance that the total repair costs for the 300 signs will exceed what value?

(b) You are about 68% certain that the total repair costs for 300 signs will fall within what interval centered around $20,400?

7-29　After several years of growth, Douglas fir trees being cultivated by a nursery currently have a mean height of 72 inches and a standard deviation of 10 inches. The heights are approximately normally distributed.

(a) What proportion of the time will random samples of 100 trees show a mean height between 70 and 75 inches?

(b) What mean heights for samples of 100 will fall more than 3 standard errors from the mean of 72 inches?

7-30　A pop-music record firm wants the distribution of lengths of cuts on its records to have an average of 2 minutes and 15 seconds (135 seconds) and a standard deviation of 10 seconds, so that disc jockeys will have plenty of time for commercials within each 5-minute period. The population of times for cuts is approximately normally distributed with only a negligible skew to the right. You have just timed the cuts on a new release and have found that the 10 cuts average 140 seconds.

(a) What percent of the time will the average be 140 seconds or longer, if the new release is randomly selected?

(b) Suppose the music firm had wanted 10 cuts to average 140 seconds less than 5% of the time, what must the population mean be given that the standard deviation remains at 10 seconds?

7-31　A manufacturer of light bulbs says that its light bulbs have a mean life of 700 hours and a standard deviation of 120 hours. You purchased 144 of these bulbs with the idea that you would purchase more if the mean life of your sample is more than 680 hours. What is the probability that you will not buy again from this manufacturer?

7-32　Take a random sample of 40 single-digit numbers from the random number table (table 1 of appendix D). Calculate the sample mean $\bar{x}$. Having calculated the sample mean, what do you believe the population mean μ to be?

Vocabulary List

Be able to define each term. In addition, describe, in your own words, and give an example of each term. Your examples should not be ones given in class or in the textbook.

The terms followed by bracketed numbers were defined in earlier chapters (the bracketed numbers are those of the earlier chapters), but we ask you to define them again to show your increased understanding of their meaning.

central limit theorem
frequency distribution [2]
probability distribution [5]
random sample [2]
repeated sampling

sampling distribution
standard error of the mean
theoretical distribution
z score [2, 6]

Quiz

Answer "True" if the statement is always true. If the statement is not always true, replace the words shown in color with words that make the statement always true.

7–1   A sampling distribution **is** a distribution listing all the sample statistics that describe a particular sample.

7–2   The histograms of **all** sampling distributions are symmetrically shaped.

7–3   The mean of the sampling distribution of $\bar{x}$'s is equal to the mean of the **sample**.

7–4   The standard error of the mean is the standard deviation of the population **from which the samples have been taken**.

7–5   The standard error of the mean **increases** as the sample size increases.

7–6   The shape of the distribution of sample means is always that of a **normal** distribution.

7–7   A **probability** distribution of a sample statistic is a distribution of many values of that statistic which were obtained by repeated sampling.

7–8   The central limit theorem provides us with a description of the three characteristics of a sampling distribution of sample **medians**.

7–9   Random digits are distributed **rectangularly**.

7–10   We **do not need** to take repeated samples in order to use the concept of the sampling distribution.

Working with Your Own Data

The central limit theorem is very important to the development of the rest of this course. Its proof, which requires the use of calculus, is beyond the intended level of this course. However, the truth of the CLT can be demonstrated both theoretically and by experimentation. The following series of questions will help to verify the central limit theorem both ways.

A The Population

Consider the theoretical population that contains the three numbers 0, 3, and 6 in equal proportions.

1. (a) Construct the theoretical probability distribution for the drawing of a single number from this population.

    (b) Draw a histogram of this probability distribution.

    (c) Calculate the mean μ and the standard deviation σ for this population.

B The Sampling Distribution, Theoretically

Let's study the theoretical sampling distribution formed by the means of all possible samples of size 3 that can be drawn from the above population.

2. Construct a list showing all the possible samples of size 3 that could be drawn from this population. (There are 27 possibilities.)

3. Find the mean for each of the 27 possible samples listed in answer to question 2.

4. Construct the probability distribution (the theoretical sampling distribution of sample means) for these 27 sample means.

5. Construct a histogram for this sampling distribution of sample means.

6. Calculate the mean $\mu_{\bar{x}}$ and the standard error of the mean $\sigma_{\bar{x}}$.

7. Show that the results found in answers 1(c), 5, and 6 support the three claims made by the central limit theorem.

C The Sampling Distribution, Empirically

Let's now see if the central limit theorem can be verified empirically; that is, does it hold when the sampling distribution is formed by the sample means that result from several random samples?

8. Draw a random sample of size 3 from the above population. List your sample of three numbers and calculate the mean for this sample.

You may take three identical "tags" numbered 0, 3, and 6, put them in a "hat", and draw your sample using replacement between each drawing. Or you may use dice; let 0 be represented by 1 and 2, let 3 be represented by 3 and 4, and 6 by 5 and 6. Or you may use random numbers to simulate the drawing of your samples. Or you may draw your sample from the list of random samples at the bottom of the page. Describe the method you decide to use. (Ask your instructor for guidance.)

9. Repeat question 8 forty-nine (49) more times so that you have a total of fifty (50) sample means that have resulted from samples of size 3.

10. Construct a frequency distribution of the 50 sample means found in answering questions 8 and 9.

11. Construct a histogram of the frequency distribution of observed sample means.

12. Calculate the mean $\bar{\bar{x}}$ and standard deviation $s_{\bar{x}}$ of the frequency distribution formed by the 50 sample means.

13. Compare the observed values of $\bar{\bar{x}}$ and $s_{\bar{x}}$ with the values of $\mu_{\bar{x}}$ and $\sigma_{\bar{x}}$. Do they agree?

The following table contains 100 samples of size 3 that were generated randomly by computer.

6 3 0	0 3 0	6 6 0	3 3 6	6 6 3	6 3 3
0 0 3	3 0 6	3 3 0	3 6 6	0 3 0	6 6 3
6 6 6	0 3 0	6 3 6	0 6 3	6 0 3	6 3 3
6 0 0	3 0 6	6 3 3	3 3 0	3 3 0	3 3 3
3 3 3	3 0 0	6 6 6	3 3 6	0 0 6	0 6 3
6 6 6	0 0 6	3 3 0	0 6 6	0 0 3	6 6 3
0 0 6	0 0 6	6 6 6	6 3 6	6 6 0	3 0 0
3 6 6	6 3 0	3 6 3	3 0 0	3 3 6	0 6 0
3 0 0	0 3 6	6 3 3	6 0 6	3 3 6	6 0 3
0 3 6	3 6 3	6 6 3	6 6 0	3 3 3	3 0 0
6 3 0	6 6 0	0 3 0	6 6 0	3 6 6	0 3 6
6 3 3	0 3 0	6 6 0	6 6 3	6 6 0	3 0 3
3 6 3	3 6 0	0 0 6	0 3 3	3 6 6	0 3 6
0 6 0	6 0 0	0 6 0	0 6 6	0 3 3	0 3 6
3 3 6	3 3 3	3 3 6	6 3 6	3 3 3	3 6 6
6 3 3	3 0 0	3 0 6	6 0 3	3 6 6	6 0 3
0 3 3	6 3 0	0 3 6	0 3 6		

The calculations required are accomplished most easily with the assistance of an electronic calculator or a canned program on a computer. Your local computer center can provide a list of the available canned programs and the necessary information and assistance to run your data on the computer.

PART 3

Inferential Statistics

The central limit theorem told us about the sampling distribution of sample means. Specifically it stated that a distribution of sample means was approximately normally distributed about the mean of the population; and it stated the relationship between the standard deviation of the population and the sampling distribution. With this information we are able to make probability statements about the likelihood of certain sample mean values occurring when samples are drawn from a population with a known mean and a known standard deviation. We are now ready to turn this situation around. We will draw one sample, calculate its mean value, and then an inference about the value of the population will be made based on the sample mean's value.

In this part of the textbook we will learn about making two types of inferences: (1) the decision-making process by use of a hypothesis test procedure, and (2) the procedures for estimating a population parameter. Specifically, we will learn about making these two types of inferences about the population mean μ and the population standard deviation for normally distributed populations, and for the probability parameter p of a binomial population.

8 Introduction to Statistical Inferences

We've matched these tires against some of the toughest roads in the world! In the Baja, East Africa, Greece, Morocco, and the United States, Sears Steel Belted Radials have proven their rugged durability. That's because they're built tough! 2 steel belts and 2 rayon radial plies team-up to give you traction, mileage, and outstanding durability. We even guarantee them for 40,000 miles. So hurry in now and save on the Sears Steel Belted Radial. The tire that proved itself on the tough roads of the world.

Radial Gas-saving Story

Independent tests prove that Sears Steel Belted Radial, when compared to our leading fiber glass belted tire, improved gas mileage by an average of 7.4%, when driven at different constant speeds. That's in both foreign and domestic categories!

Chapter Objectives

A random sample of 36 pieces of data yields a mean of 4.64. What can we deduce about the population from which the sample was taken? We will be asked to answer two types of questions:

1. *Is the sample mean significantly different in value from the hypothesized mean value of 4.5?*

2. *Based on the sample, what statement can we make about the value of the population mean?*

The first question requires us to make a decision, *while the second question requires us to make an* estimation.

In this chapter and the next two chapters we will find out how a hypothesis test is used to make a statistical decision about three basic parameters: the mean μ, the standard deviation σ, and the proportion p. We will also see how an estimation of these three parameters is made. In this chapter we will concentrate our attention on learning about the basic concepts of hypothesis testing and estimation. We will deal mostly with questions about the population mean, using two methods that assume that the population standard deviation is known. This assumption is seldom realized in real-life problems, but it will make our first look at inferences much easier.

Section 8–1   The Nature of Hypothesis Testing

hypothesis test

To illustrate the decision-making process called a **hypothesis test**, let's consider Bill's claim (see exercise 5-64), "I randomly guessed to obtain the answers." Bill had taken a ten-question multiple-choice test (five possible answers per question) and stated that he had merely guessed the answers without reading the questions or the possible answers. He answered seven questions correctly. Do you believe that he did, in fact, randomly guess?

This question calls for a decision: It is reasonably possible for Bill to obtain seven correct answers by randomly guessing, or is it unlikely that he could have answered seven questions correctly as a result of random guessing? Bill knows the answer, but it is up to us to decide which alternative we think is correct. What do you think, did he guess or not? How do you justify the conclusion you reached?

Let's investigate the statistical decision-making process, the hypothesis test procedure. The statistical hypothesis test is a five-step procedure. Each part of this procedure will be demonstrated and justified as we investigate Bill's statement.

The first two steps of the hypothesis test procedure are to formulate two hypotheses.

hypothesis

Hypothesis

A statement that something is true.

STEP 1   Formulate the null hypothesis.

null hypothesis

Null Hypothesis, H_0

The hypothesis upon which we wish to focus our attention. Generally this is a statement that a population parameter has a specified value. Often the phrase "there is no difference" is used in its interpretation—thus the name "null" hypothesis.

STEP 2   Formulate the alternative hypothesis.

alternative hypothesis

Alternative Hypothesis, H_a

A statement about the same population parameter that is used in the null hypothesis. Generally this is a statement that specifies that the population parameter has a value different from the value given in the null hypothesis. The rejection of the null hypothesis will imply the acceptance of the alternative hypothesis.

The null hypothesis and the alternative hypothesis are formulated by inspecting the problem or statement to be investigated and then forming two alternative statements. For our illustration we inspect Bill's statement and formulate the two alternatives: "Bill did randomly guess on all 10 questions," or "Bill did not randomly guess on all 10 questions." One of these statements becomes the null hypothesis; the other becomes the alternative hypothesis. The statement "Bill randomly guessed at all 10 questions" relates to the binomial probability distribution where $n = 10$ and x is the number of questions answered correctly. If Bill did randomly guess, then the probability that each answer is correct is $p = 0.2$ (only 1 of the 5 choices, or 0.2, is correct). This statement about the population parameter p becomes the null hypothesis:

$H_0: p = 0.2$ (Bill did randomly guess on each question.)

The alternative hypothesis is

$H_a: p \neq 0.2$ (Bill did not randomly guess on each question.)

From this point on in the hypothesis test procedure, we will work under the assumption that the null hypothesis is a true statement. This situation might be compared to a courtroom trial, where the accused is assumed to be innocent until sufficient evidence has been presented to show otherwise. At the conclusion of the hypothesis test, we will make one of two possible **decision**   decisions. We will decide in agreement with the null hypothesis and say that we fail to reject H_0 (this corresponds to "fail to convict" or an acquittal of the accused in a trial). Or we will decide in opposition to the **null hypothesis** and say that we reject H_0 (this corresponds to conviction of the accused in a trial).

There are four possible outcomes that could be reached as a result of the null hypothesis being either true or false and the decision being either "fail to reject" or "reject." Table 8–1 shows these four possible outcomes.

| Decision | Null Hypothesis Is: | |
	True	*False*
Fail to reject H_0	Type A Correct decision	Type II Error
Reject H_0	Type I Error	Type B Correct decision

A **type A correct decision** occurs when the null hypothesis is true and we decide in its favor. A **type B correct decision** occurs when the null hypothesis is false and our decision is in opposition to the null hypothesis. A **type I error** will be committed when a true null hypothesis is rejected, that is, when the null hypothesis is true but we decided against it. A **type II error** is committed when we decide in favor of a null hypothesis that is actually false.

type I and type II errors

When we make a decision, it would be nice if it was always a correct decision. However, this is statistically impossible, since we will be making our decision on the basis of sample information. The best we can hope for **risk** is to control the **risk**, or probability, with which an error occurs. The **probability** assigned to the type I error is called **alpha**, α. The **probability** of the **alpha, beta** type II error is called **beta**, β (β is the Greek letter beta). See table 8–2. To control these errors we will assign a small probability to them. The most frequently used probability values are 0.01 or 0.05. The probability assigned to each error will depend on the seriousness of the error. The more serious the error, the less often we will be willing to allow it to occur, and therefore a smaller probability will be assigned to it. In this text we are going to devote our attention to α, the P(type I error). Further discussion of β, the P(type II error), is beyond the scope of this book.

Error	Type Error	Probability
Rejection of a true hypothesis	I	α
Failure to reject a false null hypothesis	II	β

Let's now return to our illustration and to step 3 of the hypothesis test procedure.

STEP 3   Determine the test criteria.

test criteria

Test Criteria

Consist of (1) determining a test statistic, (2) specifying a level of significance α, and (3) determining the critical region.

test statistic

Test Statistic

A random variable whose value will be used to make the decision "fail to reject H_0" or "reject H_0."

TABLE 8–3
Probability Distribution of x

x	P(x)
0	0.107
1	0.268
2	0.302
3	0.201
4	0.088
5	0.026
6	0.006
7	0.001
8	0.0+
9	0.0+
10	0.0+

The test statistic may be a sample statistic or some other value obtained from the sample results. The probability distribution of this test statistic is determined as a result of the assumption that the null hypothesis is true. In our illustration the number of correct answers x will be used as the test statistic.

Before we discuss the level of significance and the critical region, let's look at the probability distribution of x, the number of correct answers obtained. According to the null hypothesis, that is, if in fact Bill is guessing, x is binomially distributed with $n = 10$ and $p = 0.2$. The probabilities for each x value may be found in table 4 of appendix D. These probabilities are given in table 8–3. Inspection of this distribution suggests that the probability of randomly guessing 5 or more of the answers correctly is quite remote. Thus we could say that the occurrence of values for x of 5, 6, 7, 8, 9, or 10 would not support the null hypothesis "randomly guessed." If a person were to get 0, 1, 2, 3, or 4 answers correct and say that she randomly guessed, we would feel that these were likely events and agree that she most likely *did* guess. Thus we have specified a critical region; namely, $x = 5, 6, 7, 8, 9, 10$.

critical region

Critical Region

The set of values for the test statistic that will cause us to reject the null hypothesis.

critical value

The **critical value** is the "first" value in the critical region. Thus $x = 5$ is the critical value in our illustration. That is, if we take a sample and the observed value of x is 5 or more, we will reject the null hypothesis.

level of significance

Level of Significance

The probability of committing the type I error, α.

For our example this is $P(x \geq 5)$ for the binomial random variable when $n = 10$ and $p = 0.2$.

$$P(x \geq 5) = 0.026 + 0.006 + 0.001 + 3(0+) = \mathbf{0.033}$$

That is, $\alpha = 0.033$ for our critical region. (Typically α will be determined first, and then it will dictate the critical value, as we will see in the next section.)

Now that the test criteria are determined—that is, $\alpha = 0.033$ and the critical region is $x \geq 5$ for the random variable x, number of correct answers—we are ready to proceed to step 4.

STEP 4   Obtain the value of the test statistic.

Having completed the ground rules for the test (steps 1, 2, and 3), we are ready to obtain our sample and either observe or calculate the value of the test statistic. For our illustration $x = 7$; that is, the value of the test statistic is the number of answers Bill had correct. Typically we will use a formula and calculate the value of the test statistic based on sample information.

STEP 5   Make a decision and interpret it.

The decision is made by comparing the value of the test statistic found in step 4 to the test criteria of step 3.

---

decision rule

Decision Rule

If the test statistic falls within the critical region, we will reject H_0. If the test statistic does not fall in the critical region, we will fail to reject H_0. *Note*: The set of values that are not in the critical region is called the **noncritical region** or sometimes the **acceptance region**.

acceptance region

We complete the test by interpreting the decision reached. For our illustration $x = 7$ falls in the critical region and we reject H_0. Thus we conclude that the evidence ($x = 7$) does not support Bill's claim that he randomly guessed at all 10 questions.

Note that we did not prove that Bill did not guess. If he did randomly guess on all 10, he is very fortunate to have answered 7 correctly (this would be considered a "rare event"). In retrospect, we have one of two choices to make here: (1) the null hypothesis is correct and Bill did guess (a very unlikely event since α is only 0.033), or (2) the null hypothesis is incorrect and we have sample results that are representative of some other situation. The latter choice seems to be the more likely here.

Before we leave this illustration, let's look at the other side. Suppose that Jane, who answered 3 questions correctly, also claimed to have guessed. The same test procedure (steps 1, 2, and 3) would be used to test her claim. However, in step 4 we have an observed $x = 3$, and therefore we make the decision "fail to reject H_0" in step 5. What does this imply? $x = 3$ certainly does not prove that Jane did guess; perhaps she was just not prepared for the test. **Failing to reject the null hypothesis implies only that we did not find sufficient evidence to reject the statement of the null hypothesis.**

NOTE: Some people prefer to use the phrase "accept H_0" in place of "fail to reject H_0." However, failure to reject the null hypothesis states the decision more accurately; namely, that we failed to find sufficient evidence to convict (reject) the accused (H_0) as in a courtroom trial. We acquitted the defendant; we did not prove the defendant innocent. For these reasons we will use the phrase "fail to reject H_0."

Exercises

8-1 A 14-question true-false quiz is given. Define x to be the number of questions answered correctly.

(a) x is a binomial random variable if the answers are obtained by randomly guessing. Explain why and specify the values of n and p.

(b) Construct the probability distribution for x. Let the null hypothesis be "the answers were obtained by random guessing"; that is, x is binomial with $n = 14$ and $p = 0.5$.

(c) What values of x have probabilities greater than 0.05?

(d) What values of x have probabilities less than 0.05? (Let these values form the critical region.)

(e) Find α.

(f) What decision with regard to the null hypothesis would you reach if a quiz had 2 correct answers? Six correct answers? Thirteen correct answers?

8-2 As described in this section, the hypothesis-testing procedure has many similarities to a courtroom procedure. The null hypothesis "the accused is innocent" is being tested.

(a) Describe the situation involved in each of the four possible outcomes shown in table 8-1.

(b) If the accused is acquitted, does this "prove" him innocent? (Perhaps, in this respect, the phrase "fail to reject H_0" is more accurate in expressing the situation than is "accept H_0.")

(c) If the accused is found guity, does this "prove" his guilt?

8-3 Consider the following nonmathematical situation as a hypothesis test. A medic at the scene of a serious accident tests the null hypothesis "this victim is alive."

(a) Carefully state the meaning of the four possible outcomes indicated in table 8-1.

(b) Decide on the seriousness of the two possible errors.

(c) If α and β could be controlled statistically, which set of probabilities would you prefer be used if you were the victim?

(1) $\alpha = 0.001$ and $\beta = 0.10$

(2) $\alpha = 0.05$ and $\beta = 0.05$

(3) $\alpha = 0.10$ and $\beta = 0.001$

8-4 Consider the following nonmathematical situation as a hypothesis test. You just received a parachute that was inspected by an inspector whose null hypothesis is "this parachute will open."

(a) Carefully state the meaning of the four possible outcomes indicated in table 8-1.

(b) Decide on the seriousness of the two possible errors.

(c) If α and β could be controlled statistically, which set of probabilities would you prefer be used if you were going to use the parachute?

(1) $\alpha = 0.001$ and $\beta = 0.10$

(2) $\alpha = 0.05$ and $\beta = 0.05$

(3) $\alpha = 0.10$ and $\beta = 0.001$

8-5 (a) If α is assigned the value 0.001, what are we saying about the type I error?
(b) If α is assigned the value 0.05, what are we saying about the type I error?
(c) If α is assigned the value 0.10, what are we saying about the type I error?

8-6 (a) If β is assigned the value 0.001, what are we saying about the type II error?
(b) If β is assigned the value 0.05, what are we saying about the type II error?
(c) If β is assigned the value 0.10, what are we saying about the type II error?

8-7 (a) If the value of the test statistic falls in the critical region, what decision must we make?

(b) If the value of the test statistic does not fall in the critical region (that is, it falls in the noncritical region), what decision must we make?

8-8 The director of an advertising agency is concerned with the effectiveness of a television commercial.

(a) What null hypothesis is she testing if she commits a type I error when she erroneously says that the commercial is effective?

(b) What null hypothesis is she testing if she commits a type II error when she erroneously says that the commercial is effective?

8-9 (a) If the null hypothesis is true, the probability of an error in decision is identified by what name?

(b) If the null hypothesis is false, the probability of an error in decision is identified by what name?

(c) If the test statistic falls in the crtical region, what error could be made?

(d) If the test statistic falls in the noncritical region, what error might occur?

8-10   (a) Suppose that a hypothesis test is to be carried out by using $\alpha = 0.05$. What is the probability of committing a type I error?

(b) What proportion of the probability distribution is in the noncritical region provided the null hypothesis is correct?

(c) What is a critical region?

(d) What is a critical value?

Section 8–2   The Hypothesis Test (A Classical Approach)

In section 8–1 we surveyed the steps for and some of the reasoning behind a hypothesis test in a very general manner. In this section we are going to study the hypothesis test procedure as it applies to statements concerning the mean μ of a population. To simplify the introduction of these techniques, we will impose the restriction that the population standard deviation is known. (This restriction will be removed in chapter 9.) The first three illustrations deal with the procedures for formulating the null and alternative hypotheses.

Illustration 8–1

An ecologist would like to show that Rochester has an air pollution problem. Specifically, she would like to show that the mean level of carbon monoxide in downtown Rochester air is higher than 4.9 parts per million. State the null and alternative hypotheses.

Solution   To state the hypotheses we first need to identify the population parameter in question and the value to which it is being compared. The "mean level of carbon monoxide pollution" is the parameter μ and 4.9 parts per million is the specific value. Our ecologist is questioning the value of μ. It could be related to 4.9 in any one of three ways: (1) $\mu < 4.9$, (2) $\mu = 4.9$, or (3) $\mu > 4.9$. These three statements must be arranged to form two statements, one that states what the ecologist is trying to show and the other that states the opposite. $\mu > 4.9$ represents the statement "the mean level is higher than 4.9," while $\mu < 4.9$ and $\mu = 4.9$ ($\mu \leq 4.9$) represent the opposite, "the mean level is not higher than 4.9." One of these two statements will become the null hypothesis H_0 and the other becomes the alternative hypothesis H_a.

RECALL

1. The null hypothesis states that the parameter in question has a specified value.

2. A sample mean will be used as the basis for an inference about the population mean, and sample means have an approximately normal distribution as described by the central limit theorem.

3. A normal distribution is determined when its mean and standard deviation are specified.

All this is suggesting that the statement containing the equal sign will become the null hypothesis; the other statement becomes the alternative hypothesis. Thus we have

$$H_0: \mu = 4.9 \quad \text{and} \quad H_a: \mu > 4.9$$

Recall that once the null hypothesis is stated, we proceed with the hypothesis test under the assumption that the null hypothesis is true. Thus $\mu = 4.9$ locates the center of the sampling distribution of sample means. For that reason the null hypothesis will be written with an equal sign only. If $\mu = 4.9$ and $\mu < 4.9$ are stated together as the null hypothesis, the null hypothesis will be expressed as

$$H_0: \mu = 4.9 \, (\leq)$$

where the $(\leq)$ serves as a reminder of the grouping that took place when the hypothesis was formed. □

NOTE: **The equal sign must be in the null hypothesis**, regardless of the statement of the original problem.

Illustration 8–1 expresses the viewpoint that an ecologist might take. Now let's consider how the Rochester Chamber of Commerce might view the same stuation.

Illustration 8–2

In trying to promote the city, the Chamber of Commerce would be more likely to want to conclude that the mean level of carbon monoxide in downtown Rochester is less than 4.9 parts per million. State the null and alternative hypotheses related to this viewpoint.

Solution   Again the parameter of interest is the mean level μ of carbon monoxide and 4.9 is the specified value. $\mu < 4.9$ corresponds to "the mean level is less than 4.9," while $\mu \geq 4.9$ corresponds to "the mean level is not less than 4.9." Therefore, the hypotheses are

$$H_0: \mu = 4.9 \, (\geq) \quad \text{and} \quad H_a: \mu < 4.9 \qquad □$$

A more neutral point of view is suggested by illustration 8–3.

Illustration 8–3

The "mean level of carbon monoxide in downtown Rochester is not 4.9 parts per million." State the null and alternative hypotheses that correspond to this statement.

Solution   The mean level of carbon monoxide is equal to 4.9 ($\mu = 4.9$) or the mean is not equal to 4.9 ($\mu \neq 4.9$). (*Note*: "Less than or greater than" is customarily expressed as "not equal to.") Therefore,

$$H_0: \mu = 4.9 \quad \text{and} \quad H_a: \mu \neq 4.9 \qquad \square$$

The viewpoint of the experimenter affects the way the hypotheses are formed, as we have seen in these three illustrations. Generally the experimenter is trying to show that the parameter value is different from the value specified. Thus the experimenter is usually hoping to find a rejection of the null hypothesis. Illustrations 8–1, 8–2, and 8–3 represent the three arrangements possible for the $<$, $=$, and $>$ relationships between the parameter μ and the specified value 4.9.

The courtroom trial and the hypothesis test have many similarities, as already emphasized. Let's now look at another similarity. In the courtroom there is a trial only if the prosecutor has shown the court that there is sufficient reason to make the accused stand trial. If the prosecutor did not believe that the accused was guilty, there would be no trial. Much the same is true with the hypothesis test. If the experimenter believes the null hypothesis is true, he will not challenge its truth and will not be testing it. He proceeds to test the null hypothesis only when he wishes to show that the alternative is correct. Thus the alternative hypothesis is a statement claiming that the value of the parameter is "smaller than" or "larger than" or "different from" the value claimed by the null hypothesis.

The next illustration demonstrates the complete hypothesis test procedure as it is used in questions dealing with the population mean.

Illustration 8–4

For many semesters an instructor has recorded his students' grades, and the mean μ for all these students' grades is 72. The current class of 36 students seems to be better than average in ability and the instructor wants to show that according to their average "the current class is superior to his previous classes." Does the class mean $\bar{x}$ of 75.2 present sufficient evidence to support the instructor's claim that the current class is superior? Use $\alpha = 0.05$ and $\sigma = 12.0$.

Solution   To be superior this class must have a mean grade that is higher than the mean of all the previous classes. To be "equal or less than" would not be superior.

STEP 1 $H_0: \mu = 72$ ($\leq$) (class is not superior).

STEP 2 $H_a: \mu > 72$ (class is superior).

STEP 3   The level of significance $\alpha = 0.05$ is given in the statement of the problem. The standard score z is used as the test statistic when the null hypothesis is about a population mean and the standard deviation is known. Recall that the central limit theorem tells us that the sampling distribution of sample means is approximately normally distributed. Thus the normal probability distribution will be used to complete the hypothesis test. The critical region—values of the standard score z that will cause a rejection of the null hypothesis—has an area of 0.05 and is located at the extreme right of the distribution. The critical region is on the right because large values of the sample mean suggest "superior," while values near or below 72 support the null hypothesis. See the diagram.

If random samples of size 36 are taken from a population with a mean value equal to 72, many of the sample means would have values near 72, like 71, 72, 71.8, 72.5, 73, and so on. Only sample means that are considerably larger than 72 would cause us to reject the null hypothesis. The critical value, the cutoff between "not superior" and "superior," is determined by α, the probability of the type I error. $\alpha = 0.05$ was given. Thus the critical region (the shaded region of the diagram) has an area of 0.05 and a critical value of $+1.65$. (This value is obtained by using table 5 of appendix D.)

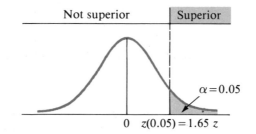

STEP 4   The value of the test statistic z will be found by using formula (7–2) and the sample information.

$$\bar{x} = 75.2 \quad \text{and} \quad n = 36$$

$$z = \frac{\bar{x} - \mu}{\sigma/\sqrt{n}}$$

RECALL: We assumed that μ was equal to 72 in the null hypothesis and that $\sigma = 12.0$ was known.

$$z = \frac{75.2 - 72}{12.0/\sqrt{36}} = \frac{3.2}{2.0} = 1.60$$

$$z^* = 1.60$$

calculated value (*)   We will use an asterisk, *, to identify the calculated value of the test statistic; the asterisk will also be used to locate its value relative to the test criteria (next step).

STEP 5   We now compare the calculated test statistic, z^*, to the test criteria set up in step 3 by locating the calculated value on the diagram and placing an asterisk at that value. Since the test statistic (calculated value) falls in the noncritical region (unshaded portion of the diagram), we must reach the following decision:

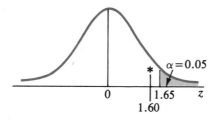

Decision: Fail to reject H_0.

conclusion   Recall that the critical region was to be shaded in the diagram, and when the test statistic falls in the critical region, we must reject H_0. Step 5 is then completed by stating a conclusion.

Conclusion: There is not sufficient evidence to show that the current class is superior.   □

Does this conclusion seem realistic? 75.2 is obviously larger than 72. Recall that 75.2 is a sample mean, and if a random sample of size 36 is drawn from a population whose mean is 72 and whose standard deviation is 12, then the probability that the sample mean is 75.2 or larger is greater than the risk, α, with which we are willing to make the type I error. [What is $P(\bar{x} > 75.2)$?]

Illustration 8–5

It has been claimed that the mean weight of women students at a college is 54.4 kilograms. Professor Schmidt does not believe the statement that the mean is 54.4 kilograms. To test the claim he collects a random sample of 100 weights from among the women students. A sample mean of 53.75 kilograms results. Is this sufficient evidence to reject the null hypothesis? Use $\alpha = 0.05$ and $\sigma = 5.4$ kilograms.

Solution   The statement suggests that the three possible relationships $(<, =, >)$ between the mean weight μ and the hypothesized 54.4 kilograms be split $(=)$ or $(<, >)$. Therefore, we have the following steps:

STEP 1 $H_0: \mu = 54.4$ (mean weight is 54.4 kilograms).

STEP 2 $H_a: \mu \neq 54.4$ (mean weight is not 54.4 kilograms).

The test statistic will be the standard score z, and the normal distribution is used since we are using the sampling distribution of sample means. The sample mean is our estimate for the population mean. Since the alternative hypothesis is "not equal to," a sample mean considerably larger than 54.4, as well as one considerably smaller than 54.4, will be in opposition to the null hypothesis. The values of the sample mean around 54.4 will support the null hypothesis. Therefore, the critical region will be split into two equal parts, one at each extreme of the normal distribution. The area of each region will be $\alpha/2$. Since $\alpha = 0.05$, each part of the critical region will have a probability of 0.025 (see the diagram). Look in table 5 of appendix D for the area 0.4750 (0.4750 = 0.5000 − 0.025). The z score of 1.96 is found and becomes 1.96 on the right of the mean and −1.96 on the left.

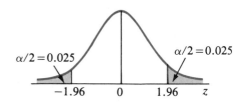

STEP 3   The test criteria are shown in the accompanying diagram.

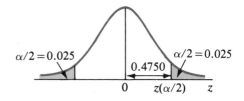

STEP 4   Calculate the test statistic z. The sample information is $\bar{x} = 53.75$ and $n = 100$. σ was given as 5.4, and the null hypothesis assumed that $\mu = 54.4$.

$$z = \frac{\bar{x} - \mu}{\sigma/\sqrt{n}} = \frac{53.75 - 54.4}{5.4\sqrt{100}} = \frac{-0.65}{0.54} = -1.204$$

$$z^* = -1.20$$

Locate z^* on the diagram constructed in step 3.

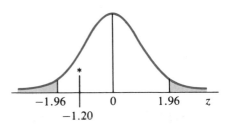

Now we are ready to identify the decision.

RECALL: If the test statistic falls in the critical region, we must reject H_0. If the test statistic falls in the noncritical region, we must fail to reject H_0.

STEP 5 The calculated value of z, z^*, falls between the critical values cited in step 3. Therefore, our decision is as follows:

Decision: Fail to reject H_0.

The interpretation of our decision is the only thing left to do. This part of the hypothesis test may very well be the most significant part, as the conclusions reached express the results found. We must be very careful to state precisely what is meant by the decision. It is often a temptation to overstate a conclusion. In general, a decision to "fail to reject H_0" may be interpreted to mean that the evidence found does not disagree with the null hypothesis. Note that this does not "prove" the truth of H_0. A decision to "reject H_0" will mean that the evidence found implies the null hypothesis is false and thus indicates the alternative hypothesis to be the case.

Conclusion: The evidence (sample mean) found does not contradict the assumption that the population mean is 54.4 kilograms. □

Before we look at another illustration, let's summarize briefly some of the details we have seen thus far.

1. The null hypothesis specifies a particular value of a population parameter.

2. The alternative hypothesis can take three forms. Each form dictates a specific location of the critical regions, as shown here:

Sign in the Alternative Hypothesis	<	≠	>
Critical Region	One region, Left side; **One-sided test**	Two regions, One on each side; **Two-sided test**	One region, Right side; **One-sided test**

one-tailed and two-tailed tests

3. For many hypothesis tests the sign in the alternative hypothesis "points" in the direction in which the critical region is located. [Think of the not equal sign ($\neq$) as being both less than (<) and greater than (>), thus pointing in both directions.]

The value assigned to α is called the significance level of the hypothesis test. Alpha cannot be interpreted to be anything other than the risk (or probability) of rejecting the null hypothesis when it is actually true. We will seldom be able to determine whether the null hypothesis is true or false; we will only decide to reject H_0 or to fail to reject H_0. The relative frequency with which we reject a true hypothesis is α, but we will never know the relative frequency with which we make an error in decision. The two ideas are actually quite different; that is, a type I error and an error in decision are two different things altogether.

Let's look at some more illustrations of the hypothesis test.

Illustration 8–6

The student body at many community colleges is considered a "commuter population." The following question was asked of the Student Affairs Office: "How far (one way) does the average community college student commute to college daily?" The office answered: "No more than 9.0 miles." The inquirer was not convinced of the truth of this and decided to test the statement. He took a sample of 50 students and found a mean commuting distance of 10.22 miles. Test the hypothesis stated above at a significance level of $\alpha = 0.05$, using $\sigma = 5$ miles.

Solution   Our first two steps follow:

STEP 1 $H_0: \mu = 9.0$ $(\leq)$

STEP 2 $H_a: \mu > 9.0$

The one-way distance traveled by the "average student" would be the same as the mean one-way distance traveled by all students. Therefore, the parameter of concern is μ. The claim "no more than 9.0 miles" implies that the three possible relationships should be grouped "no more than 9.0" $(\leq)$ versus "more than 9.0" $(>)$.

STEP 3   The critical region is on the right (since H_a contains the $>$ sign); H_0 will be rejected if it appears that the $\bar{x}$ observed is significantly greater than the 9.0 claimed. Notice that the sign in H_a $(>)$ points toward the critical region. (See the diagram.)

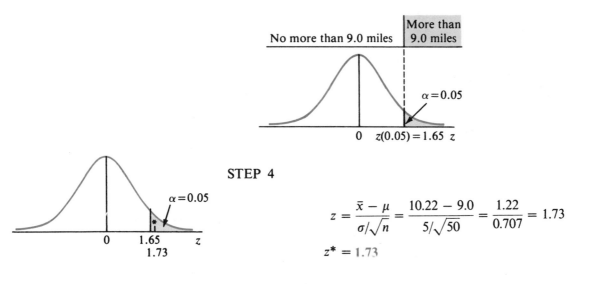

STEP 4

$$z = \frac{\bar{x} - \mu}{\sigma/\sqrt{n}} = \frac{10.22 - 9.0}{5/\sqrt{50}} = \frac{1.22}{0.707} = 1.73$$

$$z^* = 1.73$$

STEP 5 *Decision*: Reject H_0. (z^* fell in the critical region.)

Conclusion: At the 0.05 level of significance, we conclude that the average community college student probably travels more than 9.0 miles. □

Illustration 8–7

Draw a sample of 40 single-digit numbers from the random number table and test the null hypothesis $\mu = 4.5$. Use $\alpha = 0.10$ and $\sigma = 2.87$. The standard deviation of the random digits was found earlier in chapter 5. (See exercise 5–13.)

Solution

STEP 1 H_0: $\mu = 4.5$

STEP 2 H_a: $\mu \neq 4.5$

STEP 3 $\alpha = 0.10$ (See the diagram.)

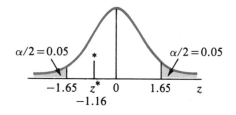

STEP 4 The following random sample was drawn from table 1 of appendix D:

2	8	2	1	5	5	4	0	9	1
0	4	6	1	5	1	1	3	8	0
3	6	8	4	8	6	8	9	5	0
1	4	1	2	1	7	1	7	9	3

$$\sum x = 159$$
$$n = 40$$
$$\bar{x} = 3.975$$

$$z = \frac{\bar{x} - \mu}{\sigma/\sqrt{n}} = \frac{3.975 - 4.50}{2.87/\sqrt{40}} = \frac{-0.525}{0.454} = -1.156$$

$$z^* = -1.16$$

z^* falls in the noncritical region, as shown in the diagram in step 3.

STEP 5 *Decision*: Fail to reject H_0.

Conclusion: The observed sample mean is not significantly different from 4.5. □

Suppose that we were to take another sample of size 40 from the table of random digits. Would we obtain the same results? Suppose that we took a third sample or a fourth? What results might we expect? What is the level of significance α? Yes, its value is 0.10, but what does it measure? Table 8–4 lists the means obtained from 10 different random samples of size 40 that were taken from table 1. The calculated value of z that corresponds to each $\bar{x}$ and the decision each would dictate are listed also. Each of the 10 calculated z scores is shown in figure 8–1. The sample number is used to identify each score. Note that one of the samples caused us to reject the null hypothesis, although we know that it is true for this situation.

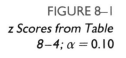

TABLE 8–4

Random Samples of Size 40 Taken from Table 1, Appendix D

Sample Number	Sample Mean, $\bar{x}$	Calculated z, z^*	Decision Reached
1	4.62	+0.26	Fail to reject H_0
2	4.55	+0.11	Fail to reject H_0
3	4.08	−0.93	Fail to reject H_0
4	5.00	+1.10	Fail to reject H_0
5	4.30	−0.44	Fail to reject H_0
6	3.65	−1.87	Reject H_0
7	4.60	+0.22	Fail to reject H_0
8	4.15	−0.77	Fail to reject H_0
9	5.05	+1.21	Fail to reject H_0
10	4.80	+0.66	Fail to reject H_0

REMEMBER: α is the probability that we reject H_0 when it is actually a true statement. Therefore, we can anticipate that a type I error will occur α of the time when testing a true null hypothesis.

FIGURE 8–1

z Scores from Table 8–4; $\alpha = 0.10$

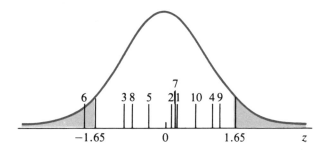

Exercises

8–11 State the null hypothesis H_0 and the alternative hypothesis H_a that would be used for a hypothesis test related to each of these statements.

(a) The mean age of the students enrolled in evening classes at our college is greater than 26 years.

(b) The mean weight of packages shipped on Air Express during the last month was less than 36.7 pounds.

(c) The mean life of fluorescent light bulbs is at least 1600 hours.

(d) The mean weight of college football players is no more than 210 pounds.

(e) The mean distance from a home in suburban Chicago to the nearest fire station is less than 4.7 miles.

(f) The mean daily high temperature during February at Daytona Beach, Florida, is not 80 degrees.

(g) The mean grade (measure of steepness) for all ski runs at Greek Peek Ski Center is no less than 20 degrees.

(h) The mean strength of welds by a new process is different from 570 pounds per unit area, the mean strength of welds by the old process.

8-12 Suppose that we wanted to test the hypothesis that the mean hourly charge for automobile repairs is at least \$23.00 per hour at the repair shops in a nearby city. Explain the conditions that would exist if we made an error in decision by committing a

(a) type I error   (b) type II error

8-13 All drugs must be approved by the Food and Drug Administration (FDA) before they can be marketed by a drug manufacturer. The FDA must weigh the error of marketing an ineffective drug with the usual risks of side effects versus the consequences of not allowing an effective drug to be sold. Suppose, using standard medical treatment, the mortality rate (r) of a certain disease is known to be a. A manufacturer submits for approval a drug that is supposed to treat this disease. The FDA sets up the hypothesis to test the mortality rate for the drug as (1) $H_0: r = a$, $H_a: r < a$, $\alpha = 0.005$; or (2) $H_0: r = a$, $H_a: r > a$, $\alpha = 0.005$.

(a) If $a = 0.95$, which test do you think the FDA would use?

(b) If $a = 0.05$, which test do you think the FDA would use? Explain.

8-14 Determine the test criteria (critical values and critical region for z) that would be used to test the null hypothesis at the given level of significance, as described in each of the following:

(a) $H_0: \mu = 20$ ($\alpha = 0.10$)
$H_a: \mu \neq 20$

(b) $H_0: \mu = 24 \, (\leq)$ ($\alpha = 0.01$)
$H_a: \mu > 24$

(c) $H_0: \mu = 10.5 \, (\geq)$ ($\alpha = 0.05$)
$H_a: \mu < 10.5$

(d) $H_0: \mu = 35$ ($\alpha = 0.01$)
$H_a: \mu \neq 35$

(e) $H_0: \mu = 14.6 \ (\geq)$ $(\alpha = 0.02)$
    $H_a: \mu < 14.6$

(f) $H_0: \mu = 6.78 \ (\leq)$ $(\alpha = 0.10)$
    $H_a: \mu > 6.78$

(g) $H_0: \mu = 21.50$ $(\alpha = 0.05)$
    $H_a: \mu \neq 21.50$

(h) $H_0: \mu = 0.034$ $(\alpha = 0.01)$
    $H_a: \mu \neq 0.034$

8-15   A sample of 36 measurements was taken to test the hypothesis stated in illustration 8-1. Does the sample mean of 5.3 provide sufficient evidence to reject the null hypothesis in the illustration? Complete the hypothesis test, using $\sigma = 1.8$ and $\alpha = 0.05$.

8-16   The ages of 50 randomly selected students enrolled in an evening class are obtained to test the claim that the mean age of students enrolled in evening classes at our college is greater than 26 years. Does a sample mean of 27.4 provide sufficient evidence to reject the null hypothesis $\mu = 26$ in favor of the alternative $\mu > 26$? Use $\sigma = 3.0$ years and $\alpha = 0.05$.

8-17   The manager at Air Express feels that the weights of packages shipped recently are less than in the past. Records show that in the past packages have had a mean weight of 36.7 pounds and a standard deviation of 14.2 pounds. A random sample of last month's shipping records yielded a mean weight of 32.1 pounds for 64 packages. Is this sufficient evidence to reject the null hypothesis in favor of the manager's claim? Use $\alpha = 0.01$.

8-18   The manufacturer of the fluorescent light bulbs used by a large industrial complex claims that the bulbs have a mean life of at least 1600 hours. The maintenance foreman had a sample of 100 bulbs randomly identified and kept a record on their use. Does a sample mean of 1562.3 support the foreman's contention that the mean life is less than 1600 hours at the 0.02 level of significance? (Use $\sigma = 150$ hours.)

8-19   A fire insurance company felt that the mean distance from a home to the nearest fire department in a suburb of Chicago was at least 4.7 miles. It set its fire insurance rates accordingly. Members of the community set out to show that the mean distance was less than 4.7 miles. This, they felt, would convince the insurance company to lower its rates. They randomly identified 64 homes and measured the distance to the nearest fire department for each. The resulting sample mean was 4.4. If $\sigma = 2.4$ miles, does the sample show sufficient evidence to support the community's claim, at the $\alpha = 0.05$ level of significance?

8-20   On a popular self-image test the mean score for public assistance recipients is 65 and the standard deviation is 5. A random sample of 42 public assistance recipients in Emerson County are given the test. They achieved a mean score of 60. Do the scores of Emerson County public assistance recipients differ from the average with respect to this variable? Use $\alpha = 0.01$.

Section 8–3 The Hypothesis Test (A Probability-Value Approach)

The classical (or traditional) hypothesis-testing procedure was described in section 8–2. An alternative approach to the decision-making process in hypothesis testing has gained popularity in recent years, largely due to the convenience and the "number crunching" ability of computers. This alternative process is to calculate and report a probability value related to the observed sample statistic. The value reported, P (capital P; not to be confused with "small p" the binomial parameter), is called the probability-value, the prob-value, or simply the P-value. The P-value is then compared to the level of significance α, the probability of the type I error, in order to make the decision of reject H_0 or fail to reject H_0.

---

prob-value

Prob-value, P

The prob-value, P, of a hypothesis test is the smallest level of significance for which the observed sample information becomes significant, provided the null hypothesis is true.

Let's return to illustration 8–4 and see how its solution would be different if solved using this "prob-value" method.

Illustration 8–8

For many semesters an instructor has recorded his students' grades, and the mean μ for all these students' grades is 72. The current class of 36 students seems to be better than average in ability and the instructor wants to show that according to their average "the current class is superior to his previous classes." Does the class mean $\bar{x}$ of 75.2 present sufficient evidence to support the instructor's claim that the current class is superior? Use $\alpha = 0.05$ and $\sigma = 12.0$.

Solution To be superior this class must have a mean grade that is higher than the mean of all the previous classes. To be "equal or less than" would not be superior.

STEP 1 $H_0: \mu = 72 \ (\leq)$ (class is not superior).

STEP 2 $H_a: \mu > 72$ (class is superior).

Notice that steps 1 and 2 are exactly the same when using either the classical or the prob-value approach to hypothesis testing.

STEP 3 Determine α, the probability of the type I error; $\alpha = 0.05$.

STEP 4   Using formula (7–2) and the sample information, we will obtain the calculated value of the test statistic $z*$.

$$z = \frac{\bar{x} - \mu}{\sigma/\sqrt{n}} \quad \text{and} \quad \bar{x} = 75.2 \quad \text{and} \quad n = 36$$

RECALL   We assume that μ was equal to 72 in the null hypothesis and that $\sigma = 12.0$ was known.

$$z = \frac{75.2 - 72.0}{12.0/\sqrt{36}} = \frac{3.2}{2.0} = 1.60$$

$$z* = 1.60$$

(REMINDER: We use an asterisk, *, to identify the calculated value of the test statistic; the asterisk will also be used to locate its value on the probability distribution in the next step.)

Notice that step 4 in the prob-value approach is exactly the same as step 4 in the classical approach.

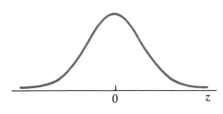

STEP 5   Draw a sketch of the probability distribution for the test statistic; z in this case. Next locate the value of $z*$ calculated in step 4 on this distribution (see diagram) and determine what part of the distribution relative to $z*$ represents the prob-value, P. Then calculate P.

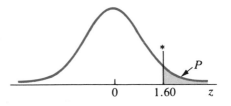

The alternative hypothesis indicates that we are interested in that part of the probability distribution that lies to the right of $z*$ since the "greater than" sign was used.

$$P = P(z > z*) = P(z > 1.60)$$

$$= 0.5000 - 0.4452 \quad \text{(Use table 5 in the Appendix to find the area under a normal curve.)}$$

$$P = 0.0548$$

The results of the statistical analysis are then reported.

STEP 6 The prob-value for this hypothesis test is 0.0548. If the decision is made using the 0.05 level of significance (or any α value less than or equal to 0.0548) then the decision will be to "fail to reject the null hypothesis" and the conclusion will be there is no evidence to show the class as superior. If, however, we had chosen to use a level of significance of 0.10 (or any α value greater than 0.0548), our decision will be to "reject H_0," and our conclusion would be that the current class is superior. □

Computer Solution

Minitab Printout for Illustration 8–8

Information given to the computer:
z-test of H_0: $\mu = 72.0$, H_a: $\mu > 72.0$, $\sigma = 12.0$, and location of sample data
Sample statistics: $n = 36$, $\bar{x} = 75.2$, $s = 11.87$

The calculated $z*$
The prob-value
The decision

```
--ZTEST OF POPULATION MU = 72.0 VS. ALTERNATIVE +1,
SIGMA = 12.0, DATA IN C1
C1        N = 36        MEAN = 75.2      ST. DEV. = 11.87

TEST OF MU = 72.0   VS.   MU G.T.   72.0
Z = 1.60
THE TEST IS STATISTICALLY SIGNIFICANT AT ALPHA = 0.0548
CANNOT REJECT AT ALPHA = 0.05
--STOP
```

□

Before looking at another illustration, let's summarize the details for this prob-value approach to the hypothesis testing.

1. The null and alternative hypotheses are formulated in the same manner as before.

2. Determine the level of significance, α, to be used.

3. The value of the test statistic is calculated in step 4 in exactly the same manner as it was calculated before in step 4.

4. The prob-value, P, is the area that represents values of z that are more extreme than $z*$ under the curve of the probability distribution. There are three separate cases: two are one-sided and one is two-sided.

    The direction (or sign) in the alternative hypothesis is the key.
    Case 1. If H_a is one-sided to the right (that is, contains " > "), then $P = P(z > z*)$, the area to the right of $z*$.

    Case 2. If H_a is one-sided to the left (that is, contains " < "), then $P = P(z < z*)$, the area to the left of $z*$.

    Case 3. If H_a is two-sided (that is, contains " ≠ "), then

    $$P = P(z < -|z*|) + P(z > |z*|),$$

    the sum of the area to the left of the negative value of $z*$ and the area to the right of the positive value of $z*$. Since both areas are equal, you will probably find one and double it. Thus $P = 2 \times P(z > |z*|)$.

5. The decision will be made by comparing P to the previously established value of α.

   (a) If the **calculated prob-value is less than or equal to the desired** α, then the decision must be **reject** H_0

   (b) If the **calculated prob-value is greater than the desired** α, then the decision must be **fail to reject** H_0.

6. Conclusions should be worded in the same manner as previously instructed.

Let's look at an illustration involving the two-tailed procedure.

Illustration 8–9

Many of the large companies in our city have for years used the Kelley Employment Agency for testing prospective employees. The employment selection test used has historically resulted in scores distributed about a mean of 82 and a standard deviation of 8. The Brown Agency has developed a new test that is quicker and easier to administer and therefore less expensive. Brown claims that their test results are the same as those obtained on the Kelley test. Many of the companies are considering a change from the Kelley Agency to the Brown Agency in order to cut costs. However, they are unwilling to make the change if the Brown test results have a different mean value.

An independent testing firm tested 36 prospective employees. A sample mean of 80 resulted. Determine the prob-value associated with this hypothesis test.

Solution   The Brown Agency's test results will be different if the mean test score is not equal to 82. They will be the same if the mean is 82. Therefore:

STEP 1 $H_0: \mu = 82$ (test results have same mean)

STEP 2 $H_a: \mu \neq 82$ (test results have different mean)

STEP 3   Step 3 is omitted when a question asks for the prob-value and not a decision.

STEP 4   The sample information $n = 36$ and $\bar{x} = 80$ and formula (7–2) are used to calculate z^*

$$z = \frac{\bar{x} - \mu}{\sigma/\sqrt{n}}$$

$$z = \frac{80 - 82}{8/\sqrt{36}} = \frac{-2}{8/6} = -1.50$$

$$z^* = -1.50$$

STEP 5   The value of z^* is located on the normal distribution. (See the diagram.) Since the alternative hypothesis indicates a two-tailed test, we must find the probability associated with two areas, namely $P(z < -|z^*|)$ and

-1.50   0 z

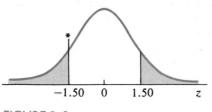

FIGURE 8–2

$P(z > |z^*|)$. And since $z^* = -1.50$, the value of $|z^*| = 1.50$. Thus $P = P(z < -1.50) + P(z > 1.50)$ as shown in figure 8–2.

$$P = P(z < -1.50) + P(z > 1.50)$$
$$= (0.5000 - 0.4332) + (0.5000 - 0.4332)$$
$$= 0.0668 + 0.0668$$
$$P = 0.1336$$

The prob-value for this hypothesis test is 0.1336. Each individual company now will make a decision whether to (a) continue to use Kelley's services or (b) change to the Brown Agency. Each will need to establish the level of significance that best fits their own situation and then make their decision using the decision rule described above. □

Exercises

8–21 Calculate the prob-value for each of the following:

(a) $H_0: \mu = 10 \quad z^* = 1.48$
   $H_a: \mu > 10$

(b) $H_0: \mu = 105 \quad z^* = -0.85$
   $H_a: \mu < 105$

(c) $H_0: \mu = 13.4 \quad z^* = 1.17$
   $H_a: \mu \neq 13.4$

(d) $H_0: \mu = 8.56 \quad z^* = -2.11$
   $H_a: \mu < 8.56$

(e) $H_0: \mu = 110 \quad z^* = -0.93$
   $H_a: \mu \neq 110$

(f) $H_0: \mu = 54.2 \quad z^* = 0.46$
   $H_a: \mu > 54.2$

8–22 The calculated prob-value for a hypothesis test is $P = 0.084$. What decision about the null hypothesis would occur:

(a) If the hypothesis test is completed at the 0.05 level of significance?

(b) If the hypothesis test is completed at the 0.10 level of significance?

 8–23 An economist claims that when the Dow-Jones average increases, the volume of shares traded on the New York Stock Exchange tends to increase. Over the past two years the daily volume on the exchange has averaged 21.5 million shares and had a standard deviation of 2.5 million. A random sample of 64 days on which the Dow-Jones average increased was selected and the average daily volume was computed. The sample average was 22 million. Calculate the prob-value for this hypothesis test.

8-24 The marketing director of A & B Cola is worried that their product is not attracting enough young consumers. To test this hypothesis she randomly surveys 100 A & B Cola consumers. The mean age of an individual in the community is 32 years and the standard deviation is 10 years. The surveyed consumers of A & B Cola have a mean age of 35. At the 0.01 level of significance, is this sufficient evidence to conclude that A & B Cola consumers are, on the average, older than the average person living in the community? Complete this hypothesis test using the prob-value approach.

Section 8-4 Estimation

Estimation is the second type of statistical inference. It is the procedure to use when answering a question that asks for the value of a population parameter. For example, "What is the mean one-way distance that Monroe Community College students commute daily?"

If you needed an answer to this question, you might take a sample from the population and calculate the sample mean $\bar{x}$. Suppose that you did draw a random sample of 100 one-way distances and a sample mean of 10.22 miles resulted. What is your estimate for the mean value of the population? If you report the sample mean $\bar{x}$ as your estimate, you will be making a point estimate.

---

point estimate

Point Estimate for a Parameter

The value of the corresponding sample statistic.

That is, the sample mean, $\bar{x} = 10.22$ miles, is the best point estimate for the mean one-way distance for this population. In other words, the mean one-way distance for all students is estimated to be 10.22 miles. We are not really implying that the population mean μ is exactly 10.22 miles. We intend this point estimate to be interpreted to say "μ is close to 10.22." From our study so far, you probably realize that when we say $\mu = 10.22$, there is very little chance that the statement is true. In previous chapters we drew samples from probability distributions where μ was known, and seldom did we obtain a sample mean exactly equal in value to μ. Why, then, should we expect a single sample to yield an $\bar{x}$ equal to μ when μ is unknown? We shouldn't.

What does it mean to say "μ is close to 10.22 miles"? Closeness is a relative term, but perhaps in this case "close" might be arbitrarily defined to be "within 1 mile" of μ. If 1 mile satisfies our intuitive idea of closeness, then to say "μ is close to 10.22" is comparable to saying "μ is between 9.22 ($10.22 - 1.0$) and 11.22 ($10.22 + 1.0$)." This suggests that we might make estimations by using intervals. The confidence interval estimate employs this interval concept and assigns a measure to the interval's reliability in estimating the parameter in question.

interval estimate	**Interval Estimate**

An interval bounded by two values and used to estimate the value of a population parameter. The values that bound this interval are statistics calculated from the sample that is being used as the basis for the estimation.

level of confidence	**Level of Confidence** $1-\alpha$

The probability that the sample selected yields boundary values that lie on opposite sides of the parameter being estimated. The level of confidence is sometimes called the **confidence coefficient**.

confidence interval	**Confidence Interval**

An interval estimate with a specified level of confidence.

The central limit theorem is the source for the information needed to construct confidence interval estimates. Our sample mean ($\bar{x} = 10.22$) is a member of the sampling distribution of sample means. The CLT describes this distribution as being approximately normally distributed with a mean $\mu_{\bar{x}} = \mu$ and a standard deviation $\sigma_{\bar{x}} = \sigma/\sqrt{n}$. The population mean is unknown; however, it exists and is constant. Let's assume that the standard deviation of the population is $\sigma = 6$.

$\sigma/\sqrt{n} = 6.0/\sqrt{100} = 0.6$

μ
(unknown)

FIGURE 8–3
Sampling Distribution, n = 100

RECALL: In this chapter we are studying the inferences about μ under the assumption that σ is known. This assumption is for convenience, and the restriction will be removed in chapter 9.

Figure 8–3 shows the sampling distribution of sample means to which our sample mean ($\bar{x} = 10.22$ and $n = 100$) belongs.

When a sample of size 100 is randomly selected from a population whose mean is μ and whose standard deviation is 6, what is the probability that the sample mean is within one unit of μ? In other words, what is $P(\mu - 1 < \bar{x} < \mu + 1)$? See figure 8–4. This probability may be found by using the normal probability distribution (table 5) and formula (7–2):

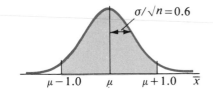

$\sigma/\sqrt{n} = 0.6$

$\mu - 1.0 \quad \mu \quad \mu + 1.0 \quad \bar{x}$

FIGURE 8–4
Probability That x̄ Is Within 1 Unit of μ

$$z = \frac{\bar{x} - \mu}{\sigma/\sqrt{n}}$$

If $\bar{x} = \mu - 1$,

$$z = \frac{(\mu - 1) - \mu}{0.6} = \frac{-1.0}{0.6} = -1.67$$

If $\bar{x} = \mu + 1$,

$$z = \frac{(\mu + 1) - \mu}{0.6} = \frac{+1.0}{0.6} = +1.67$$

Therefore,

$$P(\mu - 1 < \bar{x} < \mu + 1) = P(-1.67 < z < +1.67)$$
$$= 2 \cdot P(0 < z < 1.67)$$
$$= 2(0.4525) = \ \textbf{0.9050}$$

The probability that the mean of a random sample is within one unit of this population mean is 0.9050. Thus the probability that the population mean is within one unit of the mean of a sample is also 0.9050. Therefore, the interval 9.22 to 11.22 is a 0.9050 confidence interval estimate for the mean one-way distance commuted by the students of the college.

---

maximum error of estimate

Maximum Error of Estimate, E

One-half the width of the confidence interval. In general, E is a multiple of the standard error.

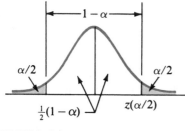

FIGURE 8–5
Each Tail Contains α/2

The preceding illustration started with the maximum error being assigned the value of 1 mile. Typically, the maximum error of estimate is determined by the level of confidence that we want our confidence interval to have. That is, $1 - \alpha$ will determine the maximum error. The level of confidence will be split so that half of $1 - \alpha$ is above the mean and half below the mean (figure 8–5). This will leave $\alpha/2$ as the probability in each of the two tails of the distribution. The z score at the boundary of the confidence interval will be z of $z(\alpha/2)$. [Recall that $\alpha/2$ is the probability (or area) under the curve to the right of this point. See chapter 6.] The standard score z is a number of standard deviations. Therefore, the maximum error of estimate E is

$$E = z(\alpha/2) \cdot \frac{\sigma}{\sqrt{n}} \tag{8-1}$$

See figure 8–6.

FIGURE 8–6
Maximum Error of Estimate

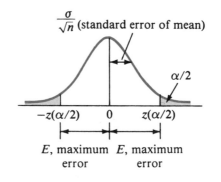

The $1 - \alpha$ confidence interval for μ is

$$\bar{x} - z(\alpha/2) \cdot \frac{\sigma}{\sqrt{n}} \quad \text{to} \quad \bar{x} + z(\alpha/2) \cdot \frac{\sigma}{\sqrt{n}} \qquad (8\text{–}2)$$

$\bar{x} - z(\alpha/2) \cdot (\sigma/\sqrt{n})$ is called the **lower confidence limit (LCL)** and $\bar{x} + z(\alpha/2) \cdot (\sigma/\sqrt{n})$ is called the **upper confidence limit (UCL)** for the confidence interval.

Illustration 8–10

Construct the 0.95 confidence interval for the estimate of the mean one-way distance that the students commute.

Solution

$$1 - \alpha = 0.95$$

$$\alpha = 0.05$$

$$\frac{\alpha}{2} = 0.025 \quad \text{(See the diagram.)}$$

z	... 0.06 ...
$\vdots$	
1.9	0.4750
$\vdots$	
	$z(0.025) = 1.96$

$$\bar{x} \pm z(\alpha/2) \cdot \frac{\sigma}{\sqrt{n}}$$

$$10.22 \pm (1.96) \cdot \frac{6}{\sqrt{100}}$$

$$10.22 \pm (1.96)(0.6)$$

$$10.22 \pm 1.176$$

$$10.22 - 1.176 = 9.044 = 9.04$$

and

$$10.22 + 1.176 = 11.396 = 11.40$$

Therefore, with 0.95 confidence we can say that the mean one-way distance is between 9.04 and 11.40, and we will write this as

(9.04, 11.40), the 0.95 confidence interval for μ ☐

NOTE: Many of the confidence interval formulas that you will be using will have a format similar to formula (8–2), namely, $\bar{x} - E$ to $\bar{x} + E$. However, perhaps the simplest way to handle the arithmetic involved is to think of the formula as $\bar{x} \pm E$, as was done in illustration 8–10. Using this format you first calculate E. Then you calculate the lower bound by subtracting and you calculate the upper bound by adding.

Illustration 8–11

To estimate the mean score on the first one-hour exam in statistics, we obtain a random sample of 38 exam scores. A sample mean of 74.3 is found. Construct the 0.98 confidence interval estimate for the mean of all first hour exam sources. Use $\sigma = 14$.

Solution

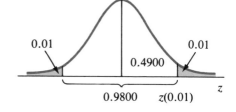

$$1 - \alpha = 0.98$$

$$\alpha = 0.02$$

$$\frac{\alpha}{2} = 0.01 \quad \text{(see the diagram.)}$$

$$\bar{x} \pm z(\alpha/2) \cdot \frac{\sigma}{\sqrt{n}}$$

$$74.3 \pm (2.33) \cdot \frac{14}{\sqrt{38}}$$

$$74.3 \pm (2.33)(2.27)$$

$$74.3 \pm 5.29$$

$$74.3 - 5.29 = \mathbf{69.01} \qquad 74.3 + 5.29 = \mathbf{79.59}$$

(69.01, 79.59) is the 0.98 confidence interval for μ. That is, with 0.98 confidence we can say that the mean exam score for all such exams is between 69.01 and 79.59. ☐

Illustration 8-12

Suppose that the sample in illustration 8-7 had been obtained for the purpose of estimating the population mean μ. The sample results ($n = 40$, $\bar{x} = 3.975$) and $\sigma = 2.87$ are used in formula (8-2) to determine the 0.90 confidence interval.

$$\bar{x} \pm z(\alpha/2) \cdot \frac{\sigma}{\sqrt{n}}$$

$$3.975 \pm 1.65 \cdot \frac{2.87}{\sqrt{40}}$$

$$3.975 \pm (1.65)(0.454)$$

$$3.975 \pm 0.749$$

$$3.975 - 0.749 = 3.226 \qquad 3.975 + 0.749 = 4.724$$
$$= 3.23 \qquad\qquad\qquad = 4.72$$

With 0.90 confidence, we think that μ is somewhere within this interval.

(3.23, 4.72) is the 0.90 confidence interval estimate for μ. (See the diagram.)

Since this sample was taken from a known population—namely, the random digits—we can look back at our answer and say that the confidence interval is correct. That is, the true value of μ, 4.5, does fall within the confidence interval. ☐

Suppose that we were to take another sample of size 40; would we obtain the same results? Suppose we took a third and a fourth. What would happen? What is the level of confidence, $1 - \alpha$? Yes, it has the value 0.90, but what does that mean? Table 8-5 lists the means obtained from 10 different random samples of size 40 taken from table 1. The 0.90 confidence interval for the estimate of μ based on each of these samples is also listed in table 8-5. Since all the interval estimates were for the mean of the random

TABLE 8-5

Samples of Size 40 Taken from Table I, Appendix D

Sample Number	Sample Mean, $\bar{x}$	0.90 Confidence Interval Estimate for μ
1	4.64	3.89 to 5.39
2	4.56	3.81 to 5.31
3	3.96	3.21 to 4.71
4	5.12	4.37 to 5.87
5	4.24	3.49 to 4.99
6	3.44	2.69 to 4.19
7	4.60	3.85 to 5.35
8	4.08	3.33 to 4.83
9	5.20	4.45 to 5.95
10	4.88	4.13 to 5.63

digits ($\mu = 4.5$), we can inspect each confidence interval and see that 9 of the 10 contain μ and would therefore be considered correct estimates. One sample (sample 6) did not yield an interval that contains μ. This is shown in figure 8–7.

FIGURE 8–7
*Interval Estimates
from Table 8–5*

Interval estimate
3.89 to 5.39
3.81 to 5.31
3.21 to 4.71
4.37 to 5.87
3.49 to 4.99
2.69 to 4.19
3.85 to 5.35
3.33 to 4.83
4.45 to 5.95
4.13 to 5.63

REMEMBER: $1 - \alpha$ is the probability that we obtain a confidence interval such that μ is contained within it.

The maximum error formula can be used to determine the sample size required to obtain a $1 - \alpha$ confidence interval estimate with a specified maximum error.

Illustration 8–13

Determine the size of the sample that will be necessary to estimate the mean weight of second-grade boys if we desire our estimate to be accurate to within 1 pound with 95% confidence. Assume that the standard deviation of such weights is 3 pounds.

Solution $1 - \alpha = 0.95$; therefore, $z(0.025) = 1.96$, as found by using table 5. The maximum error $E = 1$ and $\sigma = 3$. Use formula (8–1).

$$E = z(\alpha/2) \cdot \frac{\sigma}{\sqrt{n}}$$

$$1.0 = 1.96 \cdot \frac{3}{\sqrt{n}}$$

$$1 = \frac{5.88}{\sqrt{n}}$$

$$\sqrt{n} = 5.88$$

$$n = 34.57$$

Therefore,

$$n = 35$$

NOTE: When solving for the sample size n, all fractional (decimal) values are to be *rounded up* to the next larger integer. □

Does a sample size of 35 determine a maximum error of 1 pound in illustration 8–13?

$$E = z(\alpha/2) \cdot \frac{\sigma}{\sqrt{n}} = 1.96 \cdot \frac{3.0}{\sqrt{35}}$$

$$= (1.96)(0.51) = 0.999$$

This is a maximum error of just under 1.0.

The use of formula (8–1) can be made a little easier by rewriting the formula in a form that expresses n in terms of the other quantities:

$$n = \left[\frac{z(\alpha/2) \cdot \sigma}{E} \right]^2 \tag{8–3}$$

Illustration 8–14

A sample of what size would be needed to estimate the population mean to within $\frac{1}{5}$ of a standard deviation with 99 % confidence?

Solution $1 - \alpha = 0.99$; therefore, $\alpha/2 = 0.005$ and $z(0.005) = 2.58$, as found by using table 5. The maximum error E is to be $\frac{1}{5}$ of σ; that is, $E = \sigma/5$. Using formula (8–3), we have

$$n = \left[\frac{z(\alpha/2) \cdot \sigma}{E} \right]^2 = \left[\frac{(2.58)(\sigma)}{\sigma/5} \right]^2$$

$$= \left[\frac{(2.58\sigma)(5)}{\sigma} \right]^2 = [(2.58)(5)]^2$$

$$= (12.90)^2 = 166.41 = 167 \qquad \square$$

Exercises

8–25   A sample is taken from a population in order to estimate the mean. $\bar{x} = 35.6$ and $n = 100$.

(a) Give a point estimate for μ.

(b) If the standard deviation of this population is believed to be 16, determine a 95 % confidence interval estimate for the population mean.

8-26 A certain population of the annual incomes of unskilled laborers has a standard deviation of $1200. A random sample of 36 results in $\bar{x} = \$7280$.

(a) Give a point estimate for the population mean annual income.

(b) Estimate μ with a 95 % confidence interval.

(c) Estimate μ with a 99 % confidence interval.

8-27 A sample of 60 evening students' ages is obtained in order to estimate the mean age of evening students. $\bar{x} = 25.3$ years. The population variance is 16.

(a) Give a point estimate for μ.

(b) Find the 95 % confidence interval estimate for μ.

(c) Find the 99 % confidence interval estimate for μ.

8-28 The lengths of 200 fish caught in Cayuga Lake had a mean of 14.3 inches. The population standard deviation is 2.5 inches.

(a) Find the 90 % confidence interval for the population mean length.

(b) Find the 98 % confidence interval for the population mean length.

8-29 The average distance from 64 randomly selected homes to the nearest fire department was found to be 4.4 miles. Use a population standard deviation of 2.4 miles.

(a) Find the maximum error of estimate at 99 % confidence.

(b) Find the 99 % confidence interval.

8-30 A random group of 144 students obtained a mean score of 73.2 on a college placement exam. The population of all exam scores has a standard deviation of 8. Find the 95 % confidence interval estimate for the population mean.

8-31 A sample of what size would be needed to estimate a population mean to within 6 units with 98 % confidence if the population has a standard deviation of 16?

8-32 How large a sample should be taken if the population mean is to be estimated with 99 % confidence to within $75? The population has a standard deviation of $900.

8-33 By measuring the amount of time it takes a component of a product to move from one work station to the next, an engineer has estimated that the standard deviation is 5 seconds.

(a) How many measurements should be made in order to be 95 % certain that the maximum error of estimation will not exceed 1 second?

(b) What sample size is required for a maximum error of 2 seconds?

8-34 A medical products firm has developed a new drug that requires very strict control on the amount of the drug placed in each capsule. It can be hazardous if too much drug is used and ineffective if not enough is used. The maximum error of estimate for the mean is 2 milligrams. The standard deviation is believed to be 7 milligrams. If 99 % confidence is desired, what size sample should be taken to be certain that the mean amount of the drug per capsule is correct?

In Retrospect

Two forms of inference were studied in this chapter: hypothesis testing and estimation. They may be, and often are, used separately. However, it would seem quite natural for the rejection of a null hypothesis to be followed by a confidence interval estimate. (If the value claimed is wrong, we will usually want to know just what the true value is.)

These two forms of inference are quite different but they are related. There is a certain amount of crossover between the use of the two inferences. For example, suppose that you had sampled and calculated a 90% confidence interval for the mean of a population. The interval was 10.5 to 15.6. Following this someone claims that the true mean is 15.2. Your confidence interval estimate can be compared to this claim. If the claimed value falls within your interval estimate, you would fail to reject the null hypothesis that $\mu = 15.2$ at a 10% level of significance in a two-tailed test. If the claimed value (say 16.0) falls outside the interval, you would then reject the null hypothesis that $\mu = 16.0$ at $\alpha = 0.10$ in a two-tailed test. If a one-tailed test is required, or if you prefer a different value of α, a separate hypothesis test must be used.

The news article at the beginning of the chapter suggests that Sears' tests have proved that its steel-belted radial tire will give you 40,000 miles of rugged service. This advertisement, like many others, goes a long way in selling the consumer on a particular product. The sophistication of the presentation creates a lasting and very suggestive impression. Now examine this advertisement in the light of what you have learned in this chapter.

In this chapter we have restricted our discussion of inferences to the mean of a population for which the standard deviation is known. In the next two chapters we will discuss inferences about other population parameters and eliminate the restriction about the known value for standard deviation.

Chapter Exercises

8–35   The expected mean of a certain population is 100 and its standard deviation is 12. A sample of 50 measurements gives a sample mean of 96. Using a level of significance of 0.01, a test is to be made to see if the population mean is really 100 or if it is different from 100. State or calculate the desired answer in parts (a) through (j).

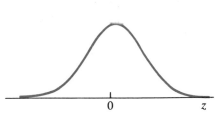

(a) H_0 (b) H_a

(c) α (d) $z(\alpha/2)$

(e) μ (based on H_0)   (f) $\bar{x}$

(g) σ (h) $\sigma_{\bar{x}}$

(i) z score for $\bar{x}$ (j) decision

(k) On the sketch locate $\alpha/2$, $z(\alpha/2)$, the rejection region for H_0, and the z score for $\bar{x}$.

8–36   A sample of 64 measurements is taken from a continuous population and the sample mean is found to be 32.0. The standard deviation of the population is known to be 2.4. An interval estimation is to be made of the mean with a level of confidence of 0.90. State or calculate the following items:

(a) $\bar{x}$ (b) σ

(c) n (d) $1 - \alpha$

(e) $z(\alpha/2)$ (f) $\sigma_{\bar{x}}$

(g) E (maximum error of estimate)   (h) upper confidence limit

(i) lower confidence limit

8–37   The college admissions office tells prospective students that the average cost of their textbooks is \$12 and the standard deviation is \$2.50. A group of students thinks that the average cost has become higher than \$12 and decides to test the admissions office's claim against their alternative, using a sample of size 45.

(a) If they use $\alpha = 0.05$, what is the cutoff value for $\bar{x}$ that will support their belief?

(b) Their sample data are summarized by $n = 45$ and $\sum x = 562.25$. Is this sufficient evidence to reject the admissions office's claim?

8–38   A rope manufacturer has conducted a large number of tests over a long period of time. They have found that the rope has a mean breaking strength of 300 pounds and a standard deviation of 24 pounds. Assume that these values are μ and σ. It is believed that by using a recently developed process, the mean breaking strength can be increased.

(a) Design a null and alternative hypothesis such that rejection of the null hypothesis will imply that the mean breaking strength has increased.

(b) If the above decision rule is used with $\alpha = 0.01$, what is the critical value for the test statistic and what value of $\bar{x}$ corresponds to it if a sample of size 45 is used?

(c) Using the decision rule established in part (a), what is the prob-value associated with rejecting the null hypothesis when 45 tests result in a sample mean of 305?

8–39   Mr. Smith, a discouraged commuter who uses the city's bus system, decided to keep a record of the amount of time he wasted as a result of the tardiness of his bus. (x is the lateness of his bus, in minutes.) After several weeks his records showed the following summary: 80 bus rides, 228 total minutes of tardiness. When Mr. Smith complained to the bus company, they replied that their buses averaged no more than 2 minutes late, with a variance of 20. Does Mr. Smith have sufficient evidence to reject the bus company's claim? Use $\alpha = 0.05$.

(a) Solve using the classical approach.

(b) Solve using the prob-value approach.

8–40   In a large supermarket the customer's waiting time to check out is approximately normally distributed with a standard deviation of 2.5 minutes. A sample of 24 customer waiting times produced a mean of 10.6 minutes. Is this evidence sufficient to reject the supermarket's claim that its customer checkout time averages no more than 9 minutes? Complete this hypothesis test using the 0.02 level of significance.

(a) Solve using the classical approach.

(b) Solve using the prob-value approach.

8-41 At a very large firm, the clerk-typists were sampled to see if the salaries differed among departments for workers in similar categories. In a sample of 50 of the firm's accounting clerks, the average annual salary was $16,010. The firm's personnel office insists that the average salary paid to all clerk-typists with the firm is $15,650 and that the standard deviation is $1,800. At the 0.05 level of significance, can we conclude that the accounting clerks receive a higher average salary than other clerk-typists do?

(a) Solve using the classical approach.

(b) Solve using the prob-value approach.

8-42 A manufacturer of a certain cigarette claims that its cigarette averages no more than 17.5 milligrams of nicotine and the standard deviation is 1.8 milligrams. A sample of 42 randomly selected cigarettes was tested and a mean of 18.3 milligrams was observed. At the 0.05 level of significance, do we have sufficient evidence to reject the manufacturer's claim?

(a) Solve using the classical approach.

(b) Solve using the prob-value approach.

8-43 The mean length of time required for returning students to register at our college has been 90 minutes, and the standard deviation is 15 minutes. A new "faster" registration procedure was introduced this semester. The students seem to think the registration people "blew it." The students collected a sample; the sample statistics are $n = 45$ and $\bar{x} = 95.6$. Does the sample show sufficient evidence, at the 0.05 level of significance, to support their contention?

(a) Solve using the classical approach.

(b) Solve using the prob-value approach.

8-44 Suppose that a confidence interval is assigned a level of confidence of $1 - \alpha = 0.95$. How is the 0.95 used in constructing the confidence interval?

8-45 A random sample of the scores of 100 applicants for clerk-typist positions at a large insurance company showed a mean score of 72.6. The preparer of the test maintained that good applicants should average 75.0.

(a) Determine the 99 % confidence interval estimate for the mean score of all applicants at the insurance company. Assume that the standard deviation of test scores is 10.5.

(b) Can the insurance company conclude that it is getting good applicants (as measured by this test)?

8-46 Waiting times (in hours) at a popular restaurant are believed to be approximately normally distributed with a variance of 2.25 hours during busy periods.

(a) A sample of 20 customers reveaied a mean waiting time of 1.52 hours. Construct the 95% confidence interval for the estimate of the population mean.

(b) Suppose the mean of 1.52 hours had resulted from a sample of 32 customers. Find the 95% confidence interval.

(c) What effect does a larger sample size have on the confidence interval?

8-47 The weights of full boxes of a certain kind of cereal are normally distributed with a standard deviation of 0.27 ounce. A sample of 18 randomly selected boxes produced a mean weight of 9.87 ounces.

(a) Find the 95% confidence interval for the true mean weight of a box of this cereal.

(b) Find the 99% confidence interval for the true mean weight of a box of this cereal.

(c) What effect did the increase in the level of confidence have on the width of the confidence interval?

8-48 An office manager wishes to estimate the mean time required to handle a customer complaint. The times are believed to be normally distributed with a standard deviation of 12 minutes. A sample of 38 complaints shows a mean of 28.7 minutes. Construct the 95% confidence interval for the true mean time required to handle customer complaints.

8-49 A large department store has 20,000 customers that use the store's charge account plan. To estimate the amount currently owed to the store, a random sample of 64 accounts was selected. The sample mean was $142, and the population standard deviation is $70.

(a) Find the 95% confidence interval estimate for the true mean account balance.

(b) Find the 95% confidence interval estimate for the total amount owed to the store through these charge accounts.

8-50 The dean of our college wishes to estimate the grade point average of all sorority members. A sample of what size should he take if he wishes to estimate their grade point average correct to within 0.1 at 98% confidence? (Assume that $\sigma = 0.5$.)

8-51 A manufacturer of automobiles wishes to estimate the mean gasoline mileage that its customers will obtain with its new compact model. How many sample runs must be performed in order that the estimate be accurate to within 0.3 mile per gallon at 95% confidence? (Assume that $\sigma = 1.5$.)

8-52 A fish hatchery manager wishes to estimate the mean length of her 3-year-old hatchery-raised trout. She wants to make a 99% confidence interval estimate accurate to within $\frac{1}{3}$ of a standard deviation. How large a sample does she need to take?

8-53 Eighty-five of the newly admitted freshmen at a college were randomly selected and asked to perform a certain physical fitness task. The times recorded had a mean value of 12.87 minutes. The time required to complete the task, x, is distributed with a standard deviation of 1.3 minutes. Construct the 90% confidence interval estimate for the mean time required by all freshmen.

8-54 The admissions office at Memorial Hospital recently stated that the mean age of its patients was 42 years. A random sample of 120 ages was obtained from the admissions office records in an attempt to disprove the claim. Is a sample mean of 44.2 years significantly larger than the claimed 42 years, at the $\alpha = 0.05$ level? Use $\sigma = 20$ years.

(a) Solve using the classical approach.

(b) Solve using the prob-value approach.

8-55 A random sample of 100 trout is taken from the vats at Happy Acres Fish Hatchery in order to estimate the length of the fish. The variance in their lengths is 4.0. Construct the 99% confidence interval estimate for the mean length of all such fish based on a sample mean of 13.15 centimeters.

8-56 A random sample of 280 rivets is selected from a very large shipment. The rivets are to have a mean diameter of no more than $\frac{5}{16}$ inch. Does a sample mean of 0.3126 show sufficient reason to reject the null hypothesis that the mean diameter is no more than $\frac{5}{16}$ inch, at the $\alpha = 0.01$ level of significance? Use $\sigma = 0.0006$.

(a) Solve using the classical approach.

(b) Solve using the prob-value approach.

Vocabulary List

Be able to define each term. In addition, describe, in your own words, and give an example of each term. Your examples should not be ones given in class or in the textbook.

The terms followed by bracketed numbers were defined in earlier chapters (the bracketed numbers are those of the earlier chapters), but we ask you to define them again to show your increased understanding of their meaning.

acceptance region
alpha (α)
alternative hypothesis
beta (β)
calculated value (*)
conclusion
confidence interval estimate
critical region
critical value
decision

estimation
hypothesis
hypothesis test
level of confidence
level of significance
lower confidence limit
maximum error of estimate
null hypothesis
one-tailed test
parameter [1]

point estimate   test criteria
prob-value   test statistic
risk   two-tailed test
sample size   type I error
sample statistic [1, 2]   type II error
standard error [7]   upper confidence limit
standard error of mean [7] $z(\alpha)$ [6]
statistic [1, 2]

Quiz

Answer " True" if the statement is always true. If the statement is not always true, replace the words shown in color with words that make the statement always true.

8–1   Beta is the probability of a type I error.

8–2 $1 - \alpha$ is known as the level of significance of a hypothesis test.

8–3   The standard error of the mean is the standard deviation of the sample taken.

8–4   The maximum error of estimate is controlled by three factors: level of confidence, sample size, and standard deviation.

8–5   Alpha is the measure of the area under the curve of the standard score that lies in the rejection region for H_0.

8–6   The risk of making a type I error is directly controlled in a hypothesis test by establishing a level for α.

8–7   Failing to reject the null hypothesis when it is false is a correct decision.

8–8   If the acceptance region in a hypothesis test is made wider (assuming σ and n remain fixed), α becomes larger.

8–9   Rejection of a null hypothesis that is false is a type II error.

8–10   To conclude that the mean is higher (or lower) than a claimed value, the value of the test statistic must fall in the acceptance region.

9 Inferences Involving One Population

Only One Voter in Three Wants Reagan to Run Again Two Years Hence But Most Think President Will Enter 1984 Race

Amid speculation about President Reagan's plans to run in 1984, a mid-March Gallup Poll shows the public leaning to the view that Reagan should not seek re-election.

A total of 52 percent would not like to see him run in 1984, while 35 percent would. Even among the President's own party members, as many as 20 percent say they would not like to see the 71-year-old Chief Executive enter the 1984 contest. . . .

While Americans lean in opposition to the President's running in 1984, they are generally convinced he will enter the race two years from now. Nationwide, 59 percent hold this view, including 62 percent of Republicans, 62 percent of Independents, and 56 percent of Democrats.

Sampling Error

In interpreting survey results, it should be remembered that all sample surveys are subject to sampling error, that is, the extent to which the results may differ from what would be obtained if the whole population surveyed had been interviewed. The size of such sampling errors depends largely on the number of interviews.

The tables . . . may be used in estimating sampling error. The completed allowances have taken into account the effect of the sample design upon sampling error. They may be interpreted as indicating the range (plus or minus the figure shown) within which the results of repeated samplings in the same time period could be expected to vary, 95 percent of the time, assuming the same sampling procedure, the same interviewers, and the same questionnaire.

Table A shows how much allowance should be made for the sampling error of a percentage.

The table would be used in the following manner: Say a reported percentage is 33 for a group which includes 1500 respondents. Go to row "percentages near 30" in the table and then to the column headed "1500". The number at this point is three, which means that the 33 percent obtained in the sample is subject to a sampling error of plus or minus three points. Another way of saying it is that very probably (95 chances out of 100) the average of repeated samplings would be somewhere between 30 and 36, with the most likely figure the 33 obtained.

Sampling Tolerances

TABLE A

Recommended Allowance for Sampling Error of a Percentage in Percentage Points (at 95 in 100 confidence level)*

	Size of Sample						
	1500	1000	750	600	400	200	100
Percentages near 10	2	2	3	4	4	5	7
Percentages near 20	2	3	4	4	5	7	9
Percentages near 30	3	4	4	5	6	8	10
Percentages near 40	3	4	4	5	6	9	11
Percentages near 50	3	4	4	5	6	9	11
Percentages near 60	3	4	4	5	6	9	11
Percentages near 70	3	4	4	5	6	8	10
Percentages near 80	2	3	4	4	5	7	9
Percentages near 90	2	2	3	4	4	5	7

Chapter Objectives

In chapter 8 we discussed two forms of statistical inference: (1) the hypothesis test and (2) the confidence interval estimation. The study of these two inferences was restricted to inferences about the population parameter μ under the restriction that σ was known. However, the population standard deviation is generally not known. Thus the first section of this chapter deals with the inferences about μ when σ is unknown. In addition, we will learn how to perform hypothesis tests and confidence interval estimations about the population parameters σ (standard deviation) and p (proportion).

Section 9–1 Inferences about the Population Mean

Until now we have made inferences about the mean in situations where the standard deviation of the population is known. Since such situations seldom occur, it is important to know what to do when σ is not given. If the sample size is large enough (generally speaking, samples of more than $n = 30$ pieces of data are considered large enough), the sample standard deviation s is a good estimate of the population standard deviation $σ$, and we may simply substitute s for $σ$ in the procedures discussed in chapter 8. [Recall that the sample standard deviation s is the value calculated by using formula (2–8) or one of its equivalent forms.] If the population that we are sampling is approximately normal and $n ≤ 30$, we will base our procedures on **Student's *t* distribution**. Student's *t* distribution (or just *t* distribution) is the distribution of the *t* **statistic**, which is defined as

$$t = \frac{\bar{x} - \mu}{s/\sqrt{n}} \tag{9-1}$$

In 1908 W. S. Gosset, an Irish brewery employee, published a paper about this *t* distribution under the pseudonym "Student." In deriving the *t* distribution, Gosset assumed that the samples were taken from normal populations. Although this might seem to be quite restrictive, satisfactory results are also obtained when sampling from many nonnormal populations.

The *t* distribution has the following properties (also see figure 9–1):

student's *t* distribution (margin note)

FIGURE 9–1

Normal and Student's t Distributions

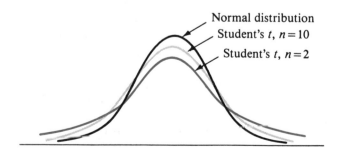

Normal distribution
Student's *t*, $n = 10$
Student's *t*, $n = 2$

304   Chapter 9   Inferences Involving One Population

Properties of the t Distribution

1. t is distributed with a mean of 0.
2. t is distributed symmetrically about its mean.
3. t is distributed with a variance greater than 1, but as the sample size n increases, the variance approaches 1.
4. t is distributed so as to be less peaked at the mean and thicker at the tails than the normal distribution.
5. t is distributed so as to form a family of distributions, a separate distribution for each sample size. The t distribution approaches the normal distribution as the sample size increases.

Although there is a separate t distribution for each sample size, $n = 2$, $n = 3, n = 4$, and so on, only certain key critical values of t will be necessary for our work. Consequently, the table for Student's t distribution (table 6 of appendix D) is a table of critical values rather than a complete table, such as that for the standard normal distribution for z. As you look at table 6, you will note that the left side of the table is identified by "df," which means "degrees of freedom." This left-hand column starts at 1 at the top and ranges to 29, then jumps to z at the bottom. As stated previously, as the sample size increases, the distribution approaches that of the normal. By inspecting table 6, you will note that as you read down any one of the columns, the entry value approaches a familiar z value at the bottom. By now you should realize that as the sample size increases, so does the number of degrees of freedom, df.

degrees of freedom

Degrees of Freedom, df

A parameter in statistics that is very difficult to define completely. Perhaps it is best thought of as an "*index number*" *used for the purpose of identifying the correct distribution to be used*. In the methods presented in this chapter, the value of df for a given situation will be $n - 1$, the sample size minus 1.

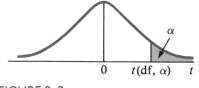

FIGURE 9–2
t Distribution Showing t (df, α)

The critical value of t to be used either in a hypothesis test or in the construction of a confidence interval will be obtained from table 6. To obtain the value of t, you will need to know two values: (1) df, the number of degrees of freedom, and (2) α, the area under the curve to the right of the right-hand critical value. A notation much like that used with z will be used to identify a critical value. $t(\text{df}, \alpha)$, read as "t of df, α," is the symbol for the value of t described in the previous sentence and shown in figure 9–2.

Illustration 9-1

Find the value of $t(10, 0.05)$ (see the accompanying diagram).

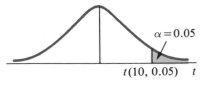

Solution   There are 10 degrees of freedom. So in table 6, appendix D, we look for the column marked $\alpha = 0.05$ and come down to df = 10.

	Amount of α in One Tail		
df	$\cdots$	0.05	$\cdots$
$\vdots$			
10		1.81	
$\vdots$			

From the table we see that $t(10, 0.05) =$ **1.81**   □

For the values of t on the left-hand side of the mean, we can use one of two notations. The t value shown in figure 9-3 could be $t(\mathrm{df}, 0.95)$, since the area to the right of it is 0.95. Or it could be identified by $-t(\mathrm{df}, 0.05)$, since the t distribution is symmetric about its mean zero.

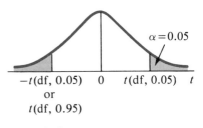

FIGURE 9-3
t Value on Left Side of Mean

Illustration 9-2

Find the value of $t(15, 0.95)$.

Solution   There are 15 degrees of freedom. In table 6 we look for the column marked $\alpha = 0.05$ and come down to df = 15. The table value is 1.75. Thus $t(15, 0.95) =$ **−1.75** (the value is negative because it is to the left of the mean; see the diagram).

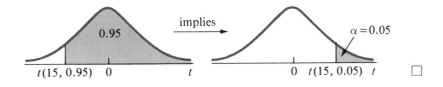

The t statistic is used in problems concerned with μ in much the same manner as z was used in chapter 8. In hypothesis-testing situations we will use formula (9-1) to calculate the test statistic value in step 4 of our procedure,

Illustration 9-3

Let's return to the hypothesis of illustration 8-1 concerning air pollution: "the mean carbon monoxide level of air pollution is no more than 4.9." Does a random sample of 25 readings (sample results: $\bar{x} = 5.1$ and $s = 2.1$) present sufficient evidence to cause us to reject this claim? Use $\alpha = 0.05$.

Solution

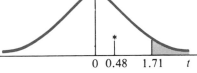

STEP 1 $H_0: \mu = 4.9 \ (\leq)$

STEP 2 $H_a: \mu > 4.9$

STEP 3 $\alpha = 0.05$, df $= 25 - 1 = 24$, and $t(24, 0.05) = 1.71$, from table 6, appendix D. (See the diagram.)

STEP 4 $t = \dfrac{\bar{x} - \mu}{s/\sqrt{n}} = \dfrac{5.1 - 4.9}{2.1/\sqrt{25}} = \dfrac{0.2}{2.1/5} = \dfrac{0.20}{0.42} = 0.476$

$t^* = \textbf{0.48}$

Comparing this value with the test criteria, we have the situation shown in the diagram.

STEP 5 *Decision*: Fail to reject H_0 (t^* is in the noncritical region).

Conclusion: We do not have enough evidence to reject the claim that the mean carbon monoxide level is no more than 4.9. □

NOTE: **If the value of df** (df $= n - 1$) in an exercise such as illustration 9-3 **is larger than 29**, then the critical value for $t(\text{df}, \alpha)$ actually becomes nearly the same as $z(\alpha)$, the z score listed at the bottom of table 6. Therefore, $t(\text{df}, \alpha)$ is not given when df > 29; $z(\alpha)$ is used instead.

Because table 6 has critical values for the Student's t distribution, the prob-value cannot be calculated for a hypothesis test that involves the use of t. However, the prob-value can be estimated.

Illustration 9-4

Let's return to illustration 9-3. Note that $t^* = 0.48$, df $= 24$ and $H_a: \mu > 4.9$. Thus for step 5 of the prob-value solution we have

$$P = P(t > 0.48, \text{knowing df} - 24). \text{ (See the diagram.)}$$

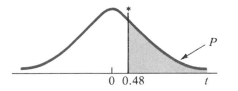

Portion of Table 6

df	0.25
⋮	
24	0.685

By inspecting the df = 24 row of table 6, you can determine that the prob-value is greater than 0.25. The 0.685 entry in the table tells us that the $P(t > 0.685) = 0.25$ as shown in the following figure.

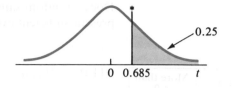

By comparing $t^* = 0.48$ (see the diagram) we see that the **prob-value P is greater than 0.25.**

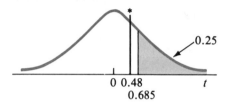

Illustration 9–5

Determine the P-value for the following hypothesis test.

$$H_0 : \mu = 55$$
$$H_a : \mu \neq 55$$

$$df = 15 \quad \text{and} \quad t^* = -1.84$$

Solution

$$P = P(t < -1.84) + P(t > 1.84) = 2P(t > 1.84)$$

Portion of Table 6

df	0.05	0.025
⋮		
15	1.75	2.13

By inspecting the df = 15 row of table 6 in Appendix D we find $P(t > 1.84)$ is between 0.025 and 0.05. Therefore P is between the double of these amounts. Thus the prob-value is between 0.05 and 0.10, **$0.05 < P < 0.10$**

The population mean may be estimated when σ is unknown in a manner similar to that used when σ is known. The difference is the use of Student's t in place of z and the use of s, the sample standard deviation, as an estimate of σ. The formula for the $1 - \alpha$ **confidence interval** of estimation then becomes

$$\bar{x} - t(df, \alpha/2) \cdot \frac{s}{\sqrt{n}} \quad \text{to} \quad \bar{x} + t(df, \alpha/2) \cdot \frac{s}{\sqrt{n}} \qquad (9\text{–}2)$$

where $df = n - 1$.

Illustration 9–6

A random sample of size 20 is taken from the weights of babies born at Northside Hospital during the year 1982. A mean of 6.87 pounds and a standard deviation of 1.76 pounds were found for the sample. Estimate, with 95% confidence, the mean weight of all babies born in this hospital in 1982.

Solution   The information we are given is $\bar{x} = 6.87$, $s = 1.76$, and $n = 20$. $1 - \alpha = 0.95$ implies that $\alpha = 0.05$; $n = 20$ implies that df $= 19$. From table 6 we get $t(19, 0.025) = 2.09$. See the diagram.

The interval is thus

$$\bar{x} \pm t(19, 0.025) \cdot \frac{s}{\sqrt{n}}$$

$$6.87 \pm 2.09 \cdot \frac{1.76}{\sqrt{20}}$$

$$6.87 \pm \frac{(2.09)(1.76)}{4.472}$$

$$6.87 \pm 0.82$$

$$6.87 - 0.82 = 6.05 \quad \text{and} \quad 6.87 + 0.82 = 7.69$$

The 0.95 confidence interval for μ is (6.05, 7.69). That is, with 95% confidence we estimate the mean weight to be between 6.05 and 7.69 pounds.   □

Exercises

9–1   Find these critical values using table 6 in appendix D.

    (a) $t(25, 0.05)$ (b) $t(10, 0.10)$

    (c) $t(15, 0.01)$ (d) $t(21, 0.025)$

    (e) $t(21, 0.95)$ (f) $t(26, 0.975)$

    (g) $t(27, 0.99)$ (h) $t(60, 0.025)$

9-2 Name (use the notation of exercise 9-1) and find the following critical values of t:

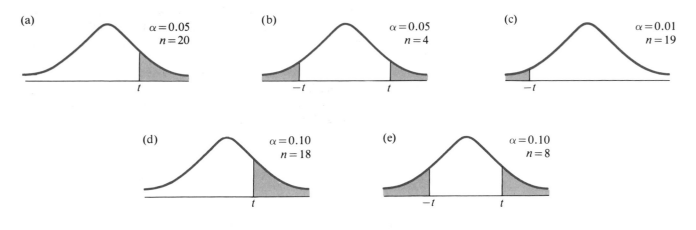

(a) $\alpha = 0.05$, $n = 20$

(b) $\alpha = 0.05$, $n = 4$

(c) $\alpha = 0.01$, $n = 19$

(d) $\alpha = 0.10$, $n = 18$

(e) $\alpha = 0.10$, $n = 8$

9-3 Ninety percent of Student's t distribution lies between $t = -1.89$ and $t = 1.89$ for how many degrees of freedom?

9-4 (a) Find the first percentile of Student's t distribution with 24 degrees of freedom.

(b) Find the 95th percentile of Student's t distribution with 24 degrees of freedom.

(c) Find the first quartile of Student's t distribution with 24 degrees of freedom.

9-5 (a) State two ways in which the normal distribution and Student's t distribution are alike.

(b) State two ways in which they are different.

9-6 A random sample of 16 one-way distances that students travel to college resulted in sample information of $\bar{x} = 10.7$ and $s = 5.3$ miles. At the 0.05 level of significance, does this show sufficient evidence to reject the claim that the average student travels no more than 8.5 miles to college?

(a) Solve using the classical approach.

(b) Solve using the prob-value approach.

9-7 A student group maintains that the average student must travel for at least 25 minutes in order to reach college each day. The college admissions office obtained a random sample of 36 one-way travel times from students. The sample had a mean of 19.4 minutes and a standard deviation of 9.6 minutes. Does the admissions office have sufficient evidence to reject the students' claim? Use $\alpha = 0.01$.

(a) Solve using the classical approach.

(b) Solve using the prob-value approach.

9-8 Homes in a nearby college town have a mean value of $58,950. It is assumed that homes in the vicinity of the college have a higher value. To test this theory, a random sample of 12 homes are chosen from the college area. Their mean valuation is $62,460 and the standard deviation is $5,200. Complete a hypothesis test using $\alpha = 0.05$.

(a) Solve using the classical approach.

(b) Solve using the prob-value approach.

9-9 Taking a random sample of 25 individuals registering for the draft in a particular town, we find that the mean waiting time in the registration line was 12.6 minutes and the standard deviation was 3.0 minutes. Estimate the mean waiting time for all individuals registering using a 90% confidence interval.

9-10 While doing an article on the high cost of college education, a reporter took a random sample of the cost of textbooks for this semester. The random variable x is the cost of one book. Her sample data can be summarized by $n = 41$, $\sum x = 550.22$, and $\sum (x - \bar{x})^2 = 1617.984$.

(a) Find the sample mean $\bar{x}$.

(b) Find the sample standard deviation s.

(c) Find the 0.90 confidence interval to estimate the true mean textbook cost for this semester, based on this sample.

(Retain your answers for use in exercise 9–33.)

9-11 Ten randomly selected shut-ins were each asked to list how many hours of television they watched per week. The results are:

$$82 \quad 66 \quad 90 \quad 84 \quad 75 \quad 88 \quad 80 \quad 94 \quad 110 \quad 91$$

Determine the 90% confidence interval estimate for the mean number of hours of television watched per week by shut-ins.

9 12 It is claimed that the students at a certain university will score an average of 35 on a given test. Is the claim reasonable if a random sample of test scores from this university yields:

$$33 \quad 42 \quad 38 \quad 37 \quad 30 \quad 42$$

Complete a hypothesis test using $\alpha = 0.05$.

(a) Solve using the classical approach.

(b) Solve using the prob-value approach.

(Retain your answers for use in exercise 9–34.)

Section 9–2 Inferences About Proportions

The proportion of or the percentage of a population and the probability associated with the occurrence of a particular event all involve the **binomial parameter** p. Recall that p was defined to be the theoretical or population probability of success on a single trial in a binomial experiment. Also, the random variable x is the number of successes that occur in a set of n trials. By combining the definition of empirical probability, $P'(A) = n(A)/n$ [formula (4–1)], with the notation of the binomial experiment, we define p', **the observed or sample binomial probability**, to be $p' = x/n$. Also recall that the mean and standard deviation of the binomial random variable x are found by using formulas (5–7) and (5–8): $\mu = np$ and $\sigma = \sqrt{npq}$, where $q = 1 - p$. This distribution of x is considered to be approximately normal if n is larger than 20 and if np and nq are both larger than 5. This commonly accepted rule of thumb allows us to use the normal distribution when making inferences concerning a binomial parameter p.

Generally it is easier to work with the distribution of p' rather than the distribution of x. Consequently, we will convert formulas (5–7) and (5–8) from the units of x to units of proportions. If we divide formulas (5–7) and (5–8) by n, we should change the units from those of x to those of proportion. The mean of x is np; thus the mean of p', $\mu_{p'}$, should be np divided by n, that is, np/n, or just p. (Does it seem reasonable that the mean of the distribution of observed values of p' should be p, the true proportion?) Further, the standard error of p' in this sampling distribution is

$$\sigma_{p'} = \sqrt{npq}/n = \sqrt{npq/n^2} = \sqrt{pq/n}$$

We summarize this information as follows:

An observed value of p' belongs to a sampling distribution that

1. is approximately normal
2. has a mean $\mu_{p'}$ equal to p
3. has a standard error $\sigma_{p'}$ equal to $\sqrt{pq/n}$

This approximation to the normal distribution is considered reasonable whenever n is greater than 20 and both np and nq are greater than 5.

RECALL: The standard deviation of a sampling distribution is called the standard error.

As a result of these new definitions for μ and σ, the calculated value of z in step 4 of a hypothesis test concerning p is obtained by using the formula

$$z = \frac{p' - p}{\sqrt{pq/n}}, \quad \text{where} \quad p' = \frac{x}{n} \tag{9–3}$$

The value of p to be used in formula (9–3) will be the value stated in the null hypothesis.

Illustration 9–7

While talking about the cars that fellow students drive, Tom made the claim that at least 15% of the students drive convertibles. Bill decided to check the validity of Tom's claim, so he took a random sample. At a level of significance of 0.10, does Bill have sufficient evidence to reject Tom's claim if there were 17 convertibles in a sample of 200 cars?

Solution

STEP 1 $H_0 : p = 0.15$ ($\geq$)   (at least 15%)

STEP 2 $H_a : p < 0.15$ (less than 15%)

STEP 3 $\alpha = 0.10$. z is found by using table 5. See the diagram.

STEP 4

$$p' = \frac{17}{200} = 0.085$$

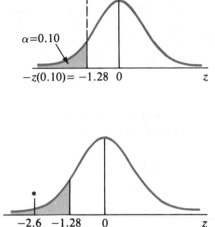

$$z = \frac{p' - p}{\sqrt{pq/n}} = \frac{0.085 - 0.150}{\sqrt{(0.15)(0.85)/200}} = \frac{-0.065}{\sqrt{0.00064}}$$

$$= \frac{-0.065}{0.025} = -2.6$$

$$z^* = -2.6$$

Comparing this value with the test criteria, we have the situation shown in the diagram.

STEP 5   *Decision*: Reject H_0 (z^* is in the critical region).

Conclusion: The evidence found contradicts Tom's claim. It appears that less than 15% of the students drive convertibles.   □

The solution to illustration 9–7 could have been carried out using the prob-value procedure. This alternative solution is shown below.

Solution

STEP 1 $H_0 : p = 0.15$ ($\geq$)   (at least 15%)

STEP 2 $H_a : p < 0.15$ (less than 15%)

STEP 3 $\alpha = 0.10$

STEP 4 $z = \dfrac{p' - p}{\sqrt{pq/n}} = \dfrac{0.085 - 0.150}{\sqrt{(0.15)(0.85)/200}}$

$z^* = -2.60$

STEP 5 $P = P(z < z^*) = P(z < -2.60) = 0.5000 - 0.4953$

$P = 0.0047$ (See the diagram.)

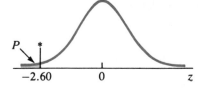

STEP 6   At the 0.10 level of significance, the sample information is significant. That is, it appears that less than 15% of the students drive convertibles.

□

When we estimate the true population proportion p, we will base our estimations on the observed value p'. The **confidence interval formula** is similar to the previous confidence interval formula.

$$p' - z(\alpha/2) \cdot \sqrt{\frac{p'q'}{n}} \quad \text{to} \quad p' + z(\alpha/2) \cdot \sqrt{\frac{p'q'}{n}} \tag{9-4}$$

where $p' = x/n$ and $q' = 1 - p'$.

Notice that the standard error, $\sqrt{pq/n}$, has been replaced by $\sqrt{p'q'/n}$. Since we do not know the value of p, we must use the best replacement available. That replacement is p', the observed value or the point estimate for p. This replacement will cause little change in the width of our confidence interval.

Illustration 9–8

Suppose that Bill has taken his sample (illustration 9–7) with the intention of estimating the value of p, the proportion of his fellow students who drive convertibles.

(a) Find the best point estimate for p that he could use.

(b) Determine the 90% confidence interval estimate for the true value of p by using formula (9–4).

Solution

(a) The best point estimate of p is 0.085, the observed value of p'.
(b) The confidence interval is $p' \pm z(\alpha/2) \cdot \sqrt{p'q'/n}$, where $q' = 1 - p'$, and $1 - \alpha = 0.90$; therefore, $z(\alpha/2) = z(0.05) = 1.65$. Thus the confidence interval is

$$0.085 \pm (1.65) \cdot \sqrt{\frac{(0.085)(0.915)}{200}}$$

$$0.085 \pm (1.65)\sqrt{0.000389}$$

$$0.085 \pm (1.65)(0.020)$$

$$0.085 \pm 0.033$$

$$0.085 - 0.033 = \textbf{0.052} \qquad 0.085 + 0.033 = \textbf{0.118}$$

$$\textbf{0.052} \quad \text{to} \quad \textbf{0.118}$$

That is, the true proportion of students who drive convertibles is between 0.052 and 0.118, with 90% confidence.

□

By using the maximum error part of the confidence interval formula, it is possible to determine the size of the sample that must be taken in order to estimate p with a desired accuracy. The **maximum error of estimate for a proportion** is

$$E = z(\alpha/2) \cdot \sqrt{\frac{pq}{n}} \qquad (9\text{–}5)$$

When using this formula, we must decide how accurate our answer must be. (Remember that we are estimating p. Therefore, E will be expressed in hundredths.) We need to establish the level of confidence with which we wish to work. If you have any indication of the value of p, use this value for $p(q = 1 - p)$. If there is no indication of an approximate value for p, then by assigning p the value 0.5, you will obtain the largest possible sample size that is required.

For ease of use, formula (9–5) can be expressed as

$$n = \frac{[z(\alpha/2)]^2 \cdot p \cdot q}{(E)^2} \qquad (9\text{–}6)$$

Illustration 9–9

Determine the sample size that is required to estimate the true proportion of blue-eyed community college students if you want your estimate to be within 0.02 with 90% confidence.

Solution

STEP 1 $1 - \alpha = 0.90$; therefore, $z(\alpha/2) = z(0.05) = 1.65$

STEP 2 $E = 0.02$

STEP 3   Use $p = 0.5$; therefore, $q = 1 - p = 0.5$.

STEP 4   Use formula (9–6) to find n:

$$n = \frac{(1.65)^2 \cdot (0.5) \cdot (0.5)}{(0.02)^2} = \frac{0.680625}{0.0004} = 1701.56$$

$$= \mathbf{1702} \qquad \qquad \square$$

Illustration 9–10

A manufacturer of automobiles purchases bolts from a supplier who claims his bolts to be approximately 5% defective. Determine the sample size that will be required to estimate the true proportion of defective bolts if we want our estimate to be within 0.02 with 90% confidence.

Solution

STEP 1 $1 - \alpha = 0.90$; therefore, $z(\alpha/2) = z(0.05) = 1.65$

STEP 2 $E = 0.02$

STEP 3   The supplier's claim is "5% defective"; thus $p = 0.05$; therefore, $q = 1 - p = 0.95$.

STEP 4   Use formula (9–6) to find n:

$$n = \frac{(1.65)^2(0.05)(0.95)}{(0.02)^2} = \frac{0.12931875}{0.0004} = 323.3$$

$$= 324 \qquad \Box$$

Notice the difference in the sample size required in illustrations 9–9 and 9–10. The only real difference between the problems is the value that was used for p. In illustration 9–9 we used $p = 0.5$, and in illustration 9–10 we used $p = 0.05$. Recall that $p = 0.5$ gives a sample of maximum size. Thus it will be of great advantage to have an indication of the value expected for p if p is much different from 0.5.

Exercises

9–13   We are testing $H_0 : p = 0.2$. We decide to reject H_0 if after 15 trials we observe more than 5 successes.

(a) State an appropriate alternative hypothesis.

(b) What is the level of significance of this test?

(c) If we observe 5 successes, do we reject H_0?

(d) If we observe 6 successes, do we reject H_0?

(e) Suppose that H_0 is $p = 0.1$ and we use the same decision rule, what happens to the level of significance?

9–14   You are testing the hypothesis $p = \frac{1}{3}$ and have decided to reject this hypothesis if after 25 trials you observe 14 or more successes.

(a) If the null hypothesis is true and you observe 13 successes, then you will do which of the following? (1) Correctly fail to reject H_0? (2) Correctly reject H_0? (3) Commit a type I error? (4) Commit a type II error?

(b) Find the significance level of your test. (Use normal approximation.)

(c) If the true probability of success is $\frac{1}{2}$ and you observe 13 successes, then you will do which of the following? (1) Correctly fail to reject H_0? (2) Correctly reject H_0? (3) Commit a type I error? (4) Commit a type II error?

(d) Calculate the prob-value for your hypothesis test after 13 successes are observed.

9–15   A poll of 1000 randomly selected voters was taken to estimate the proportion of the population that were in support of the president's current foreign aid programs. If 545 were favorable, did we find sufficient evidence to reject

the president's claim that he has the support of the majority of all voters?
Use $\alpha = 0.05$.

(a) Solve using the classical approach.

(b) Solve using the prob-value approach.

9-16 A county judge has agreed that he will give up his county judgeship and run for a state judgeship unless there is evidence that more than 25 % of his party is in opposition. A random sample of 800 party members included 217 who opposed him. Does this sample suggest that he should run for the state judgeship in accordance with his agreement? Carry out this hypothesis test by using $\alpha = 0.10$.

(a) Solve using the classical approach.

(b) Solve using the prob-value approach.

9-17 A politician claims that she will receive 60 % of the vote in an upcoming election. The results of a properly designed random sample of 100 voters showed that 50 of those sampled will vote for her. Is it likely that her assertion is correct at the 0.05 level of significance?

(a) Solve using the classical approach.

(b) Solve using the prob-value approach.

9-18 The full-time student body of our college is composed of 50 % males and 50 % females. Does a random sample of students (30 male, 20 female) from an introductory chemistry course show sufficient evidence to reject the hypothesis that the proportion of male and of female students who take this course is the same as that of the whole student body? Use $\alpha = 0.05$.

(a) Solve using the classical approach.

(b) Solve using the prob-value approach.

9-19 Construct 90 % confidence intervals for the binomial parameter p for each of the following pairs of values. Place your answers on the chart.

	Observed Proportion $p' = x/n$	Sample Size	Lower Limit	Upper Limit
(a)	$p' = 0.3$	$n = 30$		
(b)	$p' = 0.7$	$n = 30$		
(c)	$p' = 0.5$	$n = 10$		
(d)	$p' = 0.5$	$n = 100$		
(e)	$p' = 0.5$	$n = 1000$		

(f) Compare answers (a) and (b).

(g) Compare answers (c), (d), and (e).

9-20 A bank randomly selected 250 checking account customers and found that 110 of them also had savings accounts at this same bank. Construct a 0.95 confidence interval estimate for the true proportion of checking account customers that also have savings accounts.

9-21 A public opinion poll based on 400 randomly selected people yields a sample proportion, $p' = 0.37$, of the voters favoring an issue. Find a 99% confidence interval estimate for p, which is the true population proportion that favors the issue. Interpret this interval.

9-22 In a sample of 60 randomly selected students, only 22 favored the amount being budgeted for next year's intermural and interscholastic sports. Construct the 0.99 confidence interval estimate for the proportion of all students who support the proposed budget amount.

9-23 A bank believes that approximately $\frac{2}{5}$ of its checking account customers have used at least one other service provided by the bank within the last six months. How large a sample will be needed to estimate the true proportion to within 5% at the 0.98 level of confidence?

9-24 Members of the student senate are working on a proposal concerning a pub on campus. They feel that if 60% of the student body favor the proposal, it should be passed. How large a sampling needs to be taken to enable them to estimate the proportion of students who favor the proposal to within 4% with 99% confidence?

Section 9-3 Inferences about Variance and Standard Deviation

Often problems arise that require us to make inferences about variability. For example, a soft drink bottling company has a machine that fills 32-ounce bottles. They need to control the variance σ^2 (or standard deviation σ) among the amount x of soft drink put in each bottle. The mean amount placed in each bottle is important, but a correct mean amount does not ensure that the filling machine is working correctly. If the variance is too large, there could be many bottles that are overfilled and many that are underfilled. Thus this bottling company will want to maintain as small a variance (or standard deviation) as possible.

We will study two kinds of inferences in this section: (1) the hypothesis test concerning the variance (or standard deviation) of one population and (2) the estimation of the variance (or standard deviation) of one population. Often in these two inferences it is customary to talk about variance instead of the standard deviation. This is because the techniques employ the sample variance rather than the standard deviation. However, remember that the standard deviation is the square root of the variance; thus to talk about the variance of a population is comparable to talking about the standard deviation.

Let's return to our problem. Suppose the soft drink bottling company wishes to detect when the variability in the amount of soft drink placed in each bottle gets out of control. A variance of 0.0004 is considered acceptable, and the company will want to adjust the bottle-filling machine when the variance becomes larger than this value. The decision will be made by using the hypothesis test procedure. The null hypothesis is that the variance is no larger than the specified value 0.0004; the alternative hypothesis is that the variance is larger than 0.0004.

$$H_0 : \sigma^2 = 0.0004\ (\leq) \qquad \text{(variance not out of control)}$$

$$H_a : \sigma^2 > 0.0004 \qquad \text{(variance out of control)}$$

chi-square

The test statistic that will be used in making a decision about the null hypothesis is **chi-square**, χ^2 (χ is the Greek lowercase letter chi). The calculated value of chi-square will be obtained by using the formula

$$\chi^2 = \frac{(n-1)s^2}{\sigma^2} \qquad\qquad (9\text{–}7)$$

where s^2 is the sample variance, n is the sample size, and σ^2 is the value specified in the null hypothesis.

When random samples are drawn from a normal population of a known variance σ^2, the quantity $(n-1)s^2/\sigma^2$ possesses a probability distribution that is known as the **chi-square distribution**. The equations that define the chi-square distribution are not given here; they are beyond the level of this book. However, to use the chi-square distribution, we must be aware of the following properties (see also figure 9–4):

Properties of the Chi-square Distribution

1. χ^2 is nonnegative in value; it is zero or positively valued.

2. χ^2 is not symmetrical; it is skewed to the right.

3. There are many χ^2 distributions. Like the t distribution, there is a different χ^2 distribution for each degree-of-freedom value.

For the inferences discussed in this section, the number of degrees of freedom df is equal to $n-1$.

The critical values for chi-square are obtained from table 7 in appendix D. The critical values will be identified by two values: degrees of freedom df and the area under the curve to the right of the critical value being sought. Thus $\chi^2(\text{df}, \alpha)$ is the symbol used to identify the critical value of chi-square with df degrees of freedom and with α being the area to the right, as shown in

FIGURE 9–4
Various Chi-square Distributions

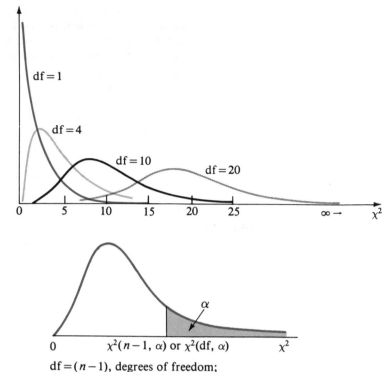

FIGURE 9–5
*Chi-square Distribution
Showing χ^2 (df, α)*

df = $(n-1)$, degrees of freedom;
α is the area under curve to the right of a particular value

figure 9–5. Since the chi-square distribution is not symmetrical, the critical values associated with both tails are given in table 7.

Illustration 9–11

Find χ^2 (20, 0.05).

Solution   In table 7 you will find the value shown in the table here. Therefore, $\chi^2(20, 0.05) =$ **31.4**. □

	Area Under Curve to the Right
df	⋯ **0.050** ⋯
⋮	
20	31.4
⋮	

Illustration 9–12

Find χ^2 (14, 0.90).

Solution   df = 14 and the area to the right of the critical value is 0.90, as shown in the figure. Therefore, χ^2 (14, 0.90) = **7.79**.

$\chi^2(14, 0.90)$ χ^2

df	Area Under Curve to the Right
	$\cdots$ 0.90 $\cdots$
$\vdots$	
14	7.79
$\vdots$	

$\square$

NOTE: When df > 2, the mean value of the chi-square distribution is df. The mean is located to the right of the mode (the value where the curve reaches its high point). By locating the value of df on your sketch of the χ^2 distribution, you will establish an approximate scale so that the values can be located in their respective positions. See figure 9–6.

mode

0   df χ^2
      mean

FIGURE 9–6
Location of Mean and Mode for χ^2 Distribution

Illustration 9–13

Recall that the soft drink bottling company wanted to control the variance by not allowing the variance to exceed 0.0004. Does a sample of size 28 with a variance of 0.0010 indicate that the bottling process is out of control (with regard to variance) at the 0.05 level?

Solution

STEP 1 $H_0: \sigma^2 = 0.0004$ ($\leq$) (not out of control)

STEP 2 $H_a: \sigma^2 > 0.0004$ (out of control)

STEP 3 $\alpha = 0.05$ and $n = 28$; therefore, df = 27. The test statistic is χ^2, and the critical region is the right tail, with an area of 0.05. $\chi^2(27, 0.05)$ is found in table 7. See the diagram.

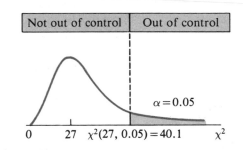

Not out of control	Out of control

$\alpha = 0.05$

0   27 $\chi^2(27, 0.05) = 40.1$ χ^2

STEP 4

$$\chi^2 = \frac{(n-1)s^2}{\sigma^2} = \frac{(28-1)(0.0010)}{0.0004}$$

$$= \frac{(27)(0.001)}{0.0004} = \frac{0.0270}{0.0004} = 67.5$$

$$\chi^{2*} = 67.5$$

See figure 9–7.

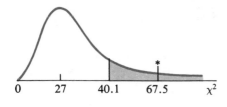

FIGURE 9–7

STEP 5 *Decision*: Reject H_0 (χ^{2*} is in the critical region).

Conclusion: The bottling process is out of control with regard to the variance.

The prob-value can be estimated for hypothesis tests using the chi-square test statistic in much the same manner as when Student's t was used. □

Illustration 9–14

Find the prob-value for the following hypothesis test.

$$H_0 : \sigma^2 = 150$$

$$H_a : \sigma^2 > 150$$

$$df = 18 \text{ and } \chi^{2*} = 32.7$$

Solution

$$P = P(\chi^2 > 32.7) \quad \text{(See the diagram.)}$$

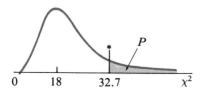

By inspecting the df $= 18$ row of table 7 we find that 32.7 is between 31.5 and 34.8. Therefore the prob-value is between 0.010 and 0.025. □

Illustration 9–15

One of the factors used in determining the usefulness of a particular exam as a measure of students' abilities is the amount of "spread" that occurs in the grades. A set of test results has little value if the range of the grades is very small. However, if the range of grades is quite large, there is a definite difference in the scores achieved by the "better" students and the scores achieved by the "poorer" students.

On an exam with a total of 100 points, it has been claimed that a standard deviation of 12 points is desirable. To determine whether or not the last one-hour exam he gave his class was a good test, a professor tested the hypothesis above at $\alpha = 0.05$ by using the exam scores of the class. There were 28 scores and the standard deviation of those 28 scores was found to be 10.5. Does the instructor have evidence, at the 0.05 level of significance, that this exam does not have the specified standard deviation?

Solution   The information given is $n = 28$, $s = 10.5$, and $\alpha = 0.05$.

STEP 1 $H_0 : \sigma = 12$

STEP 2 $H_a : \sigma \neq 12$

STEP 3 $\alpha = 0.05$; the critical values are χ^2 (27, 0.975) $= 14.6$ and χ^2 (27, 0.025) $= 43.2$. See the diagram.

STEP 4

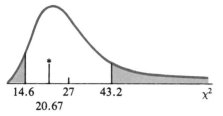

14.6   27   43.2
   20.67

$$\chi^2 = \frac{(n-1)s^2}{\sigma^2} = \frac{(27)(10.5)^2}{(12)^2} = \frac{2976.75}{144} = 20.6719$$

$$\chi^{2*} = 20.67$$

The diagram shows this value compared to the test criteria.

STEP 5   *Decision*: Fail to reject H_0.

Conclusion: The instructor does not have sufficient evidence to claim that the standard deviation is different from 12.   □

These tests for variance may be one-tailed or two-tailed tests, in accordance with the statement of the claim being tested.

The formula for chi-square may be reworked to give the values at the extremities of the confidence interval:

$$\chi^2 = \frac{(n-1)\cdot s^2}{\sigma^2} \quad \text{(solve for } \sigma^2\text{)}$$

$$\sigma^2 = \frac{(n-1)\cdot s^2}{\chi^2} \tag{9-8}$$

When constructing a $1 - \alpha$ confidence interval, the critical values of chi-square are separately substituted into formula (9–8) to obtain the two endpoints of the confidence interval of estimation. Note that χ^2 (df, $1 - \alpha/2$) is less than χ^2(df, $\alpha/2$). Therefore, after dividing, the numbers will be in the opposite order, yielding the following **confidence interval for variance**:

$$\frac{(n-1)s^2}{\chi^2(\text{df}, \alpha/2)} \quad \text{to} \quad \frac{(n-1)s^2}{\chi^2(\text{df}, 1-\alpha/2)} \tag{9-9}$$

If the confidence interval for the standard deviation is desired, we need only take the square root of each of the numbers in formula (9–9).

$$s\cdot\sqrt{\frac{(n-1)}{\chi^2(\text{df}, \alpha/2)}} \quad \text{to} \quad s\cdot\sqrt{\frac{(n-1)}{\chi^2(\text{df}, 1-\alpha/2)}} \tag{9-10}$$

Illustration 9–16

Using the professor's results from illustration 9–15 ($n = 28$, $s = 10.5$), calculate the 95% confidence interval for the estimate of the population variance and standard deviation.

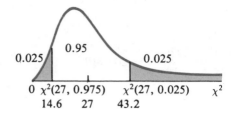

0.025 0.95 0.025

0 $\chi^2(27, 0.975)$ $\chi^2(27, 0.025)$ χ^2
     14.6   27   43.2

Solution The information given is $n = 28$ and $s = 10.5$. For a 95% confidence interval, $\alpha = 0.05$ and hence $\alpha/2 = 0.025$. The critical values for χ^2 are shown in the diagram. The confidence interval, using formula (9–9), is

$$\frac{(27)(10.5)^2}{43.2} \quad \text{to} \quad \frac{(27)(10.5)^2}{14.6}$$

$$\frac{2976.75}{43.2} \quad \text{to} \quad \frac{2976.75}{14.6}$$

$$\textbf{68.9} \quad \text{to} \quad \textbf{203.9}$$

That is, with 95% confidence we estimate the population variance to be between 68.9 and 203.9.

The confidence interval for the standard deviation can be found by taking the square root of 68.9 and of 203.9. The 95% confidence interval estimate for the standard deviation is **8.3** to **14.3**. □

Exercises

9–25 Find these critical values by using table 7 of appendix D.

(a) $\chi^2 (18, 0.01)$ (b) $\chi^2 (16, 0.025)$
(c) $\chi^2 (8, 0.10)$ (d) $\chi^2 (28, 0.01)$
(e) $\chi^2 (22, 0.95)$ (f) $\chi^2 (10, 0.975)$
(g) $\chi^2 (50, 0.90)$ (h) $\chi^2 (24, 0.99)$

9–26 Name (use the notation of exercise 9–25) and find the critical values of χ^2.

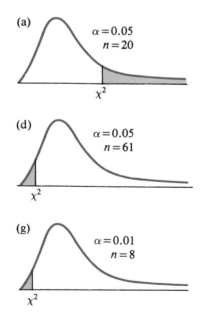

(a) $\alpha = 0.05$ $n = 20$

(b) $\alpha = 0.01$ $n = 5$

(c) $\alpha = 0.025$ $n = 18$

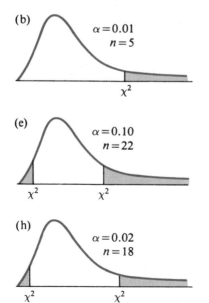

(d) $\alpha = 0.05$ $n = 61$

(e) $\alpha = 0.10$ $n = 22$

(f) $\alpha = 0.05$ $n = 7$

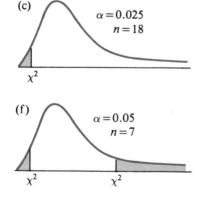

(g) $\alpha = 0.01$ $n = 8$

(h) $\alpha = 0.02$ $n = 18$

9-27 (a) What value of chi-square for 5 degrees of freedom subdivides the area under the distribution curve such that 5% is to the right and 95% is to the left?

(b) What is the value of the 95th percentile for the chi-square distribution with 5 degrees of freedom?

(c) What is the value of the 90th percentile for the chi-square distribution with 5 degrees of freedom?

9-28 (a) The central 90% of the chi-square distribution with 11 degrees of freedom lies between what values?

(b) The central 95% of the chi-square distribution with 11 degrees of freedom lies between what values?

(c) The central 99% of the chi-square distribution with 11 degrees of freedom lies between what values?

9-29 In the past the standard deviation of weights of certain 32.0-ounce packages filled by a machine was 0.25 ounce. A random sample of 20 packages showed a standard deviation of 0.35 ounce. Is the apparent increase in variability significant at the 0.10 level of significance?

(a) Solve using the classical approach.

(b) Solve using the prob-value approach.

9-30 The standard deviation of the annual temperatures of a city over a period of 100 years is 15° Fahrenheit. Using the mean temperature on the 15th day of each month during the last 12 months, a standard deviation of annual temperatures was computed as 10.5° Fahrenheit. Is the temperature in the city significantly less variable than in the past at a significance level of 0.10?

(a) Solve using the classical approach.

(b) Solve using the prob-value approach.

9-31 A commercial farmer harvests his entire field of a vegetable crop at one time. Therefore he would like to plant a variety of green beans that mature all at one time (small standard deviation between maturity times of individual plants). A seed company has developed a new hybrid strain of green beans that it believes to be better for the commercial farmer. The maturity time of the standard variety has an average of 50 days and a standard deviation of 2.1 days. A random sample of 30 plants of the new hybrid showed a standard deviation of 1.65 days. Does this sample show a significant lowering of the standard deviation at the 0.05 level of significance?

(a) Solve using the classical approach.

(b) Solve using the prob-value approach.

9-32 x is a random variable with a normal distribution. A sample of size 22 resulted in $\sum x = 397.3$ and $\sum x^2 = 7374.09$.

(a) What would be the point estimate for the population variance?

(b) What would be the 0.90 confidence interval estimate for the population variance?

(c) What would be the 0.90 confidence interval estimate for the population standard deviation?

9-33 (a) Determine the best point estimate for σ^2, the variance of the costs of textbooks in exercise 9-10.

(b) Construct the 0.95 confidence interval for σ^2 using the sample information.

9-34 Use the sample data given in exercise 9-12 to construct the 0.95 confidence interval estimate for the standard deviation of the test scores.

In Retrospect

We have studied inferences, both hypothesis testing and confidence interval estimation, for three basic population parameters: mean μ, proportion p, and standard deviation σ. When we make inferences about a single population, we are usually concerned with one of these three values. Table 9–1 identifies the formula that is used in each of the inferences for problems involving a single population.

TABLE 9–1

Formulas to Use for Inferences Involving a Single Population

Situation	Test Statistic	Formula to be Used	
		Hypothesis Test	Interval Estimate
One mean			
σ known	z	Formula (7–2)	(8–2)
σ unknown	t	Formula (9–1)	(9–2)
One proportion	z	Formula (9–3)	(9–4)
One standard deviation	χ^2	Formula (9–7)	(9–10)
One variance	χ^2	Formula (9–7)	(9–9)

In this chapter we also used the maximum error of estimate term of formula (9–6) to determine the size of sample required to make estimates about the population proportion with the desired accuracy.

The news article at the beginning of this chapter reports several observed sample percentages as point estimates of population proportions. The article gives only the point estimates. However, the sampling error explanation that accompanies the article refers to the 95 percent confidence interval estimate. By combining a point estimate with its corresponding maximum error of estimate (sampling tolerance) obtained from table A, we can construct an interval estimate. In exercise 9–51 you will do some calculating to verify the 95 percent level of confidence.

In the next chapter we will discuss inferences about two populations, whose respective means, proportions, and standard deviations are to be compared.

Chapter Exercises

9-35 Air pollution is determined by measuring several different elements that can be detected in the air. One of them is carbon monoxide. The sample of daily readings in the accompanying table was obtained from a local newspaper.

Day	1	2	3	4	5	6	7	8	9	10	11	12
Carbon Monoxide	3.5	3.9	2.8	3.1	3.1	3.4	4.8	3.2	2.5	3.5	4.4	3.1

(a) Calculate the mean and standard deviation for this sample.

(b) Carbon monoxide is measured and interpreted according to the accompanying scale. Does the sample show sufficient evidence to allow us to conclude that the carbon monoxide level is low, that is, $\mu < 4.9$, at $\alpha = 0.05$?

(c) Does the sample show sufficient evidence to allow us to reject the claim that variance in the carbon monoxide readings is no more than 0.25 at $\alpha = 0.05$?

9-36 Using the sample given in exercise 9-35 and the result found in answering part (a), construct the following confidence intervals:

(a) 90% confidence interval for estimating the mean daily level of carbon monoxide pollution

(b) 95% confidence interval for estimating the mean daily level of carbon monoxide pollution

(c) 99% confidence interval for estimating the mean daily level of carbon monoxide pollution

9-37 Using the sample given in exercise 9-35 and the result found in answering part (a), construct the following confidence intervals:

(a) 90% confidence interval for estimating the standard deviation of the carbon monoxide pollution

(b) 95% confidence interval for estimating the standard deviation of carbon monoxide pollution

(c) 99% confidence interval for estimating the standard deviation of the carbon monoxide pollution

9-38 Verify algebraically that formula (9-6) is equivalent to formula (9-5).

9-39 A random sample of size 72 is taken from very extensive records on cash awards made to company employees for helpful suggestions. If the mean of this sample is $140 and its standard deviation is $25, construct a 99% confidence interval for the true mean of all cash awards given by the company.

9-40 A manufacturer of television sets claims that the maintenance expenditures for his product will average no more than $50 during the first year following the expiration of the warranty. A consumer group has asked you to substantiate or discredit the claim. The results of a random sample of 50 owners of such television sets showed that the mean expenditure was $61.60 and the standard deviation was $32.46. At the 0.01 level of significance, should you conclude that the producer's claim is true or not likely to be true?

(a) Solve using the classical approach.

(b) Solve using the prob-value approach.

9-41 It has been suggested that abnormal human males tend to occur more in children born to older-than-average parents. Case histories of 20 abnormal males were obtained and the ages of the 20 mothers were

$$31 \quad 21 \quad 29 \quad 28 \quad 34 \quad 45 \quad 21 \quad 41 \quad 27 \quad 31$$
$$43 \quad 21 \quad 39 \quad 38 \quad 32 \quad 28 \quad 37 \quad 28 \quad 16 \quad 39$$

The mean age at which mothers in the general population give birth is 28.0 years.

(a) Calculate the sample mean and standard deviation.

(b) Does the sample give sufficient evidence to support the claim that abnormal male children have older-than-average mothers? Use $\alpha = 0.05$.

(1) Solve using the classical approach.

(2) Solve using the prob-value approach.

9-42 A natural gas utility is considering a contract for purchasing tires for its fleet of service trucks. The decision will be based on expected mileage. For a sample of 100 tires tested, the mean mileage was 36,000 and the standard deviation was 2,000 miles. Estimate the mean mileage that the utility should expect from these tires using a 96% confidence interval.

9-43 A manufacturer wishes to estimate the mean life of a new line of automobile batteries. How large a sample should it take in order to estimate the mean life to within 2 months at 99% confidence, if the lifetimes of batteries of this type typically have a standard deviation of 6 months?

9-44 A West coast radio station is promoting a popular music group named *Warren Peace and his Atom Bombs*. In the past, 60% of the listeners of the station have liked the music groups that the station promotes. Out of a randomly selected sample of 200 listeners, 102 of them like the *Warren Peace* group. At the 0.02 level of significance, test the hypothesis that there is no difference between the attitude of the listeners in this sample and listeners in the past.

(a) Solve using the classical approach.

(b) Solve using the prob-value approach.

9-45   You are interested in comparing the hypothesis $p = 0.8$ against the alternative $p < 0.8$. In 100 trials you observe 73 successes. Calculate the prob-value associated with this result.

9-46   An instructor asks each of the 54 members of his class to write down "at random" one of the numbers $1, 2, 3, \ldots, 13, 14, 15$. Since the instructor believes that students like gambling, he considers that 7 and 11 are lucky numbers. He counts the number of students, x, who selected 7 or 11. How large must x be before the hypothesis of randomness can be rejected at the 0.05 level?

9-47   Construct the 0.98 confidence interval estimate for the proportion of voters who favor the selection of a new city manager, based on a random sample of 400 voters in which 285 said that they favored the idea of a new city manager.

9-48   The marketing research department of an instant-coffee firm conducted a survey of married males to determine the proportion of married males who preferred their brand. Twenty of the 100 in the random sample preferred the company's brand. Use a 95 % confidence interval to estimate the proportion of all married males that prefer this company's brand of instant coffee. Interpret your answer.

9-49   A company is drafting an advertising campaign that will involve endorsements by noted athletes. In order for the campaign to succeed, the endorser must be both highly respected and easily recognized. A random sample of 100 prospective customers are shown photos of various athletes. If the customer recognizes the athlete then he or she is asked if they respect the athlete. In the case of a top woman golfer, 16 of the 100 respondents recognized the golfer's picture and indicated that they also respected her. At the 95 % level of confidence, what is the true proportion with which this woman golfer is both recognized and respected?

9-50   A local auto dealership advertises that 90 % of customers, whose autos were serviced by their service department, are pleased with the results. As a researcher you take exception to this statement since you are aware that many people are reluctant to express dissatisfaction even if they are not pleased. A research experiment was set up in which those in the sample had received service by this dealer within the last two weeks. During the interview, the individuals were led to believe that the interviewer was new in town and was considering taking his car to this dealer's service department. Of the 60 sampled, 14 said that they were dissatisfied and would not recommend the department.

(a) Estimate the proportion of dissatisfied customers using a 95 % confidence interval.

(b) What can be concluded about the dealer's claim knowing your answer in part (a)?

9-51 The news article at the beginning of the chapter reports the observed percentage as a point estimate of the population percentage. The sampling error explanation describes how the point estimate can be expressed as a 95% confidence interval estimate. Reread the article and the sampling error explanation.

(a) The article reports 35% of the people would like to see Reagan run again for president in 1984. If this information is based on a sample of 1500, what is the sampling tolerance (maximum error of estimate) according to table A?

(b) Using formula (9–5) calculate the maximum error of estimate for the situation in question (a).

(c) How does your answer in (b) compare to the value found in table A for (a)? Why did the maximum error get rounded up for table A?

(d) Express the estimate for the population proportion of the public, who want Reagan to run in 1984, as a 95% confidence interval.

9-52 The State Motor Vehicle Department wishes to estimate the proportion of first-year drivers that could be classified as "careless drivers." How large a sample should it take if it wishes to be within 3% at 95% confidence? (It is expected that approximately $\frac{1}{3}$ of all drivers are careless.)

9-53 A production process is considered to be out of control if the produced parts have a mean length different from 27.5 millimeters or a standard deviation that is greater than 0.5 millimeter. A sample of 30 parts yields a sample mean of 27.63 millimeters and a sample standard deviation of 0.87 millimeter.

(a) At the 0.05 level of significance, does this sample indicate that the process should be adjusted in order to correct the standard deviation of the product?

(b) At the 0.05 level of significance, does this sample indicate that the process should be adjusted in order to correct the mean value of the product?

9-54 Bright-lite claims that its 60-watt light bulb burns with a length of life that is approximately normally distributed with a standard deviation of 81 hours. A sample of 101 bulbs had a variance of 8075. Is this sufficient evidence to reject Bright-lite's claim in favor of the alternative, "the standard deviation is larger than 81 hours," at the 0.05 level of significance?

(a) Solve using the classical approach.

(b) Solve using the prob-value approach.

9-55 The weights of the drained fruit found in 21 randomly selected cans of peaches packed by Sunny Fruit Cannery were (in ounces)

11.0	11.6	10.9	12.0	11.5	12.0	11.2
10.5	12.2	11.8	12.1	11.6	11.7	11.6
11.2	12.0	11.4	10.8	11.8	10.9	11.4

(a) Calculate the sample mean and standard deviation.

(b) Construct the 0.98 confidence interval for the estimate of the mean weight of drained peaches per can.

9-56 Oranges are selected at random from a large shipment that just arrived. A sample is taken to estimate the size (circumference, in inches) of the oranges. The sample data are summarized as follows: $n = 100$, $\sum x = 878.2$, and $\sum(x - \bar{x})^2 = 49.91$.

(a) Determine the mean and standard deviation for this sample.

(b) What is the point estimate for μ, the mean circumference of oranges in this shipment?

(c) What is the 0.95 confidence interval estimate for μ?

(d) What is the point estimate for the standard deviation σ of the circumferences?

(e) What is the 0.95 confidence interval for σ?

9-57 The case histories for the 20 children in exercise 9–41 showed their fathers to have ages of:

$$32 \quad 48 \quad 33 \quad 22 \quad 29 \quad 30 \quad 45 \quad 25 \quad 26 \quad 43$$
$$36 \quad 27 \quad 34 \quad 20 \quad 35 \quad 55 \quad 52 \quad 38 \quad 34 \quad 37$$

(a) Calculate the sample mean and standard deviation.

(b) Estimate the mean age of all fathers of abnormal male children using a 0.95 confidence interval.

(c) Estimate the standard deviation of the ages of all fathers for these children using a 0.95 confidence interval.

9-58 In a large cherry orchard the average yield has been 4.35 tons per acre for the last several years. A new fertilizer was tested on 15 randomly selected 1-acre plots. The yields from these plots follow:

$$3.56 \quad 5.00 \quad 4.88 \quad 4.93 \quad 3.92$$
$$4.25 \quad 5.12 \quad 5.13 \quad 4.79 \quad 4.45$$
$$5.35 \quad 4.81 \quad 3.48 \quad 4.15 \quad 4.72$$

At the 0.05 level of significance, do we have sufficient evidence to claim that there was a significant increase in production?

(a) Solve using the classical approach.

(b) Solve using the prob-value approach.

Vocabulary List

Be able to define each term. In addition, describe, in your own words, and give an example of each term. Your examples should not be ones given in class or in the textbook.

The terms followed by bracketed numbers were defined in earlier chapters (the bracketed numbers are those of the earlier chapters), but we ask you to define them again to show your increased understanding of their meaning.

acceptance region [8]
calculated value [8]
chi-square
conclusion [8]
critical region [8]
critical value [8]
decision [8]
degrees of freedom
inference [8]
level of confidence [8]
level of significance [8]
maximum error of estimate [8]
one-tailed test [8]

parameter [1, 8]
proportion [6]
random variable [5, 6]
response variable [1]
sample size [8]
sample statistic [1, 2]
σ known
σ unknown
standard error [7, 8]
Student's t
test statistic [8]
two-tailed test [8]

Quiz

Answer "True" if the statement is always true. If the statement is not always true, replace the words shown in color with words that make the statement always true.

9–1   The Student's t distribution is an approximately normal distribution but is more **dispersed** than the normal distribution.

9–2   The **chi-square** distribution is used for inferences about the mean when σ is unknown.

9–3   The **Student's t** distribution is used for all inferences about a population's variance.

9–4   If the test statistic falls in the critical region, the null hypothesis has **been proved true**.

9–5   When the test statistic is t and the number of degrees of freedom exceeds 30, the critical value of t is very close to that of z.

9–6   When making inferences about one mean when the value of σ is not known, the **z score** is the test statistic.

9–7   The chi-square distribution is a skewed distribution whose mean is always **2**.

9–8   Often the concern with testing the variance (or standard deviation) is to keep its size under control or relatively small. Therefore, many of the hypothesis tests with chi-square will be **one-tailed**.

9–9 $\sqrt{npq}$ is the standard error of proportion.

9–10   The sampling distribution of p' is approximately distributed as **chi-square**.

10 Inferences Involving Two Populations

Study Shows Women Are "Outworking" Men

WASHINGTON—The old saw that goes "a woman's work is never done" appears to apply to the workplace as well as the home, according to a University of Michigan study showing that women take fewer breaks than men and expend more energy at lower pay on their jobs.

The study, based on personal interviews with 208 men and 168 women in a variety of occupations, found that men, on the average, took about 52 minutes off from work in scheduled and unscheduled breaks.

In contrast, women spend only 35 minutes in such formal and informal breaks, including extra time at lunch beyond the normal hour, fixed coffee breaks and unsanctioned chats around the water cooler.

When this conscientiousness was figured in, the study showed that the disparity between the wages women earn and those of men grew even larger.

Nationally, the average working man makes $7 an hour, while women earn $4.34; but when the differences in the amount of time put in on the job during the working day are calculated, men's earnings jumped to $8.48 an hour, while women's averaged out to $4.86.

Although data from a wide range of studies has shown that women make far less for comparable work than men, the University of Michigan study also found that they work harder than men.

Based on a "work effort scale" drawn from personal interviews, the study concluded that women put out 112 per cent the effort of men at their jobs in terms of the energy level expended at work.

Unmarried women received the least amount of break time and scored highest on the work effort scale.

Married men expended more energy than unmarried men, on their jobs, but also took more total break time.

Chapter Objectives

In chapters 8 and 9 we introduced the basic concepts of hypothesis testing and confidence interval estimation in connection with inferences about one population and these parameters: mean, standard deviation, and proportion. In this chapter we continue to investigate inferences about those same three parameters however, here we will use them to compare two populations.

Section 10–1 Independent and Dependent Samples

In this chapter we are going to study the procedures for making inferences about two populations. When comparing two populations we need two samples, one from each. Two basic kinds of samples can be used: independent and dependent. The dependence or independence of a sample is **source** determined by the sources used for the data. A **source** can be a person, an object, or anything that yields a piece of data. If the same set of sources is **dependent and** used to obtain the data representing both populations, we have **dependent** **independent samples** **sampling**. If two unrelated sets of sources are used, one set from each population, we have **independent sampling**. The following illustrations should amplify these ideas further.

Illustration 10–1

A test will be conducted to see if the participants in a physical fitness class actually improve in their level of fitness. It is anticipated that approximately 500 people will sign up for this course, and the instructor decides that she will give 50 of the participants a set of tests before the course actually begins (pre-test) and then will give another set of tests to 50 participants at the end of the course (post-test). The following two sampling procedures are proposed:

> Plan *A*: Randomly select 50 participants from the list of those enrolled and give them the pre-test. At the end of the course, select another random sample of size 50 and give them the post-test.

> Plan *B*: Randomly select 50 participants and give them the pre-test; give the same set of 50 the post-test upon completion of the course.

Plan *A* illustrates independent sampling—the sources (class participants) used for each sample (pre-test and post-test) were selected separately. Plan *B* illustrates dependent sampling; the sources used for both samples are the same. □

Illustration 10–2

A test is being designed to compare the wearing quality of two brands of tires. The automobiles will be selected and equipped with the new tires and then driven under "normal" conditions for one month. Then a measurement will be taken to determine how much wear took place. Two plans will be used:

Plan C: n cars will be selected randomly and equipped with brand A and driven for the month, and n other cars will be selected and equipped with brand B and driven for the month.

Plan D: n cars will be selected randomly, equipped with one tire of brand A and one tire of brand B (the other two tires are not part of the test), and driven for the month.

In this illustration we might suspect that many other factors must be taken into account when testing automobile tires—such as age, weight, and mechanical condition of the car; driving habits of drivers; location of the tire on the car; and where the car is driven. However, at this time we are only trying to illustrate dependent and independent samples. Plan C is independent (unrelated sources) and plan D is dependent (common sources). ☐

Independent and dependent cases each have their advantages; these will be emphasized later. Both methods of sampling are often used.

Exercises

10–1 The marketing division of a large company has developed a new promotion which it hopes will effectively sell a new household product. To test the effectiveness of their new promotion, a sample of 40 housewives is randomly selected from the community. Each housewife is interviewed and rated. Then each housewife is given the promotion "pitch." Next a follow-up interview takes place. Two sets of data result from the before and the after interviews. Do these sets of data represent dependent or independent samples? Explain.

10–2 In trying to estimate the amount of growth that took place in the trees planted by the County Parks Commission recently, 36 trees were randomly selected from the 4000 planted. The heights of these trees were measured and recorded. One year later another set of 42 trees was randomly selected and measured. Do the two sets of data (36 heights, 42 heights) represent dependent or independent samples? Explain.

10–3 Twenty people were selected to participate in a psychology experiment. The 20 answered a short multiple-choice quiz about their attitudes on a particular subject. The 20 people then viewed a 45-minute film. The following day the same 20 people were asked to answer a follow-up questionnaire about their attitudes. At the completion of the experiment, the experimenter will have two sets of scores. Do these two samples represent dependent or independent samples? Explain.

10–4 Patients involved in a heart research program have previously suffered a heart attack. In studying the symptoms preceding the actual attack, a researcher interviews a sample of 20 men and 15 women and determines the amount of pain experienced beforehand. After the interviewing is complete there will be two sets of data: one from the 20 men and another from the 15 women. Do these samples represent dependent or independent samples? Explain.

10–5   Let's assume that there are 400 students in our college taking elementary statistics this semester. Describe how you would obtain two independent samples of size 25 from these 400 students in order to test some precourse skill against the same skill after completing the course.

10–6   Describe how you would obtain your samples in exercise 10–5 if you were to use dependent samples.

Section 10–2   Inferences Concerning the Difference between Two Independent Means (Variances Known or Large Samples)

When comparing the means of two populations, we typically consider the difference between their means, $\mu_1 - \mu_2$. The inferences we will make about $\mu_1 - \mu_2$ will be based on the difference between the observed sample means, $\bar{x}_1 - \bar{x}_2$. This observed difference belongs to a sampling distribution, the characteristics of which are described in the following statement.

---

independent means

If independent samples of sizes n_1 and n_2 are drawn randomly from large populations with means μ_1 and μ_2 and variances σ_1^2 and σ_2^2, respectively, **the sampling distribution of $\bar{x}_1 - \bar{x}_2$**, the difference between the means,

1. is approximately normally distributed
2. has a mean of $\mu_{\bar{x}_1 - \bar{x}_2} = \mu_1 - \mu_2$
3. has a standard error of $\sigma_{\bar{x}_1 - \bar{x}_2} = \sqrt{(\sigma_1^2/n_1) + (\sigma_2^2/n_2)}$

This normal approximation is good for all sample sizes given that the populations involved are approximately normal and the population variances σ_1^2 and σ_2^2 are known quantities.

Since the sampling distribution is approximately normal, we will use the z **statistic** in our inferences. In the **hypothesis tests** z will be determined by

$$z = \frac{(\bar{x}_1 - \bar{x}_2) - (\mu_1 - \mu_2)}{\sqrt{(\sigma_1^2/n_1) + (\sigma_2^2/n_2)}} \tag{10–1}$$

if both σ_1 and σ_2 are known quantities.

Illustration 10-3

Suppose we are interested in comparing the academic success of college students who belong to fraternal organizations with the success of those who do not belong to fraternal organizations. These two populations are clearly separate, and we should take independent samples from each of the two populations. The "Greeks" claim that fraternal organization members achieve academically at a level no lower than the nonfraternal students. (Cumulative grade point average is the measure of academic success.) Samples of size 40 are taken from each population. The means obtained are 2.03 for the 40 fraternal members (g) and 2.21 for the 40 nonfraternal members (n). Assume that the standard deviation of both populations is $\sigma = 0.6$. Complete a hypothesis test of the Greeks' claim, using $\alpha = 0.05$.

Solution   The hypotheses for this test are as follows:

STEP 1 $H_0: \mu_g = \mu_n$ $(\geq)$ or $\mu_g - \mu_n = 0$ $(\geq)$ (no lower)

STEP 2 $H_a: \mu_g < \mu_n$ or $\mu_g - \mu_n < 0$ (lower)

The null hypothesis is usually interpreted as "there is no difference between the means," and, therefore, it is customary to express it by $\mu_1 - \mu_2 = 0$.

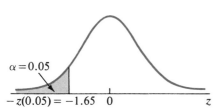

STEP 3   The test statistic used will be z. The test criteria for $\alpha = 0.05$ will be as shown in the diagram.

STEP 4   The formula for the test statistic z is formula (10-1):

$$z = \frac{(\bar{x}_g - \bar{x}_n) - (\mu_g - \mu_n)}{\sqrt{(\sigma_g^2/n_g) + (\sigma_n^2/n_n)}}$$

$$= \frac{(2.03 - 2.21) - 0}{\sqrt{[(0.6)^2/40] + [(0.6)^2/40]}} \qquad = \frac{-0.18}{\sqrt{(0.36/40) + (0.36/40)}}$$

$$= \frac{-0.18}{\sqrt{0.009 + 0.009}} = \frac{-0.18}{\sqrt{0.018}} = \frac{-0.180}{0.134} = -1.343$$

$$z^* = -1.34$$

This value is compared to the test criteria in the accompanying diagram.

STEP 5   *Decision*: Fail to reject H_0 (z^* is in the noncritical region).

Conclusion: The claim that the fraternal organization members achieve at a level no lower than nonfraternal members cannot be rejected.   ☐

We often wish to estimate the difference between the means of two different populations. When independent samples are involved, we will use the information about the sampling distribution of $\bar{x}_1 - \bar{x}_2$ and the z statistic to construct our **confidence interval estimate for $\mu_1 - \mu_2$.**

$$(\bar{x}_1 - \bar{x}_2) - z(\alpha/2) \cdot \sqrt{\frac{\sigma_1^2}{n_1} + \frac{\sigma_2^2}{n_2}}$$

to   (10–2)

$$(\bar{x}_1 - \bar{x}_2) + z(\alpha/2) \cdot \sqrt{\frac{\sigma_1^2}{n_1} + \frac{\sigma_2^2}{n_2}}$$

Illustration 10–4

Construct the 95% confidence interval estimate for the difference between the two independent means of illustration 10–3, mean grade point average of fraternal organization members and mean grade point average of non-fraternal members. Use the sample values found in illustration 10–3.

Solution The information given is $\bar{x}_g = 2.03$, $\sigma_g = 0.6$, $n_g = 40$, $\bar{x}_n = 2.21$, $\sigma_n = 0.6$, $n_n = 40$, and $1 - \alpha = 0.95$. The interval is

$$(\bar{x}_g - \bar{x}_n) \pm z(0.025) \cdot \sqrt{\frac{\sigma_g^2}{n_g} + \frac{\sigma_n^2}{n_n}}$$

$$(2.03 - 2.21) \pm (1.96) \cdot \sqrt{\frac{(0.6)^2}{40} + \frac{(0.6)^2}{40}}$$

(See illustration 10–3.)

$$(-0.18) \pm (1.96)(0.134)$$

$$-0.18 \pm 0.26$$

$$\mathbf{-0.44} \quad \text{to} \quad \mathbf{0.08}$$

is the 0.95 confidence interval for $\mu_g - \mu_n$. That is, with 95% confidence we estimate the difference between the means to be between -0.44 and $+0.08$.

           As occurred in previous chapters, the variance of a population is generally unknown when we wish to make an inference about the mean. Therefore, it is necessary to replace σ_1 and σ_2 with the best estimates available, namely, s_1 and s_2. If both samples have sizes that exceed 30, we may replace σ_1 and σ_2 in formulas (10–1) and (10–2) with s_1 and s_2, respectively, without appreciably affecting our level of significance or confidence. Thus for inferences about the difference between two population means, based on independent samples where the σ's **are unknown** and both $n_1 > 30$ and $n_2 > 30$, we will use

$$z = \frac{(\bar{x}_1 - \bar{x}_2) - (\mu_1 - \mu_2)}{\sqrt{(s_1^2/n_1) + (s_2^2/n_2)}} \qquad (10\text{--}3)$$

for the calculation of the **test statistic in the hypothesis test.** We will use

$$(\bar{x}_1 - \bar{x}_2) - z(\alpha/2) \cdot \sqrt{\frac{s_1^2}{n_1} + \frac{s_2^2}{n_2}}$$

to   (10–4)

$$(\bar{x}_1 - \bar{x}_2) + z(\alpha/2) \cdot \sqrt{\frac{s_1^2}{n_1} + \frac{s_2^2}{n_2}}$$

for calculating the endpoints of the $1 - \alpha$ **confidence interval estimate.**

Illustration 10–5

Two independent samples are taken to compare the means of two populations. The sample statistics are given in table 10–1. Can we conclude that the mean of population A is greater than the mean of population B, at the 0.02 level of significance?

TABLE 10–1
Sample Statistics for Illustration 10–5

Sample	n	$\bar{x}$	s
A	50	57.5	6.2
B	60	54.4	10.6

Solution   This problem calls for a hypothesis test for the difference of two independent means. Both n's are larger than 30; therefore, formula (10–3) will be used to calculate z.

STEP 1 $H_0: \mu_A - \mu_B = 0 \ (\le)$

STEP 2 $H_a: \mu_A - \mu_B > 0$

STEP 3 $\alpha = 0.02$. The test criteria are shown in the diagram.

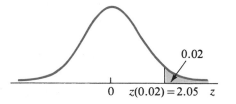

STEP 4   Using formula (10–3),

$$z = \frac{(57.5 - 54.4) - 0}{\sqrt{[(6.2)^2/50] + [(10.6)^2/60]}}$$

$$= \frac{3.1}{\sqrt{0.7688 + 1.8727}} = \frac{3.1}{1.625} = 1.908$$

$$z^* = 1.91$$

STEP 5   *Decision*: Fail to reject H_0 (z^* is in the noncritical region).

Conclusion: We do not have sufficient evidence to conclude that $\mu_A > \mu_B$.

NOTE: The null hypothesis can claim a difference, for example: the mean of A is 10 larger than the mean of B. In this situation the null hypothesis is written, $\mu_A - \mu_B = 10$ and the value 10 is used in step 4 when z is calculated.

☐

Illustration 10–6

Suppose that the samples given in illustration 10–5 were taken for the purpose of estimating the difference between the two population means. Construct the 0.99 confidence interval for the estimation of this difference.

Solution   This estimation calls for the difference between the means of two independent samples whose sizes are both greater than 30. Therefore, we will use formula (10–4).

$$(57.5 - 54.4) \pm 2.58 \sqrt{\frac{(6.2)^2}{50} + \frac{(10.6)^2}{60}}$$

$$3.1 \pm (2.58)(1.625)$$

$$3.1 \pm 4.19$$

$$-1.09 \quad \text{to} \quad 7.29$$

is the 0.99 confidence interval for $\mu_A - \mu_B$. That is, our 99% confidence interval estimate for the difference between the two population means is -1.09 to 7.29.

☐

Exercises

10–7   Small metal clips used in a furnace mechanism have a mean weight of 0.6 ounce and a variance of 0.0004. Two random samples, one of 100 and the other of 80 observations are taken on two consecutive days. Assuming that there has been no change in the production process, what is the probability that the two sample means differ by

(a) more than 0.002 ounce?   (b) less than 0.0015 ounce?

10-8 Independent samples are taken from two normal populations, each of which has a variance of 100. Do the sample means shown in the accompanying table provide sufficient evidence to reject the hypothesis that the means of these two populations are equal? Use $\alpha = 0.05$.

Sample	n	$\bar{x}$
A	30	35.6
B	35	38.3

(a) Solve using the classical approach.

(b) Solve using the prob-value approach.

10-9 Independent samples are taken from normal populations I and II with variances of 200 and 700, respectively. Do the sample data shown in the accompanying table provide sufficient evidence to reject the hypothesis that the populations have equal means, at the 0.10 level of significance?

Sample	n	$\bar{x}$
I	32	104.5
II	40	110.9

(a) Solve using the classical approach.

(b) Solve using the prob-value approach.

10-10 Use the sample information in exercise 10-8 to construct the 0.99 confidence interval estimate for the difference between the two population means.

10-11 An experiment was conducted to compare the mean absorptions of drugs in specimens of muscle tissue. Seventy-two tissue specimens were randomly divided into two equal groups. Each group was tested with one of the two drugs. Assume that the variance of absorption is 0.10 for this type of drug. The means found were $\bar{x}_A = 7.9$ and $\bar{x}_B = 8.5$. Construct the 98% confidence interval for the difference in the mean absorption rates.

10-12 The purchasing department for a regional supermarket chain is considering two sources from which to purchase 10-pound bags of potatoes. A random sample taken from each source shows the following results:

	Idaho Supers	Idaho Best
Number of Bags Weighed	100	100
Mean Weight	10.2 lbs	10.4 lbs.
Sample Variance	0.36	0.25

At the 0.05 level of significance, is there a difference between the mean weights of the 10-pound bags of potatoes?

(a) Solve using the classical approach.

(b) Solve using the prob-value approach.

10-13 A sample of the heights of female and male students participating in the college intermural program was obtained, with the results as shown in the accompanying table. Do these data provide sufficient evidence to reject the hypothesis that the average male participant is no more than 2.5 inches taller than the average female participant, at the 0.02 level of significance?

Sample	n	$\sum x$	$\sum (x - \bar{x})^2$
Female	30	1952	74.2
Male	40	2757	284.3

(a) Solve using the classical approach.

(b) Solve using the prob-value approach.

10-14 In a certain species the white-flowered plants appear to have smaller and more abundant flowers than do the red-flowered plants. The accompanying data were collected when the number of flowers on several plants were counted. Use these sample data to construct the 95% confidence interval

Flower	n	$\sum x$	$\sum (x - \bar{x})^2$
Red	32	2253	32,462.0
White	35	5157	58,600.0

estimate for the difference between the mean number of flowers on the white- and the red-flowered plants ($\mu_W - \mu_R$).

Section 10–3 Inferences Concerning Two Variances

When comparing two populations, it is quite natural that we compare their variances or standard deviations. The inferences concerning two population variances (or standard deviations) are much like those comparing means. We will study two kinds of inferences about the comparison of the variances of two populations: (1) **the hypothesis test for the equality of the two variances,** and (2) **the estimation of the ratio of the two population variances** σ_1^2/σ_2^2.

The soft drink bottling company discussed in section 9–3 is trying to decide whether to install a modern high-speed bottling machine. There are, of course, many concerns in making this decision. The variance in the amount of fill per bottle is one of them. In this respect the manufacturer of the new system contends that the variance in fills is no longer with the new machine than it was with the old. A hypothesis test for the equality of the two variances can be used to make a decision in this situation. The null hypothesis will be that the variance of the modern high-speed machine (m) is no larger than the variance of the present machine (p); that is, $\sigma_m^2 \leq \sigma_p^2$. The alternative hypothesis will then be $\sigma_m^2 > \sigma_p^2$.

$$H_0: \sigma_m^2 = \sigma_p^2 \ (\leq) \quad \text{or} \quad \frac{\sigma_m^2}{\sigma_p^2} = 1 \ (\leq)$$

$$H_a: \sigma_m^2 > \sigma_p^2 \quad \text{or} \quad \frac{\sigma_m^2}{\sigma_p^2} > 1$$

The test statistic that will be used in making a decision about the null hypothesis is F. The calculated value of F will be obtained by using the following formula:

$$F = \frac{s_1^2}{s_2^2} \qquad (10\text{--}5)$$

where s_1^2 and s_2^2 are the variances of two independent samples of sizes n_1 and n_2, respectively.

When independent random samples are drawn from normal populations with equal variances, the ratio of the sample variances, s_1^2/s_2^2, will possess a probability distribution known as the *F* **distribution** (see figure 10–1).

FIGURE 10–1
F Distribution

F distribution

Properties of the *F* Distribution

1. F is nonnegative in value; it is zero or positively valued.
2. F is nonsymmetrical; it is skewed to the right.
3. There are many F distributions, much like the t and χ^2 distributions. There is a distribution for each pair of degree-of-freedom values.

For the inferences discussed in this section, the degrees of freedom for each of the samples are $df_1 = n_1 - 1$ and $df_2 = n_2 - 1$.

The critical values for the F distribution may be obtained from tables 8a, 8b, and 8c in appendix D. Each critical value will be determined by three identification values: df_n, df_d, and the area under the curve to the right of the critical value being sought. Each F distribution has two degrees-of-freedom values: df_n, the degrees of freedom associated with the sample whose variance is in the numerator of the calculated F; and df_d, the degrees of freedom associated with the sample whose variance is in the denominator. Therefore, the symbolic name for a critical value of F will be $F(df_n, df_d, \alpha)$, as shown in figure 10–2.

FIGURE 10–2
F Distribution Showing F (df$_n$, df$_d$, α)

Table 8a in appendix D shows the critical values for $F(\text{df}_n, \text{df}_d, \alpha)$, where α is equal to 0.05; table 8b gives the critical values when $\alpha = 0.025$; table 8c gives values when $\alpha = 0.01$.

Illustration 10–7

Find $F(5, 8, 0.05)$, the critical F value for samples of size 6 and 9 with 5% of the area in the right tail.

Degrees of Freedom for Numerator

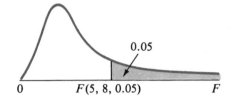

Solution From table 8a ($\alpha = 0.05$), we obtain the value shown in the table here. Therefore, $F(5, 8, 0.05) = \mathbf{3.69}$ □

Notice that $F(8, 5, 0.05)$ is 4.82. The degrees of freedom associated with the numerator and with the denominator must be kept in the correct order. (3.69 is quite different from 4.82. Check some other pairs to verify this fact.)

Tables showing the critical values for the F distribution give only the right-hand critical value. If the critical value for the left-hand tail is needed, we will obtain it by calculating the reciprocal of the related critical value obtained from the table. Expressed as a formula, this is

$$F(\text{df}_1, \text{df}_2, 1 - \alpha) = \frac{1}{F(\text{df}_2, \text{df}_1, \alpha)} \tag{10-6}$$

See figure 10–3. Notice that when the reciprocal is taken, the degrees of freedom for the numerator and the denominator are switched also. (Why do you suppose this switch is necessary?)

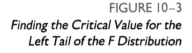

FIGURE 10–3

Finding the Critical Value for the Left Tail of the F Distribution

Illustration 10–8

Find the value for $F(10, 15, 0.99)$.

Solution   Using formula (10–6),

$$F(10, 15, 0.99) = \frac{1}{F(15, 10, 0.01)} = \frac{1}{4.56} = 0.219$$

$$= 0.22 \qquad \square$$

Illustration 10–9

Recall that our soft drink bottling company was to make a decision about the equality of the variances of amounts of fill between its present machine and a modern high-speed outfit. Does the sample information in table 10–2 present sufficient evidence to reject the manufacturer's claim that the modern, high-speed, bottle-filling machine fills bottles with no more variance than the company's present machine? Use $\alpha = 0.01$.

TABLE 10–2
Sample Results for Illustration 10–9

Sample	n	s^2
Present machine, p	22	0.0008
Modern high-speed machine, m	25	0.0018

Solution

STEP 1 $H_0: \sigma_m^2 = \sigma_p^2$ or $\sigma_m^2/\sigma_p^2 = 1$ $(\leq)$ (no more variance)

STEP 2 $H_a: \sigma_m^2 > \sigma_p^2$ or $\sigma_m^2/\sigma_p^2 > 1$ (more variance)

STEP 3 $\alpha = 0.01$. The test statistic to be used is F, since the null hypothesis is about the equality of the variances of two populations. The critical region is one-tailed and on the right because the alternative hypothesis says "greater than." $F(24, 21, 0.01)$ is the critical value. The number of degrees of freedom for the numerator is 24 (25–1) because the sample from the modern high-speed machine is associated with the numerator, as specified by the null hypothesis. $df_d = 21$ because the sample associated with the denominator has size 22. The critical value is found in table 8c and is 2.80. See the diagram.

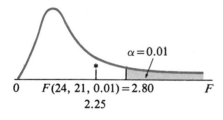

$\alpha = 0.01$

$F(24, 21, 0.01) = 2.80$

2.25

STEP 4

$$F = \frac{s_m^2}{s_p^2} = \frac{0.0018}{0.0008} = 2.25$$

$$F^* = 2.25$$

This value is shown in the diagram.

STEP 5   *Decision*: Fail to reject H_0 (F^* is in the noncritical region).

Conclusion: The samples do not present sufficient evidence to reject the manufacturer's claim. $\qquad \square$

Optional Technique

When completing a hypothesis test about the equality of two population variances, it would be convenient if we could always use a right-hand critical value for F without having to calculate the left-hand critical value. This can be accomplished by minor adjustments in the null hypothesis and in the calculation of F in step 4. The two cases we would like to change are (1) the two-tailed test and (2) the one-tailed test where the critical region is on the left. The one-tailed test with the critical region on the right already meets our criterion.

CASE 1   When a two-tailed test is to be completed, we will state the hypotheses in the normal way. The calculated value of F, F^*, will be the larger of s_1^2/s_2^2 or s_2^2/s_1^2. (One of these will have a value between zero and one; the other will be larger than one.) The critical value of F will be $F(\mathrm{df}_n, \mathrm{df}_d, \alpha/2)$, where df_n is the number of degrees of freedom for the sample whose variance is used in the numerator; df_d represents the degrees of freedom used in the denominator. Only the right-tail critical value will be needed. The test is completed in the usual fashion.

CASE 2   When a one-tailed test where the critical region is on the left is to be completed, we will interchange the position of the two variances in the statement of the hypotheses. This will reverse the direction of the alternative hypothesis and put the critical region on the right. From this point on, the procedure is the same as for the one-tailed test with a critical region on the right.

Even though the hypothesis test is the most commonly used inference about two variances, occasionally you may be asked to estimate the **ratio of two variances**, σ_A^2/σ_B^2. The best point estimate is s_A^2/s_B^2. The $1 - \alpha$ **confidence interval** for the estimate is constructed by using the following formula:

ratio of variances

$$\frac{s_A^2/s_B^2}{F(\mathrm{df}_A, \mathrm{df}_B, \alpha/2)} \quad \text{to} \quad \frac{s_A^2/s_B^2}{F(\mathrm{df}_A, \mathrm{df}_B, 1 - \alpha/2)} \qquad (10\text{--}7)$$

Recall that left-tail values of F are not in the tables. These left-tail values are found by using formula (10–6). Using formula (10–6), we can rewrite formula (10–7) as

$$\frac{s_A^2/s_B^2}{F(\mathrm{df}_A, \mathrm{df}_B, \alpha/2)} \quad \text{to} \quad \frac{s_A^2}{s_B^2} F(\mathrm{df}_B, \mathrm{df}_A, \alpha/2) \qquad (10\text{--}8)$$

(Notice that when the reciprocal is used, the degrees of freedom for the numerator and denominator are switched.)

ratio of standard deviation   If the **ratio of standard deviations** is desired, you need only take the positive square root of each of the bounds of the interval found by using formula (10–8).

NOTE: The formula for estimating the ratio of population variances or population standard deviations requires the use of the ratio of sample variances. If the standard deviation of the sample is given, it must be squared to obtain the variance.

Exercises

10–15 Using the $F(df_1, df_2, \alpha)$ notation, name each of the critical values shown on the figures. (For two-tail cases, use $\alpha/2$ in each tail.)

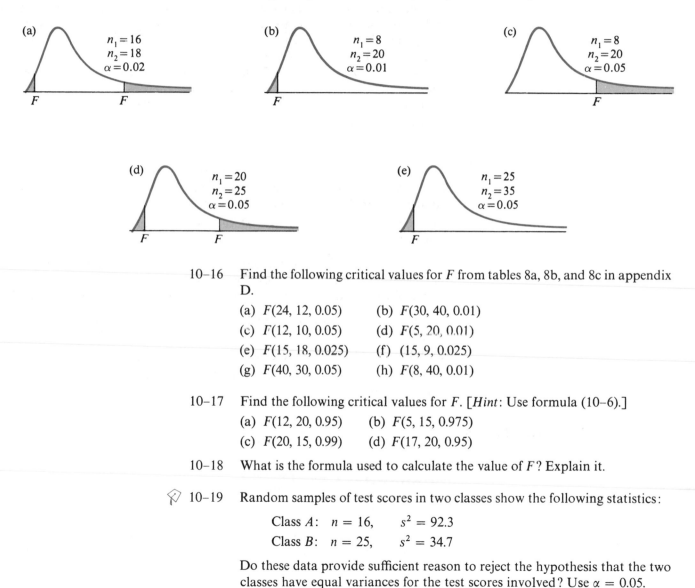

(a)
$n_1 = 16$
$n_2 = 18$
$\alpha = 0.02$
F F

(b)
$n_1 = 8$
$n_2 = 20$
$\alpha = 0.01$
F

(c)
$n_1 = 8$
$n_2 = 20$
$\alpha = 0.05$
F

(d)
$n_1 = 20$
$n_2 = 25$
$\alpha = 0.05$
F F

(e)
$n_1 = 25$
$n_2 = 35$
$\alpha = 0.05$
F

10–16 Find the following critical values for F from tables 8a, 8b, and 8c in appendix D.

(a) $F(24, 12, 0.05)$ (b) $F(30, 40, 0.01)$
(c) $F(12, 10, 0.05)$ (d) $F(5, 20, 0.01)$
(e) $F(15, 18, 0.025)$ (f) $(15, 9, 0.025)$
(g) $F(40, 30, 0.05)$ (h) $F(8, 40, 0.01)$

10–17 Find the following critical values for F. [*Hint*: Use formula (10–6).]
(a) $F(12, 20, 0.95)$ (b) $F(5, 15, 0.975)$
(c) $F(20, 15, 0.99)$ (d) $F(17, 20, 0.95)$

10–18 What is the formula used to calculate the value of F? Explain it.

10–19 Random samples of test scores in two classes show the following statistics:

Class A: $n = 16$, $s^2 = 92.3$
Class B: $n = 25$, $s^2 = 34.7$

Do these data provide sufficient reason to reject the hypothesis that the two classes have equal variances for the test scores involved? Use $\alpha = 0.05$.

10–20 To test the hypothesis that the variance among the weights of third-grade girls is greater than the variance among third-grade boys, the following statistics were obtained:

Girls: $s = 9.0$, $n = 36$

Boys: $s = 6.5$, $n = 45$

Do these statistics provide significant evidence to justify rejecting the null hypothesis in favor of the alternative $\sigma_G^2 > \sigma_B^2$, at the $\alpha = 0.05$ level?

10–21 A bakery is considering buying one of two gas ovens. The bakery requires that the temperature remain very constant during a baking operation. A study was conducted to measure the variance in temperature of the ovens during the baking process. The variance in temperature before the thermostat restarted the flame for the Monarch oven was 2.4 for 16 measurements. The variance for the Kraft oven was 3.2 for 12 measurements. Does this information provide sufficient reason to conclude that there is a difference in the variances for the two ovens? Use a 0.01 level of significance.

(a) Solve using the classical approach.

(b) Solve using the prob-value approach.

10–22 Use the sample information given in exercise 10–19.

(a) Construct the 0.95 confidence interval for the estimate of the ratio of the variance of population A to the variance of population B.

(b) Construct the 0.95 confidence interval for the estimate of the ratio of the standard deviations.

10–23 Using the sample information about the heights of students participating in the college intermural program (exercise 10–13), construct the 0.98 confidence interval for estimating the ratio of the standard deviations, male to female.

10–24 Assume that the data in exercise 10–14 were collected for the purpose of estimation.

(a) What would be your point estimate for the ratio of the two variances?

(b) Construct the 95 % confidence interval for the estimation of the ratio of the two variances.

Section 10–4 Inferences Concerning the Difference between Two Independent Means (Variances Unknown and Small Samples)

In section 10–2 we treated the cases for inferences about the difference between the means of two independent samples, where the samples were both large. We will now investigate the inference procedures to be used in situations where two independent small samples (that is, one or both samples are of size less than or equal to 30) are taken from approximately normal populations for the purpose of comparing their means. For these inferences we must use Student's t distribution. However, we must distinguish between two possible cases: (1) the variances of the two populations are equal, $\sigma_1^2 = \sigma_2^2$; or (2) the variances of the two populations are unequal, $\sigma_1^2 \neq \sigma_2^2$.

The two population variances are unknown; therefore, we will employ the F test studied in the preceding section to determine whether we have case 1 or case 2. The two sample variances will be used in a two-tailed test of the null hypothesis $H_0: \sigma_1^2 = \sigma_2^2$, or $\sigma_1^2/\sigma_2^2 = 1$. If we fail to reject H_0, we will proceed with the methods for case 1. If we reject H_0, we will proceed with case 2. Typically the same α is used for the F test as is given for the difference between the two means.

CASE 1   The procedures here are very similar to those used when the Student's t distribution was employed. The standard error of estimate must be estimated by

$$s_p \sqrt{\frac{1}{n_1} + \frac{1}{n_2}}$$

pooled estimate for the standard deviation   where s_p symbolizes the **pooled estimate for the standard deviation**. ("Pooled" means that the information from both samples is combined so as to give the best possible estimate.) The formula for s_p is

$$s_p = \sqrt{\frac{(n_1 - 1)s_1^2 + (n_2 - 1)s_2^2}{n_1 + n_2 - 2}} \qquad (10\text{-}9)$$

The number of degrees of freedom, df, is the sum of the number of degrees of freedom for the two samples, $(n_1 - 1) + (n_2 - 1)$; that is,

$$df = n_1 + n_2 - 2 \qquad (10\text{-}10)$$

With this information we can now write the formula for the **test statistic** t that will be used in a **hypothesis test**.

$$t = \frac{(\bar{x}_1 - \bar{x}_2) - (\mu_1 - \mu_2)}{s_p \sqrt{(1/n_1) + (1/n_2)}} \qquad (10\text{-}11)$$

with $n_1 + n_2 - 2$ degrees of freedom.

Illustration 10-10

In studying the nature of the student body at a community college, the following question was raised: Do male and female students tend to drive the same distance (one way) to college daily? To answer this question a random sample of size 25 was taken from each segment of the student body. The results are shown in table 10-3. Does this evidence contradict the null hypothesis that the two groups travel the same mean distances, at $\alpha = 0.10$?

TABLE 10–3
Sample Results for Illustration 10–10

Sample	n	$\bar{x}$	s^2
Male, m	25	10.22	33.95
Female, f	25	10.55	24.47

Solution Since the σ's are unknown, we must first test the variances to determine whether we have case 1 or case 2.

STEP 1 $H_0: \sigma_m^2 = \sigma_f^2$, or $\sigma_m^2/\sigma_f^2 = 1$

STEP 2 $H_a: \sigma_m^2 \neq \sigma_f^2$

STEP 3 $\alpha = 0.10$. See the diagram for the test criteria.

$\alpha/2 = 0.05$ $\qquad$ $\alpha/2 = 0.05$

$0 \qquad F(24, 24, 0.05) = 1.98 \qquad F$

STEP 4 $F^* = 33.95/24.47 = \mathbf{1.387}$ (F^* is in the noncritical region).

STEP 5 *Decision*: Fail to reject H_0.

Conclusion: Assume $\sigma_m^2 = \sigma_f^2$ and the test for the difference between means will be completed according to case 1 procedures, using t and formula (10–11).

STEP 1 $H_0: \mu_m = \mu_f$, or $\mu_m - \mu_f = 0$

STEP 2 $H_a: \mu_m \neq \mu_f$

STEP 3 $\alpha = 0.10$; df $= 25 + 25 - 2 = 48$; $t(48, 0.05) = 1.65$. See the diagram.

0.05 $\qquad$ 0.05

$-1.65 \qquad 0\,t(48, 0.05) = 1.65\ t$

STEP 4

$$t = \frac{(\bar{x}_m - \bar{x}_f) - (\mu_m - \mu_f)}{s_p\sqrt{(1/n_m) + (1/n_f)}}$$

$$= \frac{(10.22 - 10.55) - (0)}{\sqrt{[(24)(33.95) + (24)(24.47)]/48} \cdot \sqrt{(1/25) + (1/25)}}$$

$$= \frac{-0.33}{\sqrt{29.21} \cdot \sqrt{0.08}} = \frac{-0.33}{\sqrt{2.3368}}$$

$$= \frac{-0.33}{1.52866} = -0.2158$$

$$t^* = -0.22$$

STEP 5 *Decision*: Fail to reject H_0 ($t*$ is in the noncritical region).

Conclusion: There appears to be no significant difference between the mean distances taveled to college daily by male and female students. □

Having completed our discussion of case 1, let's continue on to case 2.

CASE 2 If we must assume that the two populations have **unequal variances**, then the hypothesis test for the difference between two independent means is completed by using the test statistic t. The **calculated value of t** is obtained by using

$$t = \frac{(\bar{x}_1 - \bar{x}_2) - (\mu_1 - \mu_2)}{\sqrt{(s_1^2/n_1) + (s_2^2/n_2)}} \qquad (10\text{--}12)$$

with the number of degrees of freedom for the critical value being given by the smaller of $n_1 - 1$ or $n_2 - 1$.

Illustration 10–11

Many students have complained that the soft drink vending machine A (in the student recreation room) dispenses less drink than machine B (in the faculty lounge). To test this belief several samples were taken and weighed carefully, with the results as shown in table 10–4. Does this evidence support the hypothesis that the mean amount dispensed by A is less than the mean amount dispensed by B, at the 5% level of significance?

TABLE 10–4
Sample Results for Illustration 10–11

Machine	n	$\bar{x}$	s
A	10	5.38	1.59
B	12	5.92	0.83

Solution First we test for the equality of the variances.

STEP 1 $H_0: \sigma_A^2 = \sigma_B^2$, or $\sigma_A^2/\sigma_B^2 = 1$

STEP 2 $H_a: \sigma_A^2 \neq \sigma_B^2$

STEP 3 $\alpha = 0.05$. See the diagram.

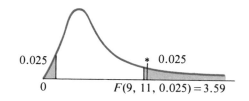

0.025   * 0.025

0        $F(9, 11, 0.025) = 3.59$

STEP 4 $F^* = (1.59)^2/(0.83)^2 = 3.67$

STEP 5   *Decision*: Reject H_0.

Conclusion: Assume that $\sigma_A^2 \neq \sigma_B^2$, with $\alpha = 0.05$.

Now we are ready to test the difference between the means by use of case 2 procedures.

STEP 1 $H_0: \mu_A = \mu_B$ or $\mu_A - \mu_B = 0$ ($\geq$)

STEP 2 $H_a: \mu_A < \mu_B$ or $\mu_A - \mu_B < 0$

STEP 3   df $= 10 - 1 = 9$ (smaller sample is of size 10); $\alpha = 0.05$. See the diagram.

STEP 4

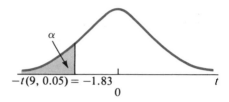

$$t = \frac{(5.38 - 5.92) - (0)}{\sqrt{[(1.59)^2/10] + [(0.83)^2/12]}}$$

$$= \frac{-0.54}{\sqrt{0.2528 + 0.0574}} = \frac{-0.54}{0.557} = -0.969$$

$$t^* = -0.97$$

STEP 5   *Decision*: Fail to reject H_0 (t^* is in the acceptance region).

Conclusion: There is not sufficient evidence shown by these samples to conclude that machine A dispenses less drink than does machine B.   □

If confidence interval estimates are to be constructed to estimate the difference between the means of two independent small samples, the F test will again be used to determine which of the following two formulas we need to use.

CASE 1   **σ's unknown but assumed equal.** The $1 - \alpha$ **confidence interval** is given by

$$(\bar{x}_1 - \bar{x}_2) - t(\text{df}, \alpha/2) \cdot s_p \cdot \sqrt{\frac{1}{n_1} + \frac{1}{n_2}}$$

to   (10–13)

$$(\bar{x}_1 - \bar{x}_2) + t(\text{df}, \alpha/2) \cdot s_p \cdot \sqrt{\frac{1}{n_1} + \frac{1}{n_2}}$$

where df $= n_1 + n_2 - 2$ and s_p is the pooled estimate for the standard deviation found by using formula (10–9).

CASE 2 **σ's unknown but assumed unequal.** The $1 - \alpha$ confidence interval is given by

$$(\bar{x}_1 - \bar{x}_2) - t(\text{df}, \alpha/2) \cdot \sqrt{\frac{s_1^2}{n_1} + \frac{s_2^2}{n_2}}$$

to

(10-14)

$$(\bar{x}_1 - \bar{x}_2) + t(\text{df}, \alpha/2) \cdot \sqrt{\frac{s_1^2}{n_1} + \frac{s_2^2}{n_2}}$$

where df is the smaller of $n_1 - 1$ or $n_2 - 1$.

Illustration 10–12

The heights of 20 randomly selected females and 30 randomly selected males were obtained from the student body of our college in order to estimate the difference in their mean heights. Find (a) a point estimate and (b) a 95% confidence interval estimate for the difference between the means, $\mu_m - \mu_f$.

Solution The sample information is given in table 10–5.

TABLE 10–5
Sample Results for Illustration 10–12

Sample	Number	Mean	Standard Deviation
Female, f	20	63.8	2.18
Male, m	30	69.8	1.92

(a) The point estimate for $\mu_m - \mu_f$ is **6.0**.
(b) First we must decide whether we should assume that $\sigma_f^2 = \sigma_m^2$ or $\sigma_f^2 \neq \sigma_m^2$.

$$H_0: \sigma_f^2 = \sigma_m^2 \quad \text{or} \quad \frac{\sigma_f^2}{\sigma_m^2} = 1$$

$$H_a: \sigma_f^2 \neq \sigma_m^2$$

$\alpha = 0.05$; see the diagram for the test criteria.

0.025   0.025

0     $F(19, 29, 0.025) \approx 2.24$

NOTE: When the correct degrees of freedom are not found in the table, the critical value is found by using the linear interpolation technique (a technique that is not discussed in this book).

$$F^* = \frac{(2.18)^2}{(1.92)^2} = 1.289$$

So we fail to reject H_0. Therefore, we assume that $\sigma_f^2 = \sigma_m^2$. The 0.95 confidence interval is then found by using formula (10–13).

$$6.0 \pm t(48, 0.025) \cdot \sqrt{\frac{19(2.18)^2 + 29(1.92)^2}{48}} \cdot \sqrt{\frac{1}{20} + \frac{1}{30}}$$

$$6.0 \pm 1.96\sqrt{4.1084}\,\sqrt{0.0833}$$

$$6.0 \pm (1.96)(2.0269)(0.2886)$$

$$6.0 \pm 1.1466$$

$$6.0 \pm 1.15$$

$$\mathbf{4.85} \quad \text{to} \quad \mathbf{7.15}$$

is the 0.95 confidence interval for $\mu_m - \mu_f$. (No specific illustration is shown for estimation where $\sigma_1^2 \neq \sigma_2^2$. All parts of such a solution can be found in various illustrations in this chapter.) ☐

Exercises

10–25 Two normal populations are sampled in order to compare their means. Do the accompanying sample data support the hypothesis that the two population means are significantly different? Use $\alpha = 0.01$. Assume that the population variances are equal.

Sample	n	$\bar{x}$	s^2
I	12	75.6	7.2
II	15	78.7	8.1

10–26 Two independent samples are drawn from normal populations, with the results as shown in the accompanying table. Does this information provide sufficient reason to reject the null hypothesis in favor of the claim that the mean of population R is significantly larger than the mean of population S? Use $\alpha = 0.05$. (Remember to perform the F test on the variance first.)

Sample	n	$\sum x$	$\sum (x - \bar{x})^2$
R	10	295	75
S	8	195	90

10-27   The two independent samples shown in the accompanying table were
         obtained in hopes of showing that the mean of population A is larger than the

Sample A	6	7	7	6	6	5	6	8	5	4
Sample B	7	2	4	3	3	5	4	6	4	2

         mean of population B. Do the samples provide significant evidence to justify
         that hope? Use $\alpha = 0.05$.

10-28   Certain "adjustments" are made between the test for the equality of two
         population means when the variances are unknown and judged to be equal
         and the test for equality of two population means when variances are un-
         known and judged to be unequal. (The t-distribution is used in both cases.)
         Explain how each of the following adjustments is accomplished.

         (a) The test statistic is modified.

         (b) The critical region is modified.

         (c) The number of degrees of freedom is modified.

10-29   Independent samples were taken from each of two normal populations in
         order to compare the two population means. The sample data are summarized
         in the accompanying table. Assume that $\sigma_1 \neq \sigma_2$.

         (a) What affect does the assumption $\sigma_1 \neq \sigma_2$ have on the situation?

         (b) Determine the number of degrees of freedom to use in comparing μ_1
         and μ_2.

         (c) Is there sufficient reason to reject the hypothesis that $\mu_1 = \mu_2$, at
         $\alpha = 0.05$?

Sample	n	$\bar{x}$	s^2
1	10	12.3	6.8
2	15	13.9	23.5

10-30   Use the information in exercise 10–25 to estimate the difference between the
         means of population I and population II. (Assume that $\sigma_I = \sigma_{II}$.) Use
         $1 - \alpha = 0.95$.

10-31   Construct the 98% confidence interval estimate for the difference between
         the means of population I and population II in exercise 10–25. (Assume
         $\sigma_I = \sigma_{II}$.)

10-32   Use the sample information in exercise 10–26 to construct the 0.95 confidence
         interval estimate for the difference between μ_R and μ_S.

10-33   Construct a 90% confidence interval estimate for the difference between μ_1
         and μ_2 based on the sample statistics given in exercise 10–29. (Assume
         that $\sigma_1 \neq \sigma_2$.)

10–34 Two samples of data were obtained to compare the means of two normal populations. The data are summarized in the accompanying table. Is there sufficient evidence to conclude that $\mu_A > \mu_B$ at $\alpha = 0.05$? (*Hint*: Does $\sigma_A = \sigma_B$?)

Sample	n	$\sum x$	$\sum x^2$
A	10	223	6735
B	10	110	1322

10–35 If a random sample of 18 homes south of Center Street in Provo, New York have a mean selling price of $15,000 and a variance of $2,400; and a random sample of 18 homes north of Center Street have a mean selling price of $16,000 and a variance of $4,800, can you conclude that there is a significant difference between the selling price of homes in these two areas of Provo at the 0.05 level?

(a) Solve using the classical approach.

(b) Solve using the prob-value approach.

10–36 Twenty laboratory mice were randomly divided into two groups of 10. Each group was fed according to two prescribed diets. At the end of three weeks the weight gained by each animal was recorded. Do the accompanying data justify the conclusion that the mean weight gained on diet B was greater than the mean weight gained on diet A, at the $\alpha = 0.05$ level of significance?

Diet A	5	14	7	9	11	7	13	14	12	8
Diet B	5	21	16	23	4	16	13	19	9	21

Section 10–5 Inferences Concerning Two Dependent Means

The procedure for comparing two dependent means is quite different from that of comparing two independent means. When comparing two independent means we make inferences about the two population means by using the difference between the two observed sample means. However, because of the relationship between two dependent samples, we will actually compare each data of the first sample with the data in the second sample that came from the same source. The two data, one from each set, that come from the same source are called **paired** data. These pairs of data are compared by using the difference in their numerical values. This difference is called a

paired difference **paired difference**. The two dependent means are then compared by using the observed mean of the resulting paired differences. The concept of using paired data for this purpose has a built-in ability to remove many otherwise uncontrollable factors. The tire wear problem (illustration 10–2) is an excellent example of such additional factors. The wearing ability of a tire is greatly affected by a multitude of factors: the size and weight of the car, the age and condition of the car, the driving habits of the driver, the number of miles driven, the condition of and types of roads driven on, the quality of the material used to make the tire, and so on.

Illustration 10–13

If we were to test the wearing quality of two tire brands by plan D, as described in illustration 10–2, all the factors above will have an equal effect on both tire brands. Table 10–6 gives the amount of wear (in thousandths of an inch)

TABLE 10–6

Sample Results for Illustration 10–13

Tire Brand	Car					
	1	*2*	*3*	*4*	*5*	*6*
A	125	64	94	38	90	106
B	133	65	103	37	102	115

that took place in such a test. One tire of each brand was placed on each of the six test cars. The position was determined with the aid of a random number table.

Since the various cars, drivers, and conditions are the same for each paired set of data, it would make sense to introduce a new measure, the paired difference d, that was observed in each pair of related data. Therefore, we add a third row to the data of table 10–6, $d =$ brand $B -$ brand A.

Car	1	2	3	4	5	6
$d = B - A$	8	1	9	-1	12	9

Do the sample data provide sufficient evidence for us to conclude that the two brands show unequal wear, at the 0.05 level of significance?

Before we can solve the problem posed in this illustration, we need some new formulas. Our two sets of sample data have been combined into one set, a set of n values of d, $d = x_1 - x_2$. The sample statistics that we will use are $\bar{d}$, the observed **mean** value of d,

$$\bar{d} = \frac{\sum d}{n} \qquad (10\text{–}15)$$

and s_d, the observed **standard deviation** of the d's:

$$s_d = \sqrt{\frac{\sum d^2 - \frac{(\sum d)^2}{n}}{n - 1}} \qquad (10\text{–}16)$$

The observed values of d in our sample are from the population of *paired differences*, whose distribution is assumed to be approximately normal with a mean value of μ_d and a standard deviation of σ_d. Since σ_d is unknown, it will be estimated by s_d. The inferences about μ_d are completed in the same manner as the inferences about μ in chapter 8 and in section 9–1 by using $\bar{d}$ as the point estimate of μ_d. The inferences are completed by using the t distribution and $s_d/\sqrt{n}$ as the approximate measure for the standard error.

$$ t = \frac{\bar{d} - \mu_d}{s_d/\sqrt{n}} \qquad (10\text{--}17) $$

with $n - 1$ degrees of freedom.

Now let's complete illustration 10–13.

Solution

mean difference   STEP 1 $H_0: \mu_d = 0$ (μ_d is **mean difference**)

STEP 2 $H_a: \mu_d \neq 0$

STEP 3 $\alpha = 0.05$. The test criteria are shown in the diagram.

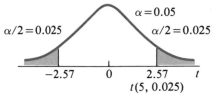

NOTE: n is the number of paired differences (d). df $= n - 1$, and $t(5, 0.025)$ is obtained from table 6, appendix D.

STEP 4

$$ t = \frac{\bar{d} - \mu_d}{s_d/\sqrt{n}} $$

First we must find $\bar{d}$ and s_d; some preliminary calculations are shown in table 10–7.

TABLE 10–7
Preliminary Calculations

d	d^2
8	64
1	1
9	81
−1	1
12	144
9	81
Total 38	372

$$\bar{d} = \frac{\sum d}{n} = \frac{38}{6} = 6.333 = \mathbf{6.3}$$

$$S_d = \sqrt{\frac{(372) - \frac{(38)^2}{6}}{5}} = \sqrt{26.27} = 5.13 = \quad \mathbf{5.1}$$

Therefore,

$$t = \frac{6.3 - 0}{5.1/\sqrt{6}} = \frac{(6.3)\sqrt{6}}{5.1} = \frac{(6.3)(2.45)}{5.1} = 3.026$$

$$t^* = \mathbf{3.03}$$

See the diagram.

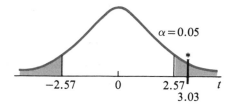

STEP 5 *Decision*: Reject H_0 (t^* is in the critical region).

Conclusion: There is a significant difference in the mean amount of wear.

Since the calculated value of t fell in the critical region on the right, we might want to conclude that one brand of these tires was better than the other. This is possible, but we must be extremely careful to interpret the conclusion correctly. Recall that the recorded data concerned the amount of wear. Therefore, the tire with the lesser amount of wear should be the better one. Our hypothesis test showed that tire brand B had significantly more wear, thereby implying that brand A is the longer-wearing tire. □

It is also possible to estimate the mean difference in paired data. The formula used for this **confidence interval** estimate is

$$\bar{d} - t(\mathrm{df}, \alpha/2) \cdot \frac{S_d}{\sqrt{n}}$$

to

(10–18)

$$\bar{d} + t(\mathrm{df}, \alpha/2) \cdot \frac{S_d}{\sqrt{n}}$$

Illustration 10–14

Construct the 95% confidence interval for the estimation of the mean difference in the paired data on tire wear, as found in illustration 10–13. The given information is $n = 6$ pieces of paired data, $\bar{d} = 6.33$, and $s_d = 5.1$.

Solution

$$\bar{d} \pm t(\text{df}, \alpha/2) \cdot \frac{s_d}{\sqrt{n}}$$

$$6.3 \pm 2.57 \cdot \frac{5.1}{\sqrt{6}}$$

$$6.3 \pm 5.4$$

$$\mathbf{0.9} \quad \text{to} \quad \mathbf{11.7}$$

is the 0.95 confidence interval for $\mu_{d = B - A}$. That is, with 95% confidence we can say that the mean difference in the amount of wear is between 0.9 and 11.7.

NOTE: This confidence interval is quite wide. This is due, in part, to the small sample size. Recall from the central limit theorem that as the sample size increases, the standard error (estimated by $s_d/\sqrt{n}$) decreases. □

Exercises

10–37 Dependent samples were taken to compare the means of two populations. The sample statistics found were

$$n = 20 \qquad \bar{d} = 3.1 \qquad s_d = 9.6$$

Do we have sufficient evidence to reject the null hypothesis that the mean difference is zero, at the $\alpha = 0.05$ level of significance?

(a) Solve using the classical approach.

(b) Solve using the prob-value approach.

10–38 The corrosive effects of various soils on coated and uncoated steel pipe were tested by using a dependent sampling plan. The data collected are summarized by

$$n = 40 \qquad \sum d = 220 \qquad \sum d^2 = 6222$$

where d is the amount of corrosion on the coated portion subtracted from the amount of corrosion on the uncoated portion. Does this sample provide sufficient reason to conclude that the coating is beneficial? Use $\alpha = 0.01$.

(a) Solve using the classical approach.

(b) Solve using the prob-value approach.

10-39   We want to know which of two types of filters should be used over an oscilloscope to help the operator pick out the image on the screen of the cathode ray tube. A test was designed in which the strength of a signal could be varied from zero to the point where the operator first detects the image. At this point, the intensity setting is read. The lower the setting the better the filter. Twenty operators were asked to make one reading for each filter. Does the data below show a significant difference in the filters at the 0.10 level?

	Filter				Filter	
Operator	1	2	Operator	1	2
1	96	92	11	88	88
2	83	84	12	89	89
3	97	92	13	85	86
4	93	90	14	94	91
5	99	93	15	90	89
6	95	91	16	92	90
7	97	92	17	91	90
8	91	90	18	78	80
9	100	93	19	77	80
10	92	90	20	93	90

(a) Solve using the classical approach.

(b) Solve using the prob-value approach.

10-40   Two men, who usually commute to work together, decide to conduct an experiment to see if one route is faster than the other. The two men feel that their driving habits are approximately the same, and therefore they decide on the following procedure. On each morning for two weeks, each will drive to work by one of the two routes; the other driver will use the second route. On the first morning A will toss a coin: if heads appear, he will use route I; if tails appear, he will use route II. On the second morning B will toss the coin: heads, route I; tails, route II. The times are recorded to the nearest minute in the accompanying table. Do these data show that one of the routes is significantly faster than the other, at $\alpha = 0.05$?

	Day									
Route	M	T	W	Th	F	M	T	W	Th	F
I	29	26	25	25	25	24	26	26	30	31
II	25	26	25	25	24	23	27	25	29	30

(a) Solve using the classical approach.

(b) Solve using the prob-value approach.

10-41 Construct the 90% confidence interval estimate for the mean difference based on the sample information given in exercise 10–37.

10-42 Construct the 0.98 confidence interval estimate for the mean difference in the amount of corrosion for the steel pipes in exercise 10–38.

10-43 A sociologist is studying the effects of a certain motion picture film upon the attitudes of the black man toward the white man. Twelve black men were randomly selected and asked to fill out a questionnaire before and after viewing the film. The scores received by the 12 men are shown in the accompanying table. Construct a 95% confidence interval estimate for the mean shift in score that takes place when this film is viewed.

Before	10	13	18	12	9	8	14	12	17	20	7	11
After	5	9	13	17	4	5	11	14	13	18	7	12

Section 10–6 Inferences Concerning Two Proportions

We are often interested in making statistical comparisons between the proportions, percentages, or probabilities associated with two populations. Such questions as the following are frequently asked: Is the proportion of homeowners who favor a certain tax proposal different from the proportion of renters who favor it? Did a larger percentage of this semester's class pass statistics than last semester's class? Do the students have a different opinion about the "new code of conduct" than does the faculty? You can probably see the many types of questions involved. In this section we will compare two population proportions by using the difference between the observed proportions, $p'_1 - p'_2$, of two independent samples.

RECALL:

1. The observed probability is $p' = x/n$, where x is the number of observed successes in n trials.
2. $q' = 1 - p'$.
3. p is the probability of success on an individual trial in a binomial probability experiment of n repeated independent trials (see page 184).

The sampling distribution of $p'_1 - p'_2$ is approximately normally distributed with a mean $\mu_{p'_1 p'_2} = p_1 - p_2$ and with a standard error of

$$\sqrt{\frac{p_1 q_1}{n_1} + \frac{p_2 q_2}{n_2}}$$

if n_1 and n_2 are sufficiently large. Therefore, when the null hypothesis that there is no difference between two proportions is to be tested, the test statistic will be z. The calculated value of z will be determined by formula (10–19):

$$z = \frac{p'_1 - p'_2}{\sqrt{pq[(1/n_1) + (1/n_2)]}}$$ (10–19)

NOTES:

1. The null hypothesis is $p_1 = p_2$, or $p_1 - p_2 = 0$.
2. The numerator of formula (10–19) written in the usual manner is $(p'_1 - p'_2) - (p_1 - p_2)$, but since the null hypothesis is assumed to be true during the test, $p_1 - p_2 = 0$ and the numerator becomes simply $p'_1 - p'_2$.
3. Since the null hypothesis is $p_1 = p_2$, the standard error of $p'_1 - p'_2$ can be written as $\sqrt{pq[(1/n_1) + (1/n_2)]}$.

Illustration 10–15

A salesman for a new manufacturer of walkie-talkies claims that the percentage of defective walkie-talkies found among his products will be no higher than the percentage of defectives found in a competitor's line. To test the salesman's statement, random samples were taken of each manufacturer's product. The sample summaries are given in table 10–8. Can we reject the salesman's claim, at the 0.05 level of significance?

TABLE 10–8
Sample Results for Illustration 10–15

Product	Sample Number	Number of Defectives	Number Checked
Salesman's	1	8	100
Competitor's	2	2	100

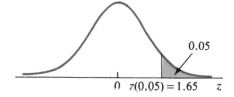

0.05

0 $z(0.05) = 1.65$ z

Solution

STEP 1 $H_0: p_1 - p_2 = 0 \ (\le)$ (no higher than)

STEP 2 $H_a: p_1 - p_2 > 0$ (higher than)

STEP 3 $\alpha = 0.05$. The test criteria are shown in the diagram.

STEP 4 $p'_1 = 8/100 = 0.08$, $p'_2 = 2/100 = 0.02$.

Since the values of p_1 and p_2 are unknown but assumed equal; (H_0), the best estimate we have for $p(p = p_1 = p_2)$ is obtained by pooling the two samples:

$$p^* = \frac{x_1 + x_2}{n_1 + n_2}$$ (10–20)

Our pooled estimate for p, p^*, then becomes

$$p^* = \frac{8 + 2}{100 + 100} = \frac{10}{200} = 0.05$$

and

$$q^* = 1 - p^* = 1 - 0.05 = 0.95$$

The value of the test statistic z is calculated by using formulas (10–19) and (10–20) together. p^* and q^* are used as estimates for p and q.

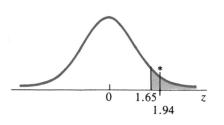

$$z = \frac{p_1' - p_2'}{\sqrt{p^* q^* [(1/n_1) + (1/n_2)]}}$$

$$= \frac{0.08 - 0.02}{\sqrt{(0.05)(0.95)[(1/100) + (1/100)]}}$$

$$= \frac{0.06}{\sqrt{(0.05)(0.95)(0.02)}} = \frac{0.06}{0.031} = 1.9354$$

$$z^* = 1.94$$

See the diagram.

STEP 5   *Decision*: Reject H_0 (z^* is in the critical region).

Conclusion: There is sufficient evidence to reject the salesman's claim.   □

Illustration 10–16

Joe claimed that the probability that a commuting college student has car trouble (of some type—car wouldn't start, flat tire, accident) on the way to college in the morning is greater than on the way home after class. He has several theories to explain his claim. The Sports Car Club thinks that the ideas of "before class" or "after class" have nothing to do with whether or not a student has car trouble. The club decides to challenge Joe's claim, and two large samples are gathered in order to test the theory. The resulting sample statistics are given in table 10–9. Complete the hypothesis test with $\alpha = 0.02$.

TABLE 10–9
Sample Results for Illustration 10–16

Time of Trouble	Number Sampled, n	Number Having Trouble
Before, b	500	30
After, a	600	28

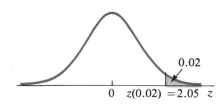

0.02

0   $z(0.02) = 2.05$ z

Solution

STEP 1 $H_0: p_b = p_a$ or $p_b - p_a = 0$ (no difference)

STEP 2 $H_a: p_b > p_a$ or $p_b - p_a > 0$ (Joe's claim)

STEP 3 $\alpha = 0.02$. The test criteria are shown in the diagram.

STEP 4

$$z = \frac{p_b' - p_a'}{\sqrt{p^* q^* [(1/n_b) + (1/n_a)]}}$$

$$p_b' = \frac{30}{500} = 0.06 \qquad p_a' = \frac{28}{600} = 0.04666 = 0.047$$

The pooled estimate for p is

$$p^* = \frac{30 + 28}{500 + 600} = \frac{58}{1100} = 0.0527 = \mathbf{0.053}$$

and thus $q^* = 1 - 0.053 = \mathbf{0.947}$. Therefore,

$$z = \frac{0.06 - 0.047}{\sqrt{(0.053)(0.947)(1/500 + 1/600)}}$$

$$= \frac{0.013}{\sqrt{0.05019(0.002 + 0.0017)}}$$

$$= \frac{0.013}{\sqrt{0.000186}} = \frac{0.013}{0.0136} = 0.956$$

$$z^* = \mathbf{0.96}$$

STEP 5   *Decision*: Fail to reject H_0 (z^* is in the acceptance region).

Conclusion: There is no evidence to indicate that the probability of car trouble is greater before class than after class.   □

To estimate the difference between the proportions of two populations, you should use the observed difference ($p_1' - p_2'$) as the point estimate. The $1 - \alpha$ **confidence interval** is given by

$$(p_1' - p_2') - z(\alpha/2) \cdot \sqrt{\frac{p_1' q_1'}{n_1} + \frac{p_2' q_2'}{n_2}}$$

to   (10–21)

$$(p_1' - p_2') + z(\alpha/2) \cdot \sqrt{\frac{p_1' q_1'}{n_1} + \frac{p_2' q_2'}{n_2}}$$

Illustration 10–17

In studying his campaign plans, Mr. Morris wishes to estimate the difference in his appeal as a candidate between men and women voters. He asks his campaign manager to take two samples and find the 99 % confidence interval estimate of the difference. A sample of 1000 voters was taken from each population, with 388 men and 459 women favoring Mr. Morris.

Solution   The campaign manager calculated the confidence interval estimate by using formula (10–21), as follows:

$$p'_m = \frac{388}{1000} = 0.388 \qquad p'_w = \frac{459}{1000} = 0.459$$

$$(0.459 - 0.388) \pm 2.58 \sqrt{\frac{(0.459)(0.541)}{1000} + \frac{(0.388)(0.612)}{1000}}$$

$$0.071 \pm 2.58\sqrt{0.000248 + 0.000237}$$

$$0.071 \pm 2.58\sqrt{0.000485}$$

$$0.071 \pm (2.58)(0.022)$$

$$0.071 \pm 0.057$$

$$\textbf{0.014} \quad \text{to} \quad \textbf{0.128}$$

is the 0.99 confidence interval for $p_w - p_m$. With 99 % confidence we can say that there is a difference of from 1.4 % to 12.8 % in Mr. Morris's voter appeal.   □

Exercises

10–44   Two manufacturing firms, producing equivalent products, claim the same rate of customer preference for their products. A random sample of customers shows 102 of 300 and 152 of 400 to prefer products A and B, respectively. Does this evidence indicate a significant difference between the proportions of preference? Use $\alpha = 0.02$.

(a) Solve using the classical approach.

(b) Solve using the prob-value approach.

10–45   Two randomly selected groups of citizens were exposed to different media campaigns, which dealt with the image of a political candidate. One week later the citizen groups were surveyed to see if they would vote for the candidate. The results were:

	Exposed to Conservative Image	Exposed to Moderate Image
Number in Sample	100	100
Proportion for the Candidate	0.40	0.50

Is there sufficient evidence to show a difference in the effectiveness of the two image campaigns at the 0.05 level of significance?

(a) Solve using the classical approach.

(b) Solve using the prob-value approach.

10-46 In a survey taken to help understand personality, 22 of 71 persons under 18 years old expressed a fear of meeting people; 23 of 91 persons 18 years old and over expressed the same fear. Should the null hypothesis "there is no difference in the true proportions" be rejected at the $\alpha = 0.10$ level of significance?

(a) Solve using the classical approach.

(b) Solve using the prob-value approach.

10-47 In a survey of 300 people from city A, 128 preferred New Spring to all other brands of deodorant soap. In city B, 149 of 400 people preferred New Spring. Find the 98% confidence interval estimate for the difference in the two proportions.

10-48 In a random sample of 40 brown-haired individuals, 22 indicated that they used hair coloring. In another random sample of 40 blonde individuals, 26 indicated that they used hair coloring. Use a 0.92 confidence interval to estimate the difference in the proportion of these groups that use hair coloring.

10-49 A test of 100 youths and 200 adults showed that 50 of the youths and 60 of the adults were careless drivers. Use the data to estimate how much larger the percentage of careless youth drivers is than the adult percentage, with a 0.95 confidence interval.

In Retrospect

In this chapter we began the comparisons of two populations by first distinguishing between independent and dependent samples, which are statistically very important and useful sampling procedures. We then proceeded to examine the inferences concerning the comparison of means, variances, and proportions for two populations.

The use of hypothesis testing and interval estimates can sometimes be interchanged; that is, the calculation of a confidence interval can often be used in place of a hypothesis test. For example, illustration 10–17 called for a confidence interval estimate. Now suppose that Mr. Morris asked: "Is there a difference in my voter appeal between men and women voters?" To answer his question you would not need to calculate a test statistic if you chose to test at $\alpha = 0.01$ using a two-tailed test. "No difference" would mean a difference of zero, which is not included in the interval from 0.014 to 0.128 (the interval determined in illustration 10–17). Therefore, a null hypothesis of "no difference" would be rejected, thereby substantiating the conclusion that there is a significant difference in voter appeal between the two groups.

The news article for this chapter shows a comparison of the amount of break time taken by men and women employees. This comparison uses the sample averages for the two groups. Note that inferences are stated about the comparison. However, the news article cites enough information only to make a point estimate for the difference in the two population means (most likely the average mentioned in the article is the mean value).

The comparison of two populations is a common occurrence. We often compare two means or two proportions and probably don't even recognize it. Look back at the Sears advertisement at the beginning of chapter 8. The statement, "Sears steel-belted Radial, when compared to our leading fiber-glass-belted tire, improved gas mileage by an average of 7.4%"

TABLE 10–10

Formulas to Use for Inferences
Involving Two Populations

Situation	Test Statistic	Formula to be Used	
		Hypothesis Test	Interval Estimate
Two independent means			
σ known	z	Formula (10–1)	(10–2)
σ unknown (large samples)	z	Formula (10–3)	(10–4)
σ unknown (equal and small samples)	t	Formulas (10–9), (10–10), (10–11)	(10–13)
σ unknown (unequal and small samples)	t	Formula (10–12)	(10–14)
Two dependent means	t	Formulas (10–15), (10–16), (10–17)	(10–15), (10–16), (10–18)
Two proportions	z	Formula (10–19)	(10–21)
Two variances	F	Formula (10–5)	(10–8)

compares the mileage obtained while using two different types of tires.

For convenience, table 10–10 identifies the formulas to use when making inferences of comparisons between two populations.

In chapters 8 through 10 we have introduced and completed hypothesis testing and confidence interval estimation for questions that involve one or two means, proportions, and variances. There is much more to be done. However, this much inferential statistics is sufficient to answer many questions. So in succeeding chapters we are ready to look at some other useful tests for other types of questions that may occur.

Chapter Exercises

10–50   Two semesters ago a sample of 150 math students showed a mean of 41.9 on the departmental final exam. Last semester another sample of 150 math students obtained an average of 44.4 on an equivalent exam. Assume that $\sigma = 12.0$ for both sets of exam grades. Should the null hypothesis "there is no real difference between these two groups" be rejected, at $\alpha = 0.05$?

(a) Solve using the classical approach.

(b) Solve using the prob-value approach.

10–51   The same achievement test is given to soldiers selected at random from two units. The scores they obtained are summarized as follows:

Unit 1: $n_1 = 70$ $\bar{x}_1 = 73.2$ $s_1 = 6.1$
Unit 2: $n_2 = 60$ $\bar{x}_2 = 70.5$ $s_2 = 5.5$

Construct a 90% confidence interval estimate for the difference in the mean level of the two units.

10–52   Out of a random sample of 100 stocks on the New York Exchange, 32 made a gain today. A random sample of 100 stocks on the American Stock Exchange showed 27 stocks making a gain.

(a) Construct the 99% confidence interval, estimating the difference in the proportion of stocks making a gain.

(b) Does the answer to part (a) suggest that there is a significant difference between the proportions of stocks making gains on the two different stock exchanges?

10–53   If 250 castings produced on mold A contained 15 defectives, and 300 castings on mold B contained 30 defectives, should the null hypothesis "there is no difference between the true proportions of defectives" be rejected, at the 0.10 level of significance?

(a) Solve using the classical approach.

(b) Solve using the prob-value approach.

10-54 Assuming that the standard deviation of the lifetime for two brands of flash-light batteries is 2.8 hours, do the sample statistics in the table substantiate the claim that brand S has a longer life than brand K, at the 0.01 level?

Brand	n	$\bar{x}$
S	25	47.6
K	30	45.2

(a) Solve using the classical approach.

(b) Solve using the prob-value approach.

10-55 A soft-drink distributor is considering two new models of dispensing machines. Both the Harvard Company machine and the Fizzit machine can be adjusted to fill the cups to a certain mean amount. However, the variation in the amount dispensed from cup to cup is a primary concern. Ten cups dispensed from the Harvard Company machine showed a variance of 0.065, while fifteen cups dispensed from the Fizzit machine showed a variance of 0.033. The factory representative from the Harvard Company maintains that his machine had no more variability than the Fizzit machine.

(a) At the 0.05 level of significance, does the sample refute the salesman's assertion?

(b) Estimate the variance for the amount of fill dispensed by the Harvard Company machine using a 90% confidence interval.

10-56 To estimate the ratio of the variances among the weights of a can of drained peaches of the leading brand to her own brand, a store manager had samples taken, with the following results:

Leading brand: $n = 16$ $s^2 = 1.968$
Her brand: $n = 25$ $s^2 = 2.834$

(a) Give a point estimate for the ratio of "leading brand" variance to "her brand" variance.

(b) Construct the 95% confidence interval estimate for the same ratio of variances.

(c) Find the 95% confidence interval for the ratio of standard deviations.

10-57 The store manager in exercise 10-56 believes that there is no difference in the drained weight of peaches between her brand and the leading brand. The sample statistics obtained to test her belief were reported as follows:

Leading brand: $\bar{x} = 9.3$ $s^2 = 1.968$ $n = 16$
Her band: $\bar{x} = 8.7$ $s^2 = 2.834$ $n = 25$

Does this sample information contradict the store manager's claim, at the 0.05 level? (See exercise 10-56.)

10-58 A survey was taken to estimate the level of sales per bookstore for a recently published technical book. In the Northeast the mean sales were 17.3 books and the variance was 41.2. Stores in the South had a mean of 20.3 books and a variance of 68.4. Forty-two stores were sampled in the Northeast while 36 were sampled in the South. At the 0.05 level of significance, is there a difference in the mean number of books sold in the two regions?

(a) Solve using the classical approach.

(b) Solve using the prob-value approach.

(c) Estimate the mean number of books sold per bookstore in the South using a 0.90 confidence interval.

10-59 Two diets will be compared. Eighty individuals are selected at random from a population of overweight musicians. Forty-five of this group are assigned diet A and the other 35 are placed on diet B. After one week the weight losses in pounds are recorded in the accompanying table.

Diet	Sample Size	Sample Mean (pounds)	Sample Variance
A	45	10.3	7
B	35	7.3	3.25

(a) Test the hypothesis that $\sigma_A = \sigma_B$ at $\alpha = 0.10$.

(b) Do the data substantiate the conclusion that the expected weight loss μ_A under diet A is greater than the expected weight loss μ_B under diet B? Test at the 0.10 level. Draw the appropriate conclusion.

(c) Construct a 90% confidence interval for $\mu_A - \mu_B$.

10 60 The score on a certain psychological test is used as an index of status frustration. The scale ranges from 0 (low frustration) to 10 (high frustration). The test was administered to independent random samples of seven radical rightists and eight Peace Corps volunteers.

Radical Rightists	6	10	3	8	8	7	9
Peace Corps Volunteers	3	5	2	0	3	1	0	4

Using the 10% level of significance, test that the mean score for both groups is the same against the alternative that it is not the same.

(a) Solve using the classical approach.

(b) Solve using the prob-value approach.

10-61 A person might use the variance or standard deviation of the daily change in the stock market price as a measure of stability. Suppose you wish to compare the stability of a company's stock this year to its stability last year. You are given the following results of taking random samples from the daily gains and losses over the last two years:

This year: $n = 25$ $s = 2.57$

Last year: $n = 25$ $s = 1.26$

(a) Construct the 90% confidence interval estimate for the ratio of last year's standard deviation to this year's.

(b) Is the daily gain more, or less, or about the same, as far as stability is concerned, for this year when compared to last year?

10–62 Show that the standard error of $p'_1 - p'_2$, which is $\sqrt{(p_1 q_1/n_1) + (p_2 q_2/n_2)}$, reduces to $\sqrt{pq[(1/n_1) + (1/n_2)]}$ when $p_1 = p_2 = p$.

10–63 (a) The advertisement used as a news article for chapter 8 states that Sears' Steel-Belted Radial tires improved gas mileage by an average of 7.4%. This 7.4% is actually an expression of a value involving two population parameters. What are the two populations? Can you explain why they reported 7.4% rather than the difference between the two mean values for the variable that measures gas mileage?

(b) The news article for this chapter states two sample averages. What are they? Assuming that the sample averages are sample means, give a point estimate for the difference between the mean amount of break time taken by men and women.

10–64 A shoe manufacturer has developed a new artificial leather heel for his shoes and would like to test the new material against leather, in hopes of showing that the new material is longer wearing than leather. Forty-eight people are outfitted with a pair of shoes that has one leather heel and one heel of the new material. After three months of normal wear, the following summary was reported for the data collected:

$$n = 48 \qquad \sum d = 88 \qquad \sum (d - \bar{d})^2 = 373.65$$

where d is defined as the amount of wear on the leather heel minus the amount of wear on the new material. Can you conclude that this sample evidence indicates that the new material does wear longer? Use $\alpha = 0.01$.

(a) Solve using the classical approach.

(b) Solve using the prob-value approach.

10–65 Two brands of paint are to be tested. Brand A is less expensive than brand B. Several chips are painted and exposed to weather conditions for a period of six months. Each chip is then judged on several qualities and a score is determined. The paint samples scored (higher scores are better) are shown in the accompanying table.

Paint A	84	86	91	93	84	88	
Paint B	90	88	92	94	84	85	92

We would like to buy the less expensive paint, unless brand B is a better paint. Do these results provide sufficient evidence to reject the null hypothesis that there is no difference? Use $\alpha = 0.05$.

(a) Solve using the classical approach.

(b) Solve using the prob-value approach.

10–66 A sample taken from insurance records shows that in 1946, 9.8% of the population of the United States was admitted into a hospital. In 1981, 14.7% was admitted. (Each sample was of size 1000.) Estimate the percentage increase of hospital admissions.

(a) Give a point estimate.

(b) Give a 95% confidence interval estimate.

(c) Does the answer to part (b) suggest that there has been an increase in the percentage of people admitted to hospitals?

10–67 To compare the merits of two short-range rockets, 8 of the first kind and 10 of the second kind are fired at a target. If the first kind have a mean target error of 36 feet and a standard deviation of 15 feet, while the second kind have a mean target error of 52 feet and a standard deviation of 18 feet, does this indicate that the second kind of rocket is less accurate than the first? Use $\alpha = 0.01$.

10–68 Two methods were used to study the latent heat of ice fusion. Both method A (an electrical method) and method B (a method of mixtures) were conducted with the specimens cooled to $-0.72°C$. The data represents the change in total heat from $-0.72°C$ to water at $0°C$, in calories per gram of mass.

Method A	Method B
79.98	80.02
80.04	79.94
80.02	79.98
80.04	79.97
80.03	79.97
80.03	80.03
80.04	79.95
79.97	79.97
80.05	
80.03	
80.02	
80.00	
80.02	

Is there a significant difference in the mean values at the 0.05 level?

(a) Solve using the classical approach.

(b) Solve using the prob-value approach.

10-69 A study was conducted to investigate the effectiveness of two different teaching methods used in physical education. The task method and the command method were both used to teach tennis to college students. A group of size 30 was used for each method. The students in both groups were instructed for the same length of time and then tested on the forehand and backhand tennis strokes.

Results	Command Group	Task Group	Calculated t
Forehand	$\bar{x} = 55.6$	$\bar{x} = 59.8$	$t^* = 2.39$
Backhand	$\bar{x} = 35.87$	$\bar{x} = 41.83$	$t^* = 3.59$

(a) Is one method more effective than the other in teaching tennis to college students?

(b) If so, for which stroke and with what level of significance?

10-70 Estimate, using a 95% confidence interval, the difference in IQ between the oldest and the youngest members (brothers and sisters) of a family based on the accompanying random sample of IQs.

Oldest	145	133	116	128	85	100	105	150	97	110	120	130
Youngest	131	119	103	93	108	100	111	130	135	113	108	125

10-71 Twelve automobiles were selected at random to test two new mixtures of unleaded gasolines. Each car was given a measured allotment of gasoline and driven; then the distance traveled was recorded. The second mixture was immediately tested in the same manner. The order in which the x and y mixtures were tested was also randomly assigned. The results are given in the accompanying table. Can you conclude that there is no real difference in

						Car						
Mixture	1	2	3	4	5	6	7	8	9	10	11	12
x	7.9	5.6	9.2	6.7	8.1	7.3	8.1	5.4	6.9	6.1	7.1	8.1
y	7.7	6.1	8.9	7.1	7.9	6.7	8.2	5.0	6.2	5.7	6.2	7.5

mileage obtained by these two gasoline mixtures, at the 0.10 level of significance?

(a) Solve using the classical approach.

(b) Solve using the prob-value approach.

10-72 To test the effect of a physical fitness course on one's physical ability, the number of sit-ups that a person could do in 1 minute, both before and after the course, was recorded. Ten randomly selected participants scored as shown in the accompanying table. Can you conclude that a significant amount of improvement took place? Use $\alpha = 0.01$.

Before	29	22	25	29	26	24	31	46	34	28
After	30	26	25	35	33	36	32	54	50	43

(a) Solve using the classical approach.

(b) Solve using the prob-value approach.

10–73 In determining the "goodness" of a test question, a teacher will often compare the percentage of better students who answer it correctly with the percentage of the poorer students who answer it correctly. One expects that the better students will answer the question correctly more frequently than the poorer students. On the last test 35 of the students with the top 60 grades and 27 with the bottom 60 answered a certain question correctly. Did the students with the top grades do significantly better on this question? Use $\alpha = 0.05$.

(a) Solve using the classical approach.

(b) Solve using the prob-value approach.

10–74 Ten soldiers were selected at random from each of two companies to participate in a rifle-shooting competition. Their scores are shown in the accompanying table. Can you conclude that company B has a higher mean score than company A? Use $\alpha = 0.05$.

Company A	72	29	62	60	68	59	61	73	38	48
Company B	75	43	63	63	61	72	73	82	47	43

10–75 Ten new recruits participated in a rifle-shooting competition at the end of their first day at training camp. The same ten competed again at the end of a full week of training and practice. Their resulting scores are shown in the accompanying table.

Time of Competition	Soldier									
	1	2	3	4	5	6	7	8	9	10
First day	72	29	62	60	68	59	61	73	38	48
One week later	75	43	63	63	61	72	73	82	47	43

(a) Does this set of 10 pairs of data show that there was a significant amount of improvement in the soldier's shooting abilities during the week? Use $\alpha = 0.05$.

(b) Notice that the data in exercises 10–74 and 10–75 are the same set of number values. Compare the types of tests and the results obtained in the solution of exercise 10–74 and part (a) of this exercise.

10–76 A group of 17 students participated in an evaluation of a special training session that claimed to improve memory. The students were randomly assigned to two groups: group A, the test group, and group B, the control group. All 17 were tested for the ability to remember certain material. Group A was given the special training; group B was not. After one month both groups were tested again, with the results as shown in the accompanying table. Do these data support the alternative hypothesis that the special training is effective, at the $\alpha = 0.01$ level of significance?

Group A

Time of Test	Student								
	1	2	3	4	5	6	7	8	9
Before	23	22	20	21	23	18	17	20	23
After	28	29	26	23	31	25	22	26	26

Group B

Time of Test	Student							
	10	11	12	13	14	15	16	17
Before	22	20	23	17	21	19	20	20
After	23	25	26	18	21	17	18	20

10-77 In a shoot-out at the OK Rifle Range, Billy and Cole each shot at 18 targets to determine who was the better shot. The scores are presented as frequency distributions.

Shooter	Score										Total
	91	92	93	94	95	96	97	98	99	100	
Billy	2	0	3	1	4	4	2	1	0	1	18
Cole	0	1	3	4	4	3	2	0	1	0	18

(a) Do these data support the claim that Cole is the more consistent point scorer, at the 0.01 level of significance?

(b) Is there sufficient evidence to reject the null hypothesis that the two shoot equally well? Use $\alpha = 0.01$.

Vocabulary List

Be able to define each term. In addition, describe, in your own words, and give an example of each term. Your examples should not be ones given in class or in the textbook.

The terms followed by bracketed numbers were defined in earlier chapters (the bracketed numbers are those of the earlier chapters), but we ask you to define them again to show your increased understanding of their meaning.

confidence interval estimate [8, 9]

dependent means

dependent samples

F distribution

hypothesis test [8, 9]

independent means

independent samples

large sample

mean difference

paired difference

pooled estimate for the standard deviation

ratio of standard deviation

ratio of variances

source (of data)

Quiz

Answer "True" if the statement is always true. If the statement is not always true, replace the words shown in color with words that make the statement always true.

10–1 When the means of two unrelated samples are used to compare two populations, we are dealing with **two dependent means**.

10–2 The use of **paired data (dependent means)** often allows for the control of unmeasurable or confounding variables because each pair is subjected to these confounding effects equally.

10–3 The **chi-square distribution** is used for making inferences about the ratio of the variances of two populations.

10–4 The **F distribution** is used when two dependent means are to be compared.

10–5 In comparing two independent means when the σ's are unknown, we need to perform an **F test on their variances** to determine the proper formula to use.

10–6 The **standard normal score** is used for all inferences concerning population proportions.

10–7 The F distribution is a **symmetric** distribution.

10–8 The number of degrees of freedom for the critical value of t is equal to **the smaller of $n_1 - 1$ or $n_2 - 1$** when making inferences about the difference between two independent means for the case where the variances are unknown and assumed to be unequal, and the sample sizes are small.

10–9 A pooled estimate for the standard deviation of two combined populations is calculated in all problems dealing with the difference between two independent means.

10–10 A **pooled estimate** for any statistic in a problem dealing with two populations is a value arrived at by combining the two separate sample statistics so as to achieve the best possible point estimate.

Working with Your Own Data

As consumers, we all purchase many bottled, boxed, canned, or packaged products. Seldom, if ever, do we question whether or not the content amount stated on the container is accurate for one of these bottles, boxes, cans, or packages.

Listed here are a few typical content listings found on various containers:

28 FL OZ (1 PT 12 OZ)
750 ml
5 FL. OZ. (148 ml)
32 FL. OZS. (1 QT.) 0.95 l
NET WT 10 OZ   283 GRAMS
NET WT. 3¾ OZ. 106 g—48 tea bags
140 1-PLY NAPKINS
77 SQ. FT.—92 TWO-PLY SHEETS—11 × 11 IN.

Have you ever thought, "I wonder if I am getting the amount that I am paying for?" And if this thought did cross your mind, did you attempt to check the validity of the content claim? The following article appeared in the *Times Union* of Rochester, New York on February 16, 1972. (This kind of situation occurs frequently.)

MILK FIRM ACCUSED OF SHORT MEASURE

The processing manager of Dairylea Cooperative, Inc., has been named in a warrant charging that the cooperative is distributing cartons of milk in the Rochester area containing less than the quantity represented.

. . . an investigator found shortages in four quarts of Dairylea milk purchased Friday.

Asst. Dist. Atty. Howard R. Relin, who issued the warrant, said the shortages ranged from 1⅛ to 1¼ ounces per quart. A quart of milk contains 32 fluid ounces.

. . . the state Agriculture and Markets Law . . . provides that a seller of a commodity shall not sell or deliver less of the commodity than the quantity represented to be sold.

. . . the purpose of the law under which . . . the dairy is charged is to ensure honest, accurate, and fair dealing with the public. There is no requirement that intent to violate the law be proved, he said.

From *The Times-Union*, Rochester, N.Y., Feb. 16, 1972.

This situation poses a very interesting legal problem: there is no need to show intent to "short-the-customer". If caught, the violators' fines are automatic and often quite severe.

A A High-Speed Filling Operation

A high-speed piston-type machine, used to fill cans with hot tomato juice, was sold to a canning company. The guarantee stated that it would fill 48-ounce cans with a mean amount of 49.5 ounces, a standard deviation of 0.072 ounce, and a maximum spread of 0.282 ounce while operating at a rate of filling 150 to 170 cans per minute. On August 12, 1982, a sample of 42 cans was gathered and the following weights were recorded. The weights, which are measured to the nearest $1/8$ oz., are recorded as variations from 49.5 oz.

$-1/8$	0	$-1/8$	0	0	0	$-1/8$	0	0	0
0	0	$-1/8$	0	$1/8$	0	$1/8$	0	$1/8$	0
0	0	$-1/8$	0	0	0	0	$-1/8$	0	0
0	0	0	0	0	0	0	0	0	0
0	0								

1. Calculate the mean $\bar{x}$, the standard deviation s, and the range of the sample data.

2. Construct a histogram picturing the sample data.

3. Does the amount of fill differ from the prescribed 49.5 ounces at the $\alpha = 0.05$ level? Test the hypothesis that $\mu = 49.5$ against an appropriate alternative.

4. Does the amount of variation, as measured by the standard deviation, satisfy the guarantee at the $\alpha = 0.05$ level?

5. Assuming that the filling machine continues to fill cans with an amount of tomato juice that is distributed normally and the mean and standard deviation are equal to the values found in question 1; what is the probability that a randomly selected can will contain less than the 48 ounces claimed on the label?

6. If the amount of fill per can is normally distributed and the standard deviation can be maintained, find the setting for the mean value that would allow only one can in every 10,000 to contain less than 48 ounces.

B Your Own Investigation

Select a "packaged" product that has a quantity of fill per package that you can and would like to investigate.

1. Describe your selected product, including the quantity per package and describe how you plan to obtain your data.

2. Collect your sample of data. (Consult your instructor for advice on size of sample.)

3. Calculate the mean $\bar{x}$ and standard deviation s for your sample data.

4. Construct a histogram picturing the sample data.

5. Does the mean amount of fill meet the amount prescribed on the label? Test using $\alpha = 0.05$.

6. Assume that the item you selected is filled continually. The amount of fill is normally distributed and the mean and standard deviation are equal to the values found in question 3. What is the probability that one randomly selected package contains less than the prescribed amount?

PART 4

More Inferential Statistics

In Part 3 we studied the two inferential statistical techniques: hypothesis testing and confidence interval estimating, in regard to one and two population settings involving the three population parameters mean, μ, standard deviation, σ, and proportion, p. In this section we will learn how to use the hypothesis test and the interval estimation techniques in other statistical situations. Some of these situations will be extensions of previous methods (for example: analysis of variance will be used to deal with more than two populations when the question involves the mean). Other situations will be alternatives to methods previously studied (such as the nonparametric techniques); while others will deal with new inferences (such as the nonparametric methods and the regression and correlation inferential techniques).

11 Additional Applications of Chi-Square

50% Opposed to Building Nuclear Plants

The New York Poll was conducted Dec. 15-18 by the National Center for Telephone Research of New York for Gannett News Service. Trained interviewers surveyed 1,000 registered voters. The calls were placed at random, but in proportion to past voter turnouts around the state.

By JOHN OMICINSKI Chief, GNS Albany Bureau

Half of all New Yorkers oppose construction of more nuclear power plants in the state, the New York State Poll shows.

And women, far more than men, account for the opposition.

Exactly 50 percent of New Yorkers surveyed said they opposed more nuclear plants in the state. That matches the percentage of voters who said they were against more plants two years ago in a New York State Poll.

The latest poll, however, shows a drop since 1977 in the percentage that favors nuclear plants—down to 32 percent from 38 percent.

There is a corresponding increase of 6 percent among those who say they're not sure.

Resistance to nuclear plants, regionally, ran highest in western New York, and lowest in the New York City north suburbs.

There is a great deal of uncertainty about the safety of nuclear plants, the poll found. But slightly more people seem to think they're safe than think they are unsafe.

Thus, it seems that most New Yorkers agree with Gov. Carey's decision last year during the campaign, to not allow more nuclear plants to be built until the radioactive waste problem is solved.

Here are the figures on the main question, "Do you favor or oppose building more nuclear power plants in New York State?":

	Yes	No	Not Sure
Total	32	50	18
Upstate	35	48	17
NY City	30	54	16
No. suburbs	32	45	23
Western NY	28	56	16

Chapter Objectives

Previously we discussed and used the chi-square distribution both to test and to estimate the value of the variance (or standard deviation) of a single population. The chi-square distribution may also be used for tests in other types of situations. In this chapter we are going to investigate some of these uses. Specifically, we will look at two tests: a multinomial experiment and the contingency table. These two types of tests will be used to compare experimental results with expected results in order to determine (1) preferences, (2) independence, and (3) homogeneity.

*The data that we will be using in these techniques will be **enumerative**, that is, the data will result from counts of occurrences.*

Section 11-1 Chi-Square Statistic

There are many problems for which the information is categorized and the results are shown by way of counts. For example, a sample of final exam scores is given in table 2–5 on page 34. In table 2–6 (page 35) the class 33–43 has a frequency of 2; the class 44–54 has a frequency of 5; and so on. The result of this tabulation is a frequency table, an idea that was first presented in chapter 2. Our discussion in this chapter differs from what was done in chapter 2, however, in that now we will work *only* with the frequencies.

cell   Suppose that we have a number of **cells** into which n observations have been sorted. (The term *cell* is synonymous with the term *class*; the terms *class* and *frequency* were defined and first used in earlier chapters. Before continuing, a brief review of sections 2–2, 2–3, and 2–4 might be beneficial.) The **observed frequencies** in each cell are denoted by $O_1, O_2, O_3, \ldots, O_k$ (see table 11–1). Note that the sum of all observed frequencies is equal to

TABLE 11–1
Observed Frequencies

	\multicolumn{5}{c}{*k Categories*}				
	1st	*2d*	*3d*	⋯	*kth*	*Total*
Observed Frequency	O_1	O_2	O_3	⋯	O_k	n

expected frequency $O_1 + O_2 + \cdots + O_k = n$. What we would like to do is to compare the observed frequencies with some **expected or theoretical frequencies**, denoted by $E_1, E_2, E_3, \ldots, E_k$, for each of these cells. Again, the sum of these expected frequencies must be exactly

$$E_1 + E_2 + \cdots + E_k = n$$

We will decide whether the observed frequencies seem to agree or seem to disagree with the expected frequencies. This will be accomplished by a hypothesis test using the **chi-square distribution, χ^2**.

The calculated value of the test statistic will be

$$\chi^2 = \sum_{\text{all cells}} \frac{(O - E)^2}{E} \tag{11-1}$$

This calculated value for chi-square will be the sum of several positive numbers, one from each cell (or category). The numerator of each term in the formula for χ^2 is the square of the difference between the values of the observed and the expected frequencies. The closer together these values are, the smaller the value of $(O - E)^2$; the farther apart, the larger the value of $(O - E)^2$. The denominator for each cell puts the size of the numerator into perspective. That is, a difference $(O - E)$ of 10 resulting from frequencies of 110 (O) and 100 (E) seems quite different from a difference of 10 resulting from 15 (O) and 5 (E).

The preceding ideas suggest that small values of chi-square indicate agreement, while larger values indicate disagreement, between the two sets of frequencies. Therefore, it is customary for these tests to be one-tailed, with the critical region on the right.

In repeated sampling the calculated value of χ^2 [formula (11-1)] will have a sampling distribution that can be approximated by the chi-square probability distribution when n is large. This approximation is generally considered adequate when all the expected frequencies are equal to or greater than 5. Recall that there is a separate chi-square distribution for each degree of freedom, df. The appropriate value of df will be described with each specific test. The critical value of chi-square, $\chi^2(\text{df}, \alpha)$, will be found in table 7 of appendix D.

In this chapter we will permit a certain amount of "liberalization" with respect to the null hypothesis and its testing. Prior to this chapter the null hypothesis has always been a statement about a population parameter (μ, σ, or p). However, there are other types of hypotheses that can be tested, such as "this die is fair," or "height and weight of individuals are independent." Notice that these hypotheses are not claims about a parameter, although sometimes they could be stated with parameter values specified.

Suppose I claimed that "this die is fair" $p = P(\text{any one number}) = \frac{1}{6}$ and you wanted to test it. What would you do? Was your answer something like this? "Roll this die, in a fair manner, many times, recording the results." Suppose you decide to roll the die 60 times. If the die is fair, what do you expect will happen? Each number, 1, 2, . . . , 6, should appear approximately $\frac{1}{6}$ of the time (that is, 10 times). If it happens that approximately 10 of each number occur, you will certainly accept the claim of fairness ($p = \frac{1}{6}$). If it happens that the die seems to favor some numbers, you will reject the claim. (The test statistic χ^2 will have a large value in that case, as we will see shortly.)

Section 11–2 Inferences Concerning Multinomial Experiments

multinomial experiment

The preceding die problem is a good illustration of a **multinomial experiment**. Let's consider this problem again. Suppose we want to test this die (at $\alpha = 0.05$) and decide whether to fail to reject or reject the claim made, "this die is fair." (The probability of each number is $\frac{1}{6}$.) It is rolled from a cup onto a smooth flat surface 60 times and the frequencies shown in the table occurred.

Number	1	2	3	4	5	6
Occurrences	7	12	10	12	8	11

The null hypothesis that the die is fair is assumed to be true. This allows us to calculate the expected frequencies. If the die was fair, we certainly would expect 10 occurrences for each number.

Let's now calculate an observed value of χ^2. These calculations are shown in table 11–2. The calculated value is $\chi^{2*} = 2.2$.

TABLE 11–2
Computations for Calculating χ^2

Number	Observed, O	Expected, E	O − E	$(O - E)^2$	$(O - E)^2/E$
1	7	10	−3	9	0.9
2	12	10	2	4	0.4
3	10	10	0	1	0.0
4	12	10	2	4	0.4
5	8	10	−2	4	0.4
6	11	10	1	1	0.1
Total	60	60	0 ⓒⓚ		**2.2**

NOTE: $\sum (O - E)$ must equal 0 since $\sum O = \sum E = n$. You can use this fact as a check, as shown in table 11–2.

Before continuing, let's set up the hypothesis-testing format.

STEP 1 H_0: The die is fair (each $p = \frac{1}{6}$).

STEP 2 H_a: The die is not fair.

STEP 3 $\alpha = 0.05$. See the figure for the test criteria. In the multinomial test, df = $k - 1$, where k is the number of cells.

STEP 4   The test statistic was calculated in table 11–2 and was found to be $\chi^{2*} = 2.2$.

STEP 5   *Decision*: Fail to reject $H_0(\chi^{2*}$ is not in the critical region).

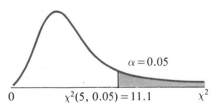

$\alpha = 0.05$

0        $\chi^2(5, 0.05) = 11.1$ χ^2

Conclusion: The observed frequencies are not significantly different from those expected of a fair die.

Before we look at other illustrations, we must define the term "multinomial experiment" and we must state the guidelines for completing the chi-square test for it.

Multinomial Experiment

An experiment with the following characteristics:

1. It consists of n identical independent trials.

2. The outcome of each trial fits into exactly one of k possible cells.

3. There is a probability associated with each particular cell, and these individual probabilities remain constant during the experiment. (It must be true that $p_1 + p_2 + \cdots + p_k = 1$.)

4. The experiment will result in a set of observed frequencies, O_1, $O_2, \ldots, O_k$, where each O_i is the number of times a trial outcome falls into that particular cell. (It must be the case that

$$O_1 + O_2 + \cdots + O_k = n.)$$

The preceding illustration with the die meets the definition of a multinomial experiment, since it has all four of the characteristics described in the definition.

1. The die was rolled n (60) times in an identical fashion, and these trials were independent of each other. (The result of each trial was unaffected by the results of other trials.)

2. Each time the die was rolled, one of six numbers resulted, and each number forms a cell.

3. The probability associated with each cell is $\frac{1}{6}$, and this was constant from trial to trial. (Six values of $\frac{1}{6}$ sum to 1.0.)

4. When the experiment was complete, we had a list of frequencies (7, 12, 10, 12, 8, and 11) which summed to 60, indicating that each of the outcomes was taken into account.

The **testing procedure** for multinomial experiments is very much as it was in previous chapters. The biggest change comes with the statement of the null hypothesis. It may be a verbal statement, such as in the die illustration: "This die is fair." Often the alternative to the null hypothesis is not stated. However, in this book the alternative hypothesis will be shown, since it seems to aid in the organization and understanding of the problem. However, it will not be used to determine the location of the critical region, as was the case previously. **For multinomial experiments we will always use a one-tailed critical region, and it will be the right-hand tail of the χ^2 distribution.**

The critical value will be determined by the level of significance assigned (α) and the number of degrees of freedom. The number of degrees of freedom (df) will be 1 less than the number of cells (k) into which the data are divided:

$$df = k - 1 \qquad\qquad (11\text{--}2)$$

Each expected frequency, E_i, will be determined by multiplying the corresponding probability (p_i) for that cell by the total number of trials n. That is,

$$E_i = n \cdot p_i \qquad\qquad (11\text{--}3)$$

One guideline should be met to ensure a good approximation to the chi-square distribution: each expected frequency should be at least 5 (that is, each $E_i \geq 5$). If this guideline cannot be met, corrective measures to ensure a good approximation should be used. These corrective measures are not covered in this book but are discussed in many other sources.

Illustration 11–1

College students have regularly insisted on freedom of choice when registering for courses. This semester there were seven sections of a particular mathematics course. They were scheduled to meet at various times with a variety of instructors. Table 11–3 shows the number of students who selected each

		Section						
	1	*2*	*3*	*4*	*5*	*6*	*7*	*Total*
Number of Students	18	12	25	23	8	19	14	119

of the seven sections. Do the data indicate that the students had a preference for certain sections? Or do the data indicate that each section was equally likely to be chosen?

Solution   If no preference was shown in the selection of sections, we would expect the 119 students to be equally distributed among the seven classes. Thus if no preference was the case, we would expect 17 students to register for each section. The test is completed as shown next, at the 5% level of significance.

STEP 1 H_0: There was no preference shown (equally distributed).

STEP 2 H_a: There was a preference shown (not equally distributed).

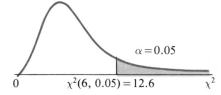

$\alpha = 0.05$

$\chi^2(6, 0.05) = 12.6$

STEP 3 $\alpha = 0.05$. See the diagram for the test criteria.

STEP 4   We use formula (11–1) to calculate the test statistic.

$$\chi^2 = \frac{(18 - 17)^2}{17} + \frac{(12 - 17)^2}{17} + \frac{(25 - 17)^2}{17}$$

$$+ \frac{(23 - 17)^2}{17} + \frac{(8 - 17)^2}{17}$$

$$+ \frac{(19 - 17)^2}{17} + \frac{(14 - 17)^2}{17}$$

$$= \frac{(1)^2 + (-5)^2 + (8)^2 + (6)^2 + (-9)^2 + (2)^2 + (-3)^2}{17}$$

$$= \frac{1 + 25 + 64 + 36 + 81 + 4 + 9}{17} = \frac{220}{17} = 12.9411$$

$$\chi^{2*} = \mathbf{12.94}$$

STEP 5   *Decision*: Reject $H_0(\chi^{2*}$ falls in the critical region).

Conclusion: There does seem to be a preference shown.

   We cannot determine, from the given information, what the preference is. It could be teacher preference, time preference, or a case of schedule conflict. Conclusions must be worded carefully to avoid suggesting conclusions that we cannot support.   □

   Not all multinomial experiments result in equal expected frequencies, as we will see in the next illustration.

Illustration 11–2

The Mendelian theory of inheritance claims that the frequencies of round and yellow, wrinkled and yellow, round and green, and wrinkled and green will occur in the ratio 9:3:3:1 when two specific varieties of peas are crossed. In testing this theory, Mendel obtained frequencies of 315, 101, 108, and 32, respectively. Do these sample data provide sufficient evidence to reject this theory, at the 0.05 level of significance?

Solution

STEP 1 H_0 : 9:3:3:1 is the ratio of inheritance.

STEP 2 H_a : 9:3:3:1 is not the ratio of inheritance.

STEP 3 $\alpha = 0.05$, $k = 4$, and df $= 3$. See the diagram.

STEP 4   The ratio 9:3:3:1 indicates probabilities of $\frac{9}{16}$, $\frac{3}{16}$, $\frac{3}{16}$, and $\frac{1}{16}$. Therefore, the expected frequencies are $9n/16$, $3n/16$, $3n/16$, and $n/16$, where

$$n = \sum O_i = 315 + 101 + 108 + 32 = 556$$

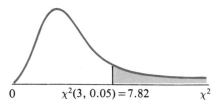

$\chi^2(3, 0.05) = 7.82$

The computations necessary for calculating χ^2 are given in table 11–4. Thus $\chi^{2*} = 0.47$.

TABLE 11–4
Computations Needed to Calculate χ^2

O	E	O − E	$(O − E)^2/E$
315	312.75	2.25	0.0162
101	104.25	−3.25	0.1013
108	104.25	3.75	0.1349
32	34.75	−2.75	0.2176
556	556.00	0 (CK)	0.4700

STEP 5   *Decision*: Fail to reject H_0.

Conclusion: There is not sufficient evidence to reject Mendel's theory.   □

Exercises

11–1   A floor polish manufacturer conducted a consumer-preference experiment to determine which of five different floor polishes was the most appealing. A sample of 100 housewives viewed five patches of flooring that had each received one of the five polishes. Each housewife indicated the patch she preferred for appearance reasons. The lighting and background were approximately the same for all patches. The results were:

Polish	A	B	C	D	E	Total
Frequency	27	17	15	22	19	100

(a) State the hypothesis for "no preference" in statistical terminology.

(b) What test statistic will be used in testing this null hypothesis?

(c) Complete the hypothesis test using $\alpha = 0.10$.

(1) Solve using the classical approach.

(2) Solve using the prob-value approach.

11–2   A certain type of flower seed will produce magenta, chartreuse, and ochre flowers in the ratio 6:3:1 (one flower per seed). A total of 100 seeds are planted and all germinate, yielding the following results:

Magenta	Chartreuse	Ochre
52	36	12

(a) If the null hypothesis (6:3:1) is true, what is the expected number of magenta flowers?

(b) How many degrees of freedom are associated with chi-square?

(c) Complete the hypothesis test using $\alpha = 0.10$.

    (1) Solve using the classical approach.

    (2) Solve using the prob-value approach.

11-3 A random sample of stocks listed on the New York Stock Exchange resulted in the frequencies shown in the accompanying table for each possible fractional price. Test the claim that all fractional prices are equally likely to occur. Use $\alpha = 0.05$.

Fractional Price	Even Dollar	$\frac{1}{8}$	$\frac{1}{4}$	$\frac{3}{8}$	$\frac{1}{2}$	$\frac{5}{8}$	$\frac{3}{4}$	$\frac{7}{8}$
Frequency	25	12	15	23	28	14	22	21

(a) Solve using the classical approach.

(b) Solve using the prob-value approach.

11-4 Previous sales records show that Inboard Custom Boats sold 20% of its boats in its Northeast sales district, and 28%, 8%, 12%, and 32%, respectively, in its Southeast, North Central, South Central, and West Coast sales districts. Of the first 500 boats sold this year, 120, 128, 43, 66, and 143 were sold in each of the five districts, respectively. Does the sales distribution so far this year seem to be the same as in previous years? Use $\alpha = 0.05$.

(a) Solve using the classical approach.

(b) Solve using the prob-value approach.

11-5 A sampling from the import distribution center for Volkswagen shows that the proportions of color choices for the Rabbit are as shown in the accompanying table. Take a sample of $n = 100$ or 200 Rabbits and compare the color preferences of local Rabbit owners to the data shown in the table.

Color	Yellow	Orange	Light Blue	Beige	Green	Red	Dark Blue
Percentage	25	17	15	12	11	10	10

(a) What null hypothesis will you test?

(b) What is the alternative hypothesis?

(c) Complete the test at $\alpha = 0.10$.

Section 11-3 Inferences Concerning Contingency Tables

contingency table A **contingency table** is an arrangement of data into a two-way classification. The data are usually frequency counts, although they need not be. The contingency table often involves two factors (or variables), and the usual question concerning such tables is whether the data indicate that the two variables are independent or dependent.

Test of Independence

To illustrate the use and analysis of a contingency table, let's consider the sex classification of liberal arts college students and their favorite academic area.

Illustration 11–3

Each person in a group of 300 students was identified as male or female and then asked whether he or she preferred taking liberal arts courses in the area of math-science, social science, or humanities. Table 11–5 is a contingency table that shows the frequencies found for these categories. Does this sample present sufficient evidence to reject the null hypothesis "preference for math-science, social science, or humanities is independent of the sex of a college student," at the 0.05 level of significance?

TABLE 11–5
Contingency Table Showing Sample Results for Illustration 11–3

Sex of Student	Favorite Subject Area			
	Math-Science, M-S	Social Science, SS	Humanities, H	Total
Male	37	41	44	122
Female	35	72	71	178
Total	72	113	115	300

Solution

STEP 1 H_0: Preference for math-science, social science, or humanities is independent of the sex of a college student.

STEP 2 H_a: Subject preference is not independent of the sex of the student.

STEP 3 To determine the critical value of chi-square, we need to know df, the number of degrees of freedom involved. In the case of contingency tables, the number of degrees of freedom is exactly the same number as the number of cells in the table that may be filled in freely when you are given the totals. The totals in this problem are shown in the following table:

			122
			178
72	113	115	300

Given these totals, you can fill in only two cells before the others are all determined. (The totals must, of course, be the same.) For example, once we pick two arbitrary values (say 50 and 60) for the first two cells of the first row (see the next table), the other four cell values are fixed.

50	60	C	122
D	E	F	178
72	113	115	300

They have to be $C = 12$, $D = 22$, $E = 53$, and $F = 103$. Otherwise the totals will not be correct. Therefore, for this problem there are two free choices. Each free choice corresponds to one degree of freedom. Hence the number of degrees of freedom for our example is 2 (df = 2). Thus if $\alpha = 0.05$ is used, the critical value is $\chi^2(2, 0.05) = 6.00$. See the diagram. (Following the conclusion of this illustration, a formula for finding df will be discussed.)

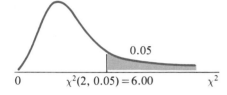

STEP 4 Before the calculated value of chi-square can be found, we need to determine the expected values E for each cell. To do this we must recall the null hypothesis, which asserts that these factors are independent. Therefore, we would expect the values to be distributed in proportion to the **marginal totals**. There are 122 males; we would expect them to be distributed among M-S, SS, and H proportionally to the 72, 113, and 115 totals. Thus the expected cell counts for males are

marginal totals

$$\frac{72}{300} \cdot 122 \qquad \frac{113}{300} \cdot 122 \qquad \frac{115}{300} \cdot 122$$

Similarly for the females, we would expect

$$\frac{72}{300} \cdot 178 \qquad \frac{113}{300} \cdot 178 \qquad \frac{115}{300} \cdot 178$$

Thus the expected values are as shown in table 11–6. (Always check the new totals with the old totals.)

TABLE 11–6
Expected Values

	M-S	SS	H	Total
	29.28	45.95	46.77	122.00
	42.72	67.05	68.23	178.00
Total	72.00	113.00	115.00	300.00

NOTE: Recall that we assume the null hypothesis to be true until there is evidence to reject it. Having made this assumption in our example, in effect we are saying that the event that a student picked at random is male and the event that a student picked at random prefers math-science courses are independent. Our point estimate for the probability that a student is male is 122/300 and the point estimate for the probability that the student prefers the math-science courses is 72/300. Therefore, the probability that both events occur is the product of the probabilities. [Refer to formula (4–8a).] Thus (122/300) · (72/300) is the probability of a selected student being male and preferring math-science. Therefore, the number of students out of 300 that are expected to be male and prefer math-science is found by multiplying the probability (or proportion) by the total number of students, 300. Thus the expected number of males who prefer math-science is (122/300) (72/300) (300) = (122/300) (72) = 29.28. The other expected values can be determined in the same manner.

Typically the contingency table is written so that it contains all this information; see table 11–7.

TABLE 11–7

Contingency Table Showing Sample Results and Expected Values

Sex of Student	Favorite Subject Area			
	Math-Science, M-S	Social Science, SS	Humanities, H	Total
Male	37 (29.28)	41 (45.95)	44 (46.77)	122
Female	35 (42.72)	72 (67.05)	71 (68.23)	178
Total	72	113	115	300

The calculated chi-square is

$$\chi^2 = \sum \frac{(O - E)^2}{E}$$

$$= \frac{(37 - 29.28)^2}{29.28} + \frac{(41 - 45.95)^2}{45.95} + \frac{(44 - 46.77)^2}{46.77}$$

$$+ \frac{(35 - 42.72)^2}{42.72} + \frac{(72 - 67.05)^2}{67.05} + \frac{(71 - 68.23)^2}{68.23}$$

$$= 2.035 + 0.533 + 0.164 + 1.395 + 0.365 + 0.112 = 4.604$$

$$\chi^{2*} = \mathbf{4.604}$$

STEP 5 *Decision*: Fail to reject $H_0(\chi^{2*}$ did not fall in the critical region.)

Conclusion: The evidence does not allow us to reject the idea of independence between the sex of a student and the student's preferred academic subject area. ☐

In general, the $r \times c$ **contingency table** (r is the number of rows; c is the number of columns) will be used to test the independence of the row factor and the column factor. The number of degrees of freedom will be determined by

$$\text{df} = (r - 1) \cdot (c - 1) \qquad (11\text{--}4)$$

where r and c are both greater than 1.

(This value for df should agree with the number of cells counted according to the general description on pages 394 and 395.)

NOTE: If either r or c is 1, then $\text{df} = k - 1$, where k represents the number of cells [formula (11–2)].

The expected values for an $r \times c$ contingency table will be found by means of the formulas found in each cell in table 11–8, where $n = $ grand

TABLE 11–8
*Expected Values for an $r \times c$
Contingency Table*

		Columns					
Rows	1	2	$\cdots$	j	$\cdots$	c	Total
1	$\dfrac{R_1 \times C_1}{n}$	$\dfrac{R_1 \times C_2}{n}$		$\dfrac{R_1 \times C_j}{n}$		$\dfrac{R_1 \times C_c}{n}$	R_1
2	$\dfrac{R_2 \times C_1}{n}$			$\cdot$			R_2
$\vdots$	$\vdots$		$\cdot$	$\cdot$		$\cdot$	$\vdots$
i	$\dfrac{R_i \times C_1}{n}$		$\cdot$	$\dfrac{R_i \times C_j}{n}$		$\cdot$	R_i
$\vdots$	$\vdots$			$\cdot$			$\vdots$
r	$\dfrac{R_r \times C_1}{n}$			$\cdot$		$\dfrac{R_r \times C_c}{n}$	R_r
Total	C_1	C_2	$\cdots$	C_j	$\cdots$	C_c	n

total. In general, the **expected value** at the intersection of the **ith row and the jth column** is given by

$$E_{i,j} = \frac{R_i \times C_j}{n} \qquad (11\text{--}5)$$

We should again observe the previously mentioned guideline: each $E_{i,j}$ should be at least 5.

NOTE: The notation used in table 11–8 and formula (11–5) may be unfamiliar to you. For convenience in referring to cells or entries in a table, $E_{i,j}$ (or E_{ij}) can be used to denote the entry in the ith row and the jth column.

That is, the first letter in the subscript corresponds to the row number and the second letter corresponds to the column number. Thus $E_{1,2}$ is the entry in the first row, second column, while $E_{2,1}$ is the entry in the second row, first column. Referring to table 11–6, $E_{1,2}$ for that table is 45.95 and $E_{2,1}$ is 42.72. The notation used in table 11–8 is interpreted in a similar manner, R_1 corresponds to the total from row 1, while C_1 corresponds to the total from column 1.

Test of Homogeneity

homogeneity

There is another type of contingency table problem. It is called a **test of homogeneity**. This test is used when one of the two variables is controlled by the experimenter so that the row (or column) totals are predetermined.

For example, suppose that we were to poll registered voters in reference to a piece of legislation proposed by the governor. In the poll 200 urban, 200 suburban, and 100 rural residents will be randomly selected and asked if they favor or oppose the governor's proposal. A total of 500 voters are to be polled. But notice that is has been predetermined (before the sample is taken) just how many are to fall within each row category, as shown in table 11–9.

TABLE 11–9
Registered Voter Poll with Predetermined Row Totals

Type of Residence	Governor's Proposal		Total
	Favor	Oppose	
Urban			200
Suburban			200
Rural			100
Total			500

In a test of this nature, we are actually testing the hypothesis "the distribution of proportions within the rows is the same for all rows." That is, the distribution of proportions in row 1 is the same as in row 2, is the same as in row 3, and so on. The alternative to this is that the distribution of proportions within the rows is not the same for all rows. This type of example may be thought of as a comparison of several multinomial experiments.

Beyond this conceptual difference, the actual testing for independence and homogeneity with contingency tables is the same. Let's demonstrate this by completing the polling illustration.

Illustration 11–4

Each person in a random sample of 500 registered voters (200 urban, 200 suburban, and 100 rural residents) was asked his or her opinion about the governor's proposed legislation. Does the sample evidence shown in table 11–10 support the hypothesis that voters within the different residence groups have different opinions about the governor's proposal? Use $\alpha = 0.05$.

TABLE 11–10

Sample Results for Illustration 11–4

Type of Residence	Governor's Proposal		
	Favor	Oppose	Total
Urban	143	57	200
Suburban	98	102	200
Rural	13	87	100
Total	254	246	500

Solution

STEP 1 H_0: The proportion of voters favoring the proposed legislation is the same in all three groups.

STEP 2 H_a: The proportion of voters favoring the proposed legislation is not the same in all three groups. (That is, in at least one group the proportions are different from the others.)

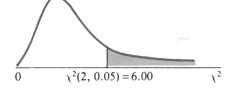

0     $\chi^2(2, 0.05) = 6.00$ χ^2

STEP 3 $\alpha = 0.05$ and df $= (3 - 1)(2 - 1) = 2$. See the diagram.

STEP 4 The expected values are found by using formula (11–5) and are given in table 11–11.

TABLE 11–11

Sample Results and Expected Values

Type of Residence	Governor's Proposal		
	Favor	Oppose	Total
Urban	143	57	200
	(101.6)	(98.4)	
Suburban	98	102	200
	(101.6)	(98.4)	
Rural	13	87	100
	(50.8)	(49.2)	
Total	254	246	500

NOTE: Each expected value is used twice in the calculation of χ^{2*}; therefore, it is a good idea to keep extra decimal places while doing the calculations.

$$\chi^2 = \frac{(143 - 101.6)^2}{101.6} + \frac{(57 - 98.4)^2}{98.4} + \frac{(98 - 101.6)^2}{101.6}$$

$$+ \frac{(102 - 98.4)^2}{98.4} + \frac{(13 - 50.8)^2}{50.8} + \frac{(87 - 49.2)^2}{49.2}$$

$$= 16.87 + 17.42 + 0.13 + 0.13 + 28.13 + 29.04 = 91.72$$

$$\chi^{2*} = \mathbf{91.72}$$

STEP 5 *Decision*: Reject H_0.

Conclusion: The three groups of voters do not all have the same proportions favoring the proposed legislation. □

Computer Solution

Minitab Printout for Illustration 11–4

Information given to computer

```
--READ TABLE INTO C1-C2
    COLUMN C1              C2
    COUNT    3              3
ROW
    1          143.        57.
    2           98.       102.
    3           13.        87.
```

Contingency table with expected values, compare to table 11–11

```
--CHISQUARE FOR DATA IN C1-C2
EXPECTED FREQUENCIES ARE PRINTED BELOW OBSERVED
FREQUENCIES
          I   C1   I   C2   I   TOTALS
    _____I_____I_____I_____
    1   I   143 I     57 I     200
        I   101.6I   98.4I
    _____I_____I_____I_____
    2   I    98 I    102 I     200
        I   101.6I   98.4I
    _____I_____I_____I_____
    3   I    13 I     87 I     100
        I    50.8I   49.2I
    _____I_____I_____I_____
    TOTALS I    254 I    246 I     500
    TOTAL CHI SQUARE =
```

The calculated value of chi-square, cell by cell, compare to solution above

```
       16.87 + 17.42 +
       0.13 +  0.13 +
      28.13 + 29.04 +
```

The calculated chi-square value, $\chi^{2}*$ ————————————→ = 91.72

The number of degrees of freedom

```
DEGREES OF FREEDOM = ( 3 - 1) × ( 2 - 1) =   2
--STOP
```

□

Exercises

11–6 A student survey was conducted to investigate opinions about a certain attitude concerning sex. The accompanying statistics were obtained. Test the claim that responses obtained are independent of the sex of the respondents. Use $\alpha = 0.05$.

Sex	Response to Question Asked	
	Approve	Disapprove
Male	60	75
Female	90	75

(a) Solve using the classical approach.

(b) Solve using the prob-value approach.

11-7 A psychologist is investigating how a person reacts to a certain situation. He feels the reaction may be influenced by how ethically pure the person's neighborhood is. He collects data on 500 people and finds:

Ethically Pure Neighborhood	Reaction			Total
	Mild	Medium	Strong	
Yes	170	100	30	300
No	70	100	30	200
Total	240	200	60	500

Does there appear to be a relationship between neighborhood and reaction at the 0.10 level of significance?

(a) Solve using the classical approach.

(b) Solve using the prob-value approach.

11-8 A survey of travelers who visited the comfort facilities of service stations for one of the very large United States petroleum distributors showed the following results:

Sex of Respondent	Quality of Restroom Facilities			Total
	Above Average	Average	Below Average	
Female	7	24	28	59
Male	8	26	7	41
Total	15	50	35	100

Using $\alpha = 0.05$, test to see if the null hypothesis, "quality of responses are independent of the sex of the respondent," can be rejected.

(a) Solve using the classical approach.

(b) Solve using the prob-value approach.

11-9   A survey of college students revealed the accompanying results on a question. Students were asked to identify the item that bothered them the most (the question asked about being late for something). Do the data collected show sufficient evidence to reject the null hypothesis that the item that bothers the student the most is independent of the sex of that student? Use $\alpha = 0.10$.

		Item		
Sex	To Class	For a Date	To Church	To Dinner
Male	18	24	15	21
Female	27	17	21	27

(a) Solve using the classical approach.

(b) Solve using the prob-value approach.

11-10 Given the following four instructors' grade distributions, determine whether or not it can be concluded that the grade distributions are different for these instructors at the $\alpha = 0.05$ level of significance.

		Grades			
Instructor	A	B	C	D or below	Totals
Smith	33	89	66	22	210
Jones	13	50	66	9	138
Johnson	18	38	31	8	95
Anderson	16	24	13	0	53
Totals	80	201	176	39	496

(a) State the null hypothesis.
(b) Solve using the classical approach.
(c) Solve using the prob-value approach.

11-11 "Fear of the dentist" is an emotion felt by many people of all age groups. A survey of 100 individuals in each of five age groups was conducted about this fear, and the results were as shown in the accompanying table.

		Age Group			
Response	Elementary	Jr. High	Sr. High	College	Adult
Fear	36	25	29	32	18
Do not fear	64	75	71	68	82

At $\alpha = 0.05$, do these data provide sufficient reason to reject the hypothesis that all age groups fear the dentist equally?

(a) Solve using the classical approach.

(b) Solve using the prob-value approach.

ψ 11-12 "Fear of darkness" is not an uncommon emotion. The accompanying data were obtained by asking 200 individuals in each group if they had serious fears of darkness. At $\alpha = 0.01$, do we have sufficient evidence to reject the hypothesis that the same proportion of each age group has serious fears of darkness? (*Hint*: The contingency table must account for all 1000 people.)

Age Group	Elementary	Jr. high	Sr. high	College	Adult
No. Who Fear	83	72	49	36	114

(a) Solve using the classical approach.

(b) Solve using the prob-value approach.

In Retrospect

In this chapter we have been concerned with tests of hypotheses using chi-square, with the cell probabilities associated with the multinomial experiment, and with the simple contingency table. In each case the basic assumptions are that a large number of observations have been made and that the resulting test statistic, $\sum [(O - E)^2/E]$, is approximately distributed as chi-square. In general, if n is large and the minimum allowable expected cell size is 5, this assumption is satisfied.

The news article at the beginning of this chapter shows a contingency table with the results listed as relative frequencies. This is a common practice when reporting the results of public opinion polls. However, to statistically interpret the results, the actual frequencies must be used. (See exercise 11-14.)

The contingency table can also be used to test homogeneity. The test for homogeneity and the test for independence look very similar and, in fact, are carried out in exactly the same way. The concepts being tested are quite different, however: equal distributions and independence. The two tests are easily distinguished from one another, for the test of homogeneity possesses predetermined marginal totals in one direction in the table. That is, before the data are collected, the experimenter determines how many subjects will be observed in each category. The only predetermined number in the test of independence is the grand total.

A few words of caution: The correct number of degrees of freedom is critical if the test results are to be meaningful. The degrees of freedom determine, in part, the critical region, and its size is important. Like other tests of hypothesis, failure to reject H_0 does not mean outright acceptance of the null hypothesis.

Chapter Exercises

11-13   The psychology department claims that the grades in its introductory course are distributed as follows: 10% A's, 20% B's, 40% C's, 20% D's, and 10% F's. In a poll of 200 randomly selected students who had previously completed this course, it was found that 16 had received A's, 43 B's, 65 C's, 48 D's, and 28 F's. Does this sample contradict the department's claim, at the 0.05 level?

(a) Solve using the classical approach.

(b) Solve using the prob-value approach.

11-14   (a) The news article at the beginning of this chapter reports the findings of a sample of 1000 people. Suppose that the percentages reported in this news article had come from the accompanying contingency table. Can the null hypothesis "the distribution of yes, no, and not sure opinions is the same for all four geographic areas of New York State" be true? Use $\alpha = 0.05$.

	Yes	No	Not Sure
Upstate	96	150	54
NY City	105	144	51
No. Suburbs	64	90	46
Western NY	56	112	32

(b) Suppose there had been only 400 people in the sample, 100 from each geographic location. Calculate the observed value of chi-square, χ^2*. Compare this value to that found in part (a). Explain the difference.

11-15   A hospital surveyed its employees to see if staff communications were satisfactory. In response to the statement, "Communication with other shifts is good," the different shifts responded as follows:

Shift Worked at the Hospital	Communication with Other Shifts is Good			
	Agree	No Opinion	Disagree	Total
7 A.M.–3 P.M.	36	8	6	50
3 P.M.–11 P.M.	32	12	6	50
11 P.M.–7 A.M.	18	12	20	50
Total	86	32	32	150

Test the homogeneity of the different shifts in regard to the quality of communication. Is there sufficient evidence to conclude that there are different opinions among the different shifts? Use $\alpha = 0.05$.

(a) Solve using the classical approach.

(b) Solve using the prob-value approach.

11-16   Four hundred people responded to a survey in which the respondent could choose any one of four alternative solutions to a problem. One hundred and twenty chose solution A, 80 chose B, 90 chose C, and 110 chose D. Do the results show that the opinions on the best solution to the problem are not equally divided between the alternatives given? Use $\alpha = 0.01$.

(a) Solve using the classical approach.

(b) Solve using the prob-value approach.

11-17 A survey of employees at an insurance firm was concerned with worker and supervisor relationships. One statement for evaluation was "I am not sure what my supervisor expects." The results of the survey are found in the following contingency table.

Years of Employment	I Am Not Sure What My Supervisor Expects		
	True	Not True	Totals
Less than 1 year	18	13	31
1–3 years	20	8	28
3–10 years	28	9	37
10 years or more	26	8	34
Total	92	38	130

Can we reject the null hypothesis that the responses to the statement and years of employment are independent at the 0.10 level of significance?

(a) Solve using the classical approach.

(b) Solve using the prob-value approach.

11-18 A random sample of 500 married men from all over the country was taken and each person was cross-classified as to the size of community that he was presently residing in and the size of community that he was reared in. The results are shown in the accompanying table. Does this sample contradict the claim of independence, at the 0.01 level of significance?

(a) Solve using the classical approach.

(b) Solve using the prob-value approach.

Size of Community Reared In	Size of Community Residing In			
	Under 10,000	10,000 to 49,999	50,000 or Over	Total
Under 10,000	24	45	45	114
10,000 to 49,999	18	64	70	152
50,000 or over	21	54	159	234
Total	63	163	274	500

11–19 Three groups of women were asked their preference in face powder from among three brands, which were called *X*, *Y*, and *Z*. The results were:

Groups of Women	Number Preferring Each Brand Brand X	Brand Y	Brand Z	Total
Inner City	45	40	35	120
Suburban	60	30	30	120
Rural	45	10	5	60
Total	150	80	70	300

Is there sufficient reason to reject the null hypothesis: "the proportion of women who prefer each face powder is the same in each population," at the 0.01 level of significance?

(a) Solve using the classical approach.

(b) Solve using the prob-value approach.

11–20 Four brands of popcorn were tested for popping. One hundred kernels of each brand were popped and the number of kernels not popped was recorded in each test; see the accompanying table. Can we reject the null hypothesis that all four brands pop equally? Test at $\alpha = 0.05$.

Brand	A	B	C	D
No. Not Popped	14	8	11	15

(a) Solve using the classical approach.

(b) Solve using the prob-value approach.

11–21 When interbreeding two strains of roses, we expect the hybrid to appear in three genetic classes in the ratio 1:3:4. If the results of an experiment yield 80 hybrids of the first type, 340 of the second type, and 380 of the third type; do we have sufficient evidence to reject the hypothesized genetic ratio at the 0.05 level of significance?

(a) Solve using the classical approach.

(b) Solve using the prob-value approach.

11–22 In order to determine what improvements are needed, the patrons of a shopping mall were asked a variety of questions in regard to the conditions at the mall. An appraisal of the aesthetics of the mall resulted in the following contingency table.

Sex of the	How Are the Aesthetics of the Mall?				
Respondent	Excellent	Fair	Good	Poor	Total
Female	6	16	21	7	50
Male	8	11	10	21	50
Totals	14	27	31	28	100

It had been decided in advance to structure the sample so that exactly 50 of each sex responded to the survey. Are the attitudes towards the aesthetics of the mall equally distributed for both sexes? Use $\alpha = 0.05$.

(a) Solve using the classical approach.

(b) Solve using the prob-value approach.

11-23 Last year's work record for the amounts of absenteeism in each of four categories for 100 randomly selected employees is compiled in the accompanying table. Do these data provide sufficient evidence to reject the hypothesis that the rate of absenteeism is the same for all categories of employees? Use $\alpha = 0.01$ and 240 work days for the year.

	Married Male	Single Male	Married Female	Single Female
Number of Employees	40	14	16	30
Days of Absence	180	110	75	135

(a) Solve using the classical approach.

(b) Solve using the prob-value approach.

11 24 The medics at a large ski center claim that the probability of a broken leg is the same for all skiers. A sampling of last month's records showed that 13 of 400 beginners sustained a broken leg, while 8 of 600 intermediates and 5 of 300 advanced skiers sustained a broken leg. At the 0.10 level, do these data show sufficient evidence to reject the medics' claim?

11-25 The sales records of two manufacturers of 40,000-mile-guarantee tires were sampled and they showed that 59 of 200 tires from one manufacturer failed within the 40,000 miles. The sample from the second manufacturer revealed 63 failures among 300 tires. Use $\alpha = 0.05$ and determine whether there is sufficient evidence to reject the hypothesis that there is no difference in the failure rates of the two brands.

11-26 The accompanying table shows the number of reported crimes committed last year in the inner part of a large city. The crimes were classified in two ways: type of crime and district of the inner city where it occurred. Do these data show sufficient evidence to reject the hypothesis that the type of crime and the district in which it occurred are independent? Use $\alpha = 0.01$.

District	Crime				
	Robbery	Assault	Burglary	Larceny	Stolen Vehicle
1	54	331	227	1090	41
2	42	274	220	488	71
3	50	306	206	422	83
4	48	184	148	480	42
5	31	102	94	596	56
6	10	53	92	236	45

Vocabulary List

Be able to define each term. In addition, describe, in your own words, and give an example of each term. Your examples should not be ones given in class or in the textbook.

The terms followed by bracketed numbers were defined in earlier chapters (the bracketed numbers are those of the earlier chapters), but we ask you to define them again to show your increased understanding of their meaning.

cell
chi-square [9]
column
contingency table
degrees of freedom [9, 10]
enumerative data
expected frequency
frequency distribution [2]

homogeneity
independence [4]
marginal totals
multinomial experiment
observed frequency [2, 4]
rows
statistic [1, 2, 8]

Quiz

Answer "True" if the statement is always true. If the statement is not always true, replace the words shown in color with words that make the statement always true.

11–1 The number of degrees of freedom for a test of a multinomial experiment is **equal to** the number of cells in the experimental data.

11–2 The **expected frequency** in a chi-square test is found by multiplying the hypothesized probability of a cell by the number of pieces of data in the sample.

11–3 The **observed** frequency of a cell is not allowed to be smaller than 5 when a chi-square test is being conducted.

11–4 In the **multinomial experiment** we have $(r - 1)$ times $(c - 1)$ degrees of freedom (r is the number of rows and c is the number of columns).

11–5 A multinomial experiment consists of n identical, **independent trials.**

11–6 A **multinomial experiment** arranges the data into a two-way classification such that the totals in one direction are predetermined.

11–7 The charts for both the multinomial experiment and the contingency table **must** be set up in such a way that each piece of data will fall into exactly one of the categories.

11–8 The test statistic $\sum [(O - E)^2/E]$ has a distribution that is **approximately normal.**

11–9 The data used in a chi-square multinomial test are always **enumerative** in nature.

11–10 The null hypothesis being tested by a test of **homogeneity** is that the distribution of proportions is the same for each of the subpopulations.

12 Analysis of Variance

Chapter
Outline

Tillage test plots
REVISITED

By **Larry Reichenberger**, Associate Machinery Editor

All tillage tools are not created equal although the weather can certainly make them look that way.

That's our analysis after putting six of your favorite primary tillage tools through the paces last season. As we told you in our 1978 Planting Issue, we conducted those tests because we felt conservation tillage had a place on the fine textured, poorly drained soils long sacred to the moldboard plow. Despite a photo finish among yields, our test plots reinforced that opinion. Here's a wrap-up of the things that shape that optimism.

We staked out 18 plots on a gently sloping field with a Chalmers silty clay loam soil type. The Chalmers is similar to the Drummer, Brookston and Webster soils common to the Corn Belt. With last fall's soils wet underfoot we used the six primary tillage tools . . . .

By the second day of May we were ready for the planter and the temperature and growth measurements that would tell us if our theories about heavy-duty conservation tillage held water.

How good is a tie. The yields from individual tillage treatments showed less variability than we had hoped, probably due to the late planting. That's why university researchers conduct these studies over many years. The only yield differences worth noting are the 7.4 bu/A spreads between the standard chisel and light disk at 115.2 bu. and the moldboard plow at 122.6 bu. The coulter-chisel and heavy disk, and surprisingly enough the V-chisel, yielded virtually the same as the moldboard plow. However, we think there are other factors that add even more merit to heavy-duty conservation tillage.

TILLAGE PLOT YIELDS—1978
Bu/A @ 15.5% Moisture

Tillage	Rep I	Rep II	Rep III	Tillage Average
Plow	118.3	125.6	123.8	122.6
V chisel	115.8	122.5	118.9	119.1
Coulter chisel	124.1	118.5	113.3	118.6
Std. chisel	109.2	114.0	122.5	115.2
Hvy. disk	118.1	117.5	121.4	119.0
Lt. disk	118.3	113.7	113.7	115.2

Chapter Objectives

Previously we have tested hypotheses about two means. In this chapter we are concerned with testing a hypothesis about several means. The **analysis of variance technique (ANOVA),** *which we are about to explore, will be used to test a hypothesis about several means: for example,*

$$H_0: \mu_1 = \mu_2 = \mu_3 = \mu_4 = \mu_5$$

By using our former technique for hypotheses about two means, we could test several hypotheses if each stated a comparison of two means. For example, we could test

$$H_1: \mu_1 = \mu_2 \qquad H_2: \mu_1 = \mu_3$$
$$H_3: \mu_1 = \mu_4 \qquad H_4: \mu_1 = \mu_5$$
$$H_5: \mu_2 = \mu_3 \qquad H_6: \mu_2 = \mu_4$$
$$H_7: \mu_2 = \mu_5 \qquad H_8: \mu_3 = \mu_4$$
$$H_9: \mu_3 = \mu_5 \qquad H_{10}: \mu_4 = \mu_5$$

In order to test the null hypothesis, H_0, that all five means are equal, we would have to test each of these ten hypotheses using our former technique for two means. Rejection of any one of the ten hypotheses about two means would cause us to reject the null hypotheses that all five means are equal. If we failed to reject all ten hypotheses about the means, we would fail to reject the main null hypothesis. Suppose we tested a null hypothesis that dealt with several means by testing all the possible pairs of two means; the overall type I error rate would become much larger than the value of α associated with a single test. The ANOVA techniques allow us to test the null hypothesis (all means are equal) against the alternative hypothesis (at least one mean value is different) with a specified value of α.

In this chapter we will introduce ANOVA. ANOVA experiments can be very complex, depending on the situation. We will restrict our discussion to the most basic experimental design, the single-factor ANOVA.

Section 12–1 Introduction to the Analysis of Variance Technique

We will begin our discussion of the analysis of variance technique by looking at an illustration.

Illustration 12–1

The temperature at which a plant is maintained is believed to affect the rate of production in the plant. The data in table 12–1 are the number, x, of units produced in one hour for randomly selected one-hour periods when the production process in the plant was operating at each of three temperature levels. The data values from repeated samplings are called replicates. Four replicates, or data values, were obtained for two of the temperatures and five were obtained for the third temperature. Does this data suggest that temperature has a significant effect on the production level at the 0.05 level?

TABLE 12–1
Sample Results for Illustration 12–1

	Temperature Levels		
	Sample from 68°F	Sample from 72°F	Sample from 76°F
	10	7	3
	12	6	3
	10	7	5
	9	8	4
		7	
Column Totals	$C_1 = 41$ $\bar{x}_1 = 10.25$	$C_2 = 35$ $\bar{x}_2 = 7.0$	$C_3 = 15$ $\bar{x}_3 = 3.75$

The level of production is measured by the mean value; $\bar{x}_i$ indicates the observed production mean at level i, where $i = 1, 2,$ and 3 corresponds to temperatures, 68°F, 72°F, and 76°F, respectively. There is a certain amount of variation among these means. Since sample means do not necessarily repeat when repeated samples are taken from a population, some variation can be expected. We will next pursue the question "Is this variation among the $\bar{x}$'s due to chance or is it due to the effect that temperature has on the production rate?"

Solution The null hypothesis that we will test is

$$H_0: \mu_{68} = \mu_{72} = \mu_{76}$$

That is, the true production mean is the same at each temperature level tested. In other words, the temperature does not have a significant effect on the production rate. The alternative to the null hypothesis is:

$$H_a: \text{not all temperature level means are equal}$$

Thus we will want to reject the null hypothesis if the data show that one or more of the means are significantly different from the others.

We will make the decision to reject H_0 or fail to reject H_0 by using the F distribution and the F test statistic. Recall from chapter 10 that the calculated value of F is the ratio of two variances. The analysis of variance procedure will separate the variation among the entire set of data into two categories.

To accomplish this separation first we work with the numerator of the fraction used to define sample variance, formula (2–7):

$$s^2 = \frac{\sum (x - \bar{x})^2}{n - 1}$$

sum of squares   The numerator of the above fraction is called the **sum of squares**:

$$\text{sum of squares} = \sum (x - \bar{x})^2 \qquad (12\text{–}1)$$

We calculate the **total sum of squares**, **SS(total)**, for the total set of data by using a formula that is equivalent to formula (12–1) but does not require the use of $\bar{x}$. This equivalent formula is

$$\text{SS(total)} = \sum (x^2) - \frac{(\sum x)^2}{n} \qquad (12\text{–}2)$$

Now we can find the SS(total) for our illustration by using formula (12–2):

$$\sum (x^2) = 10^2 + 12^2 + 10^2 + 9^2 + 7^2 + 6^2 + 7^2$$
$$+ 8^2 + 7^2 + 3^2 + 3^2 + 5^2 + 4^2$$
$$= 731$$

$$\sum x = 10 + 12 + 10 + 9 + 7 + 6 + 7$$
$$+ 8 + 7 + 3 + 3 + 5 + 4$$
$$= 91$$

$$\text{SS(total)} = 731 - \frac{(91)^2}{13}$$

$$= 731 - 637 = \mathbf{94}$$

Next, 94, the SS(total) must be separated into two parts: the sum of squares, SS(temperature), due to temperature levels and the sum of squares, SS(error), due to experimental error of replication. This splitting is often called

partitioning   **partitioning**, since SS(temperature) + SS(error) = SS(total); that is, in our illustration SS(temperature) + SS(error) = 94.

The sum of squares **SS(factor)**, [SS(temperature) for our illustration]

variation between   that measures the **variation between the factor levels** (temperatures) is found
the factor levels   by using formula (12–3):

$$\text{SS(factor)} = \left(\frac{C_1^2}{k_1} + \frac{C_2^2}{k_2} + \frac{C_3^2}{k_3} + \cdots \right) - \frac{(\sum x)^2}{n} \qquad (12\text{–}3)$$

in which C_i represents the column total, k_i represents the number of replicates at each level of the factor, and n represents the total sample size ($n = \sum k_i$).

NOTE: The data has been arranged so that each column represents a different level of the factor being tested.

Now we can find the SS(temperature) for our illustration by using formula (12–3):

$$SS(\text{temperature}) = \left(\frac{41^2}{4} + \frac{35^2}{5} + \frac{15^2}{4}\right) - \frac{(91)^2}{13}$$

$$= (420.25 + 245.00 + 56.25) - 637.0$$

$$= 721.5 - 637.0 = \mathbf{84.5}$$

variation within rows The sum of squares **SS(error)** that measures the **variation within the rows** is found by using formula (12–4):

$$SS(\text{error}) = \sum (x^2) - \left(\frac{C_1^2}{k_1} + \frac{C_2^2}{k_2} + \frac{C_3^2}{k_3} + \cdots\right) \qquad (12\text{--}4)$$

The SS(error) for our illustration can now be found by using formula (12–4):

$$\sum (x^2) = 731 \quad \text{(found previously)}$$

$$\left(\frac{C_1^2}{k_1} + \frac{C_2^2}{k_2} + \frac{C_3^2}{k_3}\right) = 721.5 \quad \text{(found previously)}$$

$$SS(\text{error}) = 731.0 - 721.5 = \mathbf{9.5}$$

NOTE: SS(total) = SS(factor) + SS(error). Inspection of formulas (12–2), (12–3), and (12–4) will verify this.

For convenience we will use an **ANOVA table** to record the sums of squares and to organize the rest of the calculations. The format of an ANOVA table is shown in table 12–2.

TABLE 12–2
Format for ANOVA Table

Source	SS	df	MS
Factor			
Error			
Total			

We have calculated the three sums of squares for our illustration. The degrees of freedom, df, associated with each of the three sources are determined as follows:

1. df(factor) is one less than the number of levels (columns) for which the factor is tested:

$$df(factor) = c - 1 \qquad (12\text{--}5)$$

where c represents the number of levels for which the factor is being tested (number of columns on the data table).

2. df(total) is one less than the total number of data:

$$df(total) = n - 1 \qquad (12\text{--}6)$$

where n represents the number of data in the total sample (that is, $n = k_1 + k_2 + k_3 + \cdots$, where k_i is the number of replicates at each level tested).

3. df(error) is the sum of the degrees of freedom for all the levels tested (columns in the data table). Each column has $k_i - 1$ degrees of freedom; therefore,

$$df(error) = (k_1 - 1) + (k_2 - 1) + (k_3 - 1) + \cdots$$

or

$$df(error) = n - c \qquad (12\text{--}7)$$

The degrees of freedom for our illustration are

$$df(temperature) = c - 1 = \ 3 - 1 = \ \mathbf{2}$$
$$df(total) = n - 1 = 13 - 1 = \mathbf{12}$$
$$df(error) = n - c = 13 - 3 = \mathbf{10}$$

The sums of squares and the degrees of freedom must check. That is,

$$SS(factor) + SS(error) = SS(total) \qquad (12\text{--}8)$$
$$df(factor) + df(error) = \ df(total) \qquad (12\text{--}9)$$

mean square The **mean square** for the factor being tested, the **MS(factor)**, and for error, the **MS(error)**, will be obtained by dividing the sum-of-square value by the corresponding number of degrees of freedom. That is,

$$MS(factor) = \frac{SS(factor)}{df(factor)} \qquad (12\text{--}10)$$

$$MS(error) = \frac{SS(error)}{df(error)} \qquad (12\text{--}11)$$

The mean squares for our illustration are

$$MS(temperature) = \frac{SS(temperature)}{df(temperature)} = \frac{84.5}{2}$$

$$= 42.25$$

$$MS(error) = \frac{SS(error)}{df(error)} = \frac{9.5}{10}$$

$$= 0.95$$

The completed ANOVA table now appears as shown in table 12–3. The

TABLE 12–3
ANOVA Table for Illustration 12–1

Source	SS	df	MS
Temperature	84.5	2	42.25
Error	9.5	10	0.95
Total	94.0	12	

hypothesis test is now completed using the two mean squares as measures of variance. The calculated value of the **test statistic, F^*,** is found by dividing the MS(factor) by the MS(error):

$$F = \frac{MS(factor)}{MS(error)} \tag{12–12}$$

The calculated value of F for our illustration is found by using formula (12–12):

$$F = \frac{MS(temperature)}{MS(error)} = \frac{42.25}{0.95}$$

$$F^* = 44.47$$

$\alpha = 0.05$

$0 \qquad F(2, 10, 0.05) = 4.10 \qquad F$

The decision to reject H_0 or fail to reject H_0 will be made by comparing the calculated value of F, $F^* = 44.47$, to a one-tailed critical value of F obtained from table 8a of appendix D. See the figure.

NOTE: Since the calculated value of F, F^*, is found by dividing MS(temperature) by MS(error), the number of degrees of freedom for the numerator is df(temperature) $= 2$ and the number of degrees of freedom for the denominator is df(error) $= 10$.

We rejected H_0 since the value F^* fell in the critical region. Therefore we conclude that room temperature does have a significant effect on the production rate. The differences in the mean production rates at the tested temperature levels were found to be significant. □

In this section we saw how the ANOVA technique separated the variance among the sample data into two measures of variance: (1) MS(factor), the measure of variance between the levels tested, and (2) MS(error), the measure of variance within the levels being tested. Then these measures of variance can be compared. For our illustration the between-level variance was found to be significantly larger than the within-level variance (experimental error). This led us to the conclusion that temperature did have a significant effect on the variable x, the number of units of production completed per hour.

In the next section we will use several illustrations to demonstrate the logic of the analysis of variance technique.

Section 12–2 The Logic behind ANOVA

Many experiments are conducted to determine the effect that some test factor has on a response variable. The test factor may be temperature (as in illustration 12–1), the manufacturer of a product, the day of the week, or any number of other things. In this chapter we are investigating the **single-factor analysis of variance**. Basically, the design for the single-factor ANOVA is to obtain independent random samples at each of the several levels of the factor being tested. We will then make a statistical decision concerning the effect that the levels of the test factor have on the response (observed) variable.

Illustrations 12–2 and 12–3 demonstrate the logic of the analysis of variance technique. Briefly, the reasoning behind the technique proceeds like this: In order to compare the means of the levels of the test factor, a measure of the **variation between the levels** (between columns on the data table), the **MS(factor)**, will be compared to a measure of the **variation within the levels** (within the columns on the data table), the **MS(error)**. If the MS(factor) is significantly larger than the MS(error), then we will conclude that the means for each of the factor levels being tested are not all the same. This implies that the factor being tested does have a significant effect on the response variable. If, however, the MS(factor) is not significantly larger than the MS(error), we will not be able to reject the null hypothesis that all means are equal.

variation between levels

variation within levels

Illustration 12–2

Do the data in table 12–4 show sufficient evidence to conclude that there is a difference in the three population means, μ_F, μ_G, and μ_H?

TABLE 12–4
Sample Results for Illustration 12–2

	Factor Levels		
	Sample from Level F	Sample from Level G	Sample from Level H
	3	5	8
	2	6	7
	3	5	7
	4	5	8
Column Totals	$C_F = 12$	$C_G = 21$	$C_H = 30$
	$\bar{x} = 3.00$	$\bar{x} = 5.25$	$\bar{x} = 7.50$

The accompanying diagram, figure 12–1, shows the relative relationship among the three samples. A quick look at the diagram suggests that the three sample means are different from each other, implying that the sampled populations have different mean values.

FIGURE 12–1

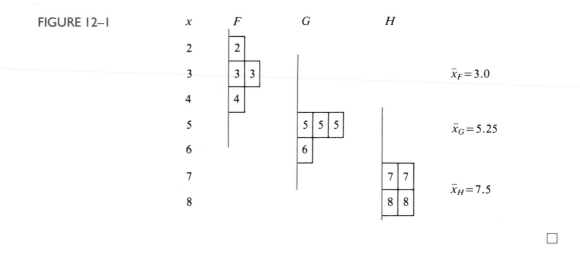

These three samples demonstrate relatively little **within-sample variation**, although there is a relatively large amount of **between-sample variation** Let's look at another illustration.

Illustration 12–3

Do the data in table 12–5 show sufficient evidence to conclude that there is a difference in the three population means, μ_J, μ_K, and μ_L?

TABLE 12–5
Sample Results for Illustration 12–3

	Factor Levels		
	Sample from Level J	Sample from Level K	Sample from Level L
	3	5	6
	8	4	2
	6	3	7
	4	7	5
Column Totals	$C_J = 21$ $\bar{x}_J = 5.25$	$C_K = 19$ $\bar{x}_K = 4.75$	$C_L = 20$ $\bar{x}_L = 5.00$

The accompanying diagram, figure 12–2, shows the relative relationship among the three samples. A quick look at the diagram **does not suggest** that the three sample means are different from each other.

FIGURE 12–2

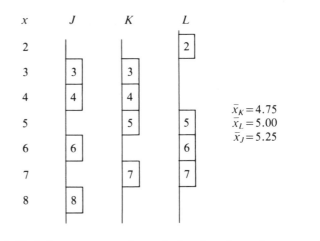

$\bar{x}_K = 4.75$
$\bar{x}_L = 5.00$
$\bar{x}_J = 5.25$

There is little between-sample variation for these three samples (that is, the sample means are relatively close in value), whereas the within-sample variation is relatively large (that is, the data values within each sample cover a relatively wide range of values).

To complete a hypothesis test for analysis of variance, we must agree on some ground rules or assumptions. In this chapter we will use the following three basic assumptions:

1. Our goal is to investigate the effect that various levels of the factor under test have on the response variable. Typically we want to find the level that yields the most advantageous values of the response variable. This, of course, means that we will probably want to reject the null hypothesis in favor of the alternative. Then a follow-up study could determine the "best" level of the factor.

2. We must assume that the effects due to chance and due to untested factors are normally distributed and that the variance caused by these effects is constant throughout the experiment.

3. We must assume independence among all observations of the experiment. (Recall that independence means that the results of one observation of the experiment do not affect the results of any other observation.) We will usually conduct the tests in a randomly assigned order to ensure independence. This technique also helps to avoid data contamination.

Section 12–3 Applications of Single-Factor ANOVA

Before continuing our ANOVA discussion, let's identify the notation, particularly the subscripts that are used (see table 12–6). Notice that each piece of data is double subscripted; the first subscript indicates the column number (test factor level) and the second subscript identifies the replicate (row) number. The column totals, C_i, are listed across the bottom of the table. The grand total, T, is equal to the sum of all x and is found by adding the column totals. Row totals can be used as a cross check, but serve no other purpose.

TABLE 12–6
Notation Used in ANOVA

	Factor Levels				
Replication	Sample from Level 1	Sample from Level 2	Sample from Level 3	...	Sample from Level C
$k = 1$	$x_{1,1}$	$x_{2,1}$	$x_{3,1}$		$x_{c,1}$
$k = 2$	$x_{1,2}$	$x_{2,2}$	$x_{3,2}$		$x_{c,2}$
$k = 3$	$x_{1,3}$	$x_{2,3}$	$x_{3,3}$		$x_{c,3}$
$\vdots$					
Column Totals	C_1	C_2	C_3	...	C_c | T

$$T = \text{grand total} = \text{sum of all } x\text{'s} = \sum x = \sum C_i$$

A mathematical model (equation) is often used to express a particular situation. In chapter 3 we used a mathematical model to help explain the relationship between the values of bivariate data. The equation $\hat{y} = b_0 + b_1 x$ served as the model when we believed that a straight line relationship existed. The probability functions studied in chapter 5 are also examples of **mathematical model** mathematical models. For the single-factor ANOVA, the **mathematical** model, formula (12–13), is an expression of the composition of each piece of data entered in our data table.

$$x_{c,k} = \mu + F_c + \varepsilon_{k(c)} \tag{12–13}$$

We interpret each term of this model as follows:

1. μ is the mean value for all the data without respect to the test factor.
2. F_c is the effect that the factor being tested has on the response variable at each different level c.

experimental error

3. $\varepsilon_{k(c)}$ (ε is the greek lower-case letter epsilon) is the **experimental error** that occurs among the k replicates in each of the c columns.

Let's look at another hypothesis test concerning an analysis of variance.

Illustration 12–4

A rifle club performed an experiment on a randomly selected group of beginning shooters. The purpose of the experiment was to determine whether shooting accuracy is affected by the method of sighting used: only the right eye open, only the left eye open, or both eyes open. Fifteen beginning shooters were selected and split into three groups. Each group experienced the same training and practicing procedures with one exception: the method of sighting used. After completing training, each student was given the same number of rounds and asked to shoot at a target. Their scores appear in table 12–7.

TABLE 12–7
Sample Results for Illustration 12–4

Method of Sighting		
Right Eye	Left Eye	Both Eyes
12	10	16
10	17	14
18	16	16
12	13	11
14		20
		21

At the 0.05 level of significance, is there sufficient evidence to reject the claim that these methods of sighting are equally effective?

Solution   In this experiment the factor is "method of sighting" and the levels are the three different methods of sighting (right eye, left eye, and both eyes open). The replicates are the scores received by each of the different students using each of the different sighting methods. The null hypothesis to

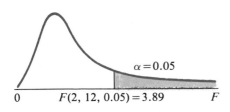

$$\alpha = 0.05$$

$$0 \qquad F(2, 12, 0.05) = 3.89 \qquad\qquad F$$

be tested is "the three methods of sighting are equally effective"; or the mean scores attained using each of the three methods are the same.

STEP 1 $H_0: \mu_R = \mu_L = \mu_B$

STEP 2 H_a: The means are not all equal (that is, at least one mean is different).

STEP 3 $\alpha = 0.05$. The test criteria are shown in the diagram.

$$df(\text{numerator}) = df(\text{method}) = 3 - 1 = 2 \quad [\text{using formula (12–5)}]$$

$$df(\text{denominator}) = df(\text{error}) = 15 - 3 = 12 \quad [\text{using formula (12–7)}]$$

STEP 4   The calculation of the test statistic F^*. Table 12–8 is used to find column totals.

TABLE 12–8
Sample Results for Illustration 12–4

| | Factor Levels: Method of Sighting | | |
Replicates	Right Eye	Left Eye	Both Eyes
$k = 1$	12	10	16
$k = 2$	10	17	14
$k = 3$	18	16	16
$k = 4$	12	13	11
$k = 5$	14		20
$k = 6$			21
Totals	$C_R = 66$	$C_L = 56$	$C_B = 98$

First the summations $\sum x$ and $\sum x^2$ need to be calculated:

$$\sum x = 12 + 10 + 18 + 12 + 14 + 10 + 17 + \cdots + 21$$
$$= 220$$

$$\sum x^2 = 12^2 + 10^2 + 18^2 + 12^2 + 14^2 + 10^2 + \cdots + 21^2$$
$$= 3392$$

Using formula (12–2), we find

$$SS(\text{total} = 3392 - \frac{(220)^2}{15}$$
$$= 3392 - 3226.67$$
$$= 165.33$$

Using formula (12–3), we find

$$SS(\text{method}) = \left(\frac{66^2}{5} + \frac{56^2}{4} + \frac{98^2}{6}\right) - 3226.67$$

$$= 3255.87 - 3226.67$$

$$= \mathbf{29.20}$$

Using formula (12–4), we find

$$SS(\text{error}) = 3392 - 3255.87 = \mathbf{136.13}$$

Use formula (12–8) to check the sum of squares.

$$SS(\text{method}) + SS(\text{error}) = SS(\text{total})$$

$$29.20 + 136.13 = 165.33$$

The number of degrees of freedom are found using formulas (12–5), (12–6), and (12–7):

$$df(\text{method}) = 3 - 1 = \mathbf{2}$$

$$df(\text{total}) = 15 - 1 = \mathbf{14}$$

$$df(\text{error}) = 15 - 3 = \mathbf{12}$$

Using formulas (12–10) and (12–11) we find,

$$MS(\text{method}) = \frac{29.20}{2} = \mathbf{14.60}$$

$$MS(\text{error}) = \frac{136.13}{12} = \mathbf{11.34}$$

The results of these computations are combined in the ANOVA table shown in table 12–9.

TABLE 12–9

ANOVA Table for Illustration 12–4

Source	SS	df	MS
Method	29.20	2	14.60
Error	136.13	12	11.34
Total	165.33	14	

The calculated value of the test statistic is then found using formula (12–12).

$$F^* = \frac{14.60}{11.34} = \mathbf{1.287}$$

STEP 5   The decision to fail to reject or reject the null hypothesis is made by comparing F^* to the critical value shown in the test criteria (step 3 above).

Decision: Fail to reject H_0.

Conclusion: The data show no evidence that would give reason to reject the null hypothesis that the three methods are equally effective. □

Computer Solution

Minitab Printout for Illustration 12–4

Information given to computer

```
--READ DATA INTO C1, LEVELS INTO C2
    COLUMN  C1            C2
    COUNT   15            15
ROW
    1       12.           1.
    2       10.           1.
    3       18.           1.
    4       12.           1.
    5       14.           1.
    6       10.           2.
    7       17.           2.
    8       16.           2.
    9       13.           2.
   10       16.           3.
   11       14.           3.
   12       16.           3.
   13       11.           3.
   14       20.           3.
   15       21.           3.
```

`--ONEWAY ANALYSIS OF VARIANCE, DATA IN C1, LEVELS IN C2`

ANALYSIS OF VARIANCE

The ANOVA table, compare to table 12–9

SOURCE	DF	SS	MS = SS/DF	F–RATIO
FACTOR	2	29.2	14.6	1.29
ERROR	12	136.1	11.3	
TOTAL	14	165.3		

The calculated value of F, $F*$

STATISTICS

LEVEL	N	MEAN	ST. DEV.
1	5	13.20	3.03
2	4	14.00	3.16
3	6	16.33	3.72

(continued)

```
POOLED ST. DEV. =  3.37
INDIVIDUAL 95 PERCENT C. I. FOR LEVEL MEANS
     (BASED ON POOLED STANDARD DEVIATION)
      +------+------+------+------+------+------+
   1  |**********|**********|
   2  |***********|**********|
   3       |*********|**********|
      +------+------+------+------+------+------+
     10.0          14.0         18.0          22.0
          12.0          16.0         20.0
--STOP
```

Recall the null hypothesis, that there is no difference between the levels of the factor being tested. A "fail to reject H_0" decision must be interpreted as the conclusion that there is no evidence of a difference due to the levels of the tested factor, whereas the rejection of H_0 implies that there is a difference between the levels. If there is a difference, the next problem is to locate the level or levels that are different. To locate this difference may be the main object of the analysis. In order to find the difference, the only method that is appropriate at this stage of our development is the inspect the data. It may be obvious which level(s) caused the rejection of H_0. In illustration 12–1 it seems quite obvious that each level is different from the other two. If the higher values are more desirable for finding the "best" level to use, we would choose that corresponding level of the factor.

To this point we have discussed analysis of variance for data dealing with one factor. It is not unusual for problems to have several factors of concern. The ANOVA techniques presented in this chapter can be developed further and applied to more complex cases.

Exercises

12–1   Suppose that an F test (as described in this chapter) has a critical value of 2.2, as shown in the diagram.

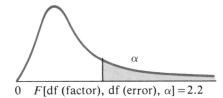

$$0 \quad F[\text{df (factor), df (error)}, \alpha] = 2.2$$

(a) What is the interpretation of a calculated value of F larger than 2.2?

(b) What is the interpretation of a calculated value of F smaller than 2.2?

(c) What is the interpretation if the calculated F were 0.1? 0.01?

12-2 Why does df(factor) the number of degrees of freedom always appear first in the critical value notation F [df(factor), df(error), α]?

12-3 Consider the accompanying table for a single-factor ANOVA. Find the following:

(a) $x_{1,2}$ (b) $x_{2,4}$ (c) C_1

(d) $\sum x$ (e) $\sum (C_i)^2$

Replicates	Level of Factor		
	1	2	3
1	3	2	7
2	0	5	4
3	1	4	5
4	5	3	6
5	7	6	2

12-4 (a) State the null and alternative hypotheses that would be under test in exercise 12-3.

(b) How would a "reject the null hypothesis" decision be interpreted?

(c) How would a "fail to reject the null hypothesis" decision be interpreted?

(d) How is such a decision reached?

12-5 Recently a new operator was assigned to a crew of workers who perform a certain job. From the records that show the number of units of work completed by each worker each day last month, a sample of size five was randomly selected for each of the two experienced workers and the new worker. At the 0.05 level of significance, does the evidence provide sufficient reason to reject the claim that there is no difference in the amount of work done by the three workers?

Units of Work (Replicates)	Workers		
	New	A	B
	8	11	10
	10	12	13
	9	10	9
	11	12	12
	8	13	13

12-6 A cookie salesman interested in increasing the sales volume of cookies in a large supermarket arranged with the manager to display his product in three different locations in the market over a 15-week period. Weekly he moved the display between the three locations. The following sales figures resulted.

Amount of Cookie Sales per Week

| | Location of Display | |
By Meat Counter	By Checkout	In Cookie Section
32	45	30
44	63	28
37	44	16
39	59	21
23	34	22

At the 0.05 level of significance, do the sample data provide sufficient reason to reject the null hypothesis that all locations are equally desirable for selling cookies?

12-7 A certain vending company's soft drink-dispensing machines are supposed to serve six ounces of beverage. Various machines were sampled and the resulting amounts of dispensed drink recorded. Does this sample evidence provide sufficient reason to reject the null hypothesis that all five machines dispense the same average amount of soft drink? Use $\alpha = 0.01$.

Amounts of Soft Drink Dispensed

| | | Machines | | |
A	B	C	D	E
3.8	6.8	4.4	6.5	6.2
4.2	7.1	4.1	6.4	4.5
4.1	6.7	3.9	6.2	5.3
4.4		4.5		5.8

In Retrospect

In this chapter we have presented an introduction to the statistical techniques known as analysis of variance. The techniques studied here were restricted to the test of a hypothesis that dealt with questions about multiple means from one given population. We were restricted to normal populations and populations with homogeneous (equal) variances. The test of multiple means is accomplished by partitioning the sum of squares into two assignable segments: (1) the sum of squares due to variation between the levels of the factor being tested, and (2) the sum of squares due to the variation between the replicates within each level. The null hypothesis about means is then tested by using the appropriate variance measurements.

Note that we restricted our development to one-factor experiments. This one-factor technique represents only a beginning to the study of analysis of variance techniques.

Refer back to the news article at the beginning of the chapter and you will see that the data given use the method of tillage (plow, etc.) as the factor and the field plots as the replicates of a one-factor analysis of variance experiment. (See exercise 12–11.)

Chapter Exercises

12–8 An experiment compared typing speeds for clerks-typists on a standard electric typewriter and on a Teletype Model 43 computer terminal. Twelve typists were randomly assigned to type on the two instruments. The scores follow:

Standard Electric	Terminal
62	52
78	60
48	47
63	48
55	52
51	40

Is there sufficient evidence to conclude that there is a difference in the population means for the two types of machines? Use $\alpha = 0.05$.

12–9 An employment agency is interested to see which of three types of ads in the help-wanted section of local newspapers is the most effective. Three types of ads (big headline, straightforward and bold print) were randomly alternated over a period of weeks and the number of people responding to the ads were noted each week. Do these data support the null hypothesis that there is no difference in the effectiveness of the ads, as measured by the mean number responding, at the 0.01 level of significance?

Number of Responses (Replicates)

Type of Advertisement

Big Headline	Straightforward	Bold Print
23	19	28
42	31	33
36	18	46
48	24	29
33	26	34
26		34

12–10 Test this hypothesis: "The day of the week does have an effect upon the amount of carbon monoxide pollution in the air in our city." See the accompanying table. At the 0.05 level, is the evidence presented by the data sufficient to reject the null hypothesis that the mean amount of carbon monoxide pollution is the same on all days?

Amount of Pollution (Replicates)	Sun	Mon	Tues	Wed	Thurs	Fri	Sat
	0.8	0.3	1.8	1.7	1.6	0.8	0.3
	0.4	1.2	1.5	1.1	0.8	1.7	1.7
	0.2	1.0	1.1	1.2	1.4	1.1	0.5
	0.7	1.4	0.9	1.5	1.2	1.3	0.9

12-11 Do the sample results given in the news article at the beginning of this chapter show sufficient evidence to support the claim that the tillage tool used to prepare the ground for planting does affect the mean yield? Complete the hypothesis test at the 0.05 level of significance.

12-12 Dividend increases during the past year for three stock groups are presented below.

Stock Groups

Natural Gas Utility	Electric Utility	Gas and Electric Utility
0.00	0.08	0.16
0.28	0.14	0.20
0.16	0.06	0.14
0.06	0.10	0.10
0.08	0.04	0.12

Do the data provide sufficient evidence to conclude that there is a difference in mean dividend increases for the three groups of stocks at the 0.01 level of significance?

12-13 The accompanying table displays the number of arrests made last year for violations of the narcotic drug laws in 24 communities. Entries are rates of arrest per 10,000 inhabitants. At $\alpha = 0.05$, is there sufficient evidence to reject the hypothesis that the mean rates of arrests are the same in all four sizes of communities?

Cities (over 250,000)	Cities (under 250,000)	Suburban Community	Rural Community
45	23	25	8
34	18	17	16
41	27	19	14
42	21	28	17
37	26	31	10
28	34	37	23

12-14 On thirty-two different days, during rush hours on a heavily traveled freeway, cars that needed assistance as a result of an accident or engine failure were counted. The accompanying table shows the number of cars that needed assistance during each observation period. At the 0.01 level of significance, do the data contradict the hypothesis that the mean number of cars that needed assistance is the same for all four categories?

Number of Cars Needing Assistance

Northbound, A.M.	Southbound, A.M.	Northbound, P.M.	Southbound, P.M.
12	9	11	9
15	7	13	8
11	10	9	12
12	16	9	10
10	8	12	13
13	11	11	9
10	9	17	11
8	8	8	7

12-15 The distance required to stop a vehicle on wet pavement was measured to compare the stopping power of four major brands of tires. A tire of each brand was tested on the same vehicle on a controlled wet pavement. The resulting distances are shown in the accompanying table. At $\alpha = 0.05$, is there sufficient evidence to conclude that there is a difference in the mean stopping distance?

Distance (Replicate)

Brand of Tire

A	B	C	D
37	37	33	41
34	40	34	41
38	37	38	40
36	42	35	39
40	38	42	41
32		34	43

12-16 To compare the effectiveness of three distinct methods of teaching reading, 26 children of equal reading aptitude were divided into three groups. Each group was instructed for a given period of time using one of the three methods. After completing the instruction period, all students were tested. The test results are shown in the accompanying table. Is the evidence sufficient to reject the hypothesis that all three instruction methods are equally effective? Use $\alpha = 0.05$.

Test Scores (Replicates)	Method I	Method II	Method III
	45	45	44
	51	44	50
	48	46	45
	50	44	55
	46	41	51
	48	43	51
	45	46	45
	48	49	47
	47	44	

12–17 The amount of taxes collected per person were analyzed on a regional basis for the state senate committee on financial affairs. At the 0.05 level, should one conclude that the means for the regions in the study differ significantly?

Amounts Collected by a State in the Region	South	West	Midwest	Northeast
	815	1004	1096	1370
	623	1058	1021	1176
	751	1126	998	975
	636	900	894	921
	584	805	768	1013
		767		798

12–18 Seven golf balls from each of six manufacturers were randomly selected and tested for durability. Each ball was hit 300 times or until failure occurred, whichever came first. Do these sample data show sufficient reason to reject the null hypothesis that the six different brands tested withstood the durability test equally well? Use $\alpha = 0.05$.

Number of Hits (Replicate)	Manufacturer Brand					
	A	B	C	D	E	F
	300	190	228	276	162	264
	300	164	300	296	175	168
	300	238	268	62	157	254
	260	200	280	300	262	216
	300	221	300	230	200	257
	261	132	300	175	256	183
	300	156	300	211	92	93

Be able to define each term. In addition, describe in your own words and give an example of each term. Your examples should not be ones given in class or in the textbook.

The terms followed by bracketed numbers were defined in earlier chapters (the bracketed numbers are those of the earlier chapters) but we ask you to define them again to show your increased understanding of their meaning.

analysis of variance	randomize [2]
ANOVA	replicate
degrees of freedom [9, 10, 11]	response variable [1]
experimental error	sum of squares
level	variance [2, 9, 10]
mathematical model	variation between levels
mean square	variation within a level
partition	variation within rows

Quiz

Answer "True" if the statement is always true. If the statement is not always true, replace the words shown in color with words that make the statement always true.

12-1 To partition the sum of squares for the total is to separate the numerical value of SS(total) into two values such that the **sum** of these two values is equal to SS(total).

12-2 A **sum of squares** is actually a measure of variance.

12-3 **Experimental error** is the name given to the variability that takes place between the levels of the test factor.

12-4 **Experimental error** is the name given to the variability that takes place among the replicates of an experiment as it is repeated under constant conditions.

12-5 **Fail to reject H_0** is the desired decision when the means for the levels of the factor being tested are all different.

12-6 The **mathematical model** for a particular problem is an equational statement showing the anticipated makeup of an individual piece of data.

12-7 The degrees of freedom for the factor are equal to **the number** of factors tested.

12-8 The measure of a specific level of a factor being tested in an ANOVA is the **variance** of that factor level.

12-9 We **need not** assume that the observations are independent to do analysis of variance.

12-10 The rejection of H_0 **indicates** that you have identified the level(s) of the factor that is (are) different from the others.

13 Linear Correlation and Regression Analysis

Nutritional Outcome of 207 Very Low-Birth-Weight Infants in an Intensive Care Unit[1]

Karyl A. Rickard, PH.D., R.D., Judith A. Ernst, R.D.,
Mary Sue Brady, R.D., Lemuel Moye, M.D.,
Richard L. Schreiner, M.D., Edwin L. Gresham, M.D.,
and James A. Lemons, M.D.[2]
*James Whitcomb Riley Hospital for Children,
Indiana University School of Medicine, Indianapolis*

*Feeding and growth outcome for 207 critically ill, very
low-birth-weight infants were highly correlated with severity
of illness.*

During the past decade, intensive care centers for newborns
throughout the country have witnessed a dramatic increase
in the number of admissions and in the survival rate of
very low-birth-weight (VLBW) infants. The impact of this
change in the patient population in the intensive care unit
for newborns can be seen in Table 1. From 1974 to 1977,
the number of babies admitted to the James Whitcomb
Riley Hospital for Children who weighed from 501 to
1,000 gm. doubled, and the number who weighed from
1,001 to 1,500 gm. tripled, while total admissions
remained unchanged. Over the same period, survival of
the 1,001 to 1,500 gm. infants reached 80 percent, while
that of the 501 to 1,000 gm. babies approached 50 percent.

Provision of adequate nutrition to the VLBW infant
is a major challenge because of the extreme physiological
immaturity and severe medical complications often
encountered. A nutrition care protocol and arbitrary
guidelines for growth were developed (1) to provide a
more uniform approach to the nutrition support of these
extremely high-risk infants. Data for the present
investigation were collected in a prospective fashion over a
two-year period to: (a) evaluate the early feeding and
growth outcome of a large population of VLBW infants
managed with the same general nutrition guidelines and
(b) assess the impact of several factors, including sex,
appropriateness of size for gestational age, and degree of
illness, on feeding and growth outcome during the first
weeks of life.

Procedures and methods

Population The population consisted of 207 VLBW
infants less than 48 hours old who were sequentially
admitted to the tertiary (level III) intensive care unit for
newborns at the James Whitcomb Riley Hospital for
Children during 1978 and 1979. All infants were 35 weeks'
gestation or less, with birth weights of less than 1,500 gm.
Eighty-two percent of the infants were outborn, requiring
transport to the James Whitcomb Riley Hospital for
Children. Eighty-two percent of the babies required
ventilator support, reflecting the intensity of care required
by this group of infants.

Complete data were obtained prospectively for each
infant. The number of infants within subgroups of birth
weight, appropriateness of weight for gestational age, and
sex are presented in Tables 2 and 3. During the initial
hospitalization, 167 babies survived. The babies who
survived and their respective weights and gestational ages
are indicated in Figure 1.

Figure 1.

The weights and gestational ages of very-low-birth-weight infants who survived (•) or
died (○) during 1978 and 1979. The reference weight percentiles by gestational age are
those of Lubchenco *et al.* (14).

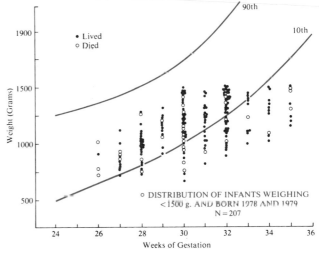

Chapter Objectives

In chapter 3 the basic ideas of regression and linear correlation analysis were introduced. (If these concepts are not fresh in your mind, review chapter 3 before beginning this chapter.) Chapter 3 was only a first look—a presentation of the basic graphic and descriptive statistical aspects of linear correlation and regression analysis. In this chapter we will take a second, more detailed look at linear correlation and regression analysis.

Previously we used the linear correlation coefficient to determine whether there was a linear relationship between two variables. Now we will determine if there is a linear relationship by use of a hypothesis test, where the probability of a type I error is fixed by the value assigned to α. In chapter 3 we introduced a set of formulas for finding the equation of the straight line of best fit. Now we wish to ask, Is the linear equation of any real use? Previously we used the equation of the line of best fit to make point predictions. Now we will make confidence interval estimations. In short, this second look at linear correlation and regression analysis is to be a much more complete presentation than the first.

*Recall that **bivariate data** are ordered pairs of numerical values. The two values are each response variables. They are paired with each other as a result of a common bond (see page 86).*

Section 13–1 Linear Correlation Analysis

In chapter 3 the linear correlation coefficient was presented as a quantity that measures the strength of a linear relationship (dependency). Now let's take a second look at this concept and see how r, the coefficient of linear correlation, works. Intuitively, we want to think about how to measure the mathematical linear dependency of one variable on another. As x increases, does y tend to increase (decrease)? How strong (consistent) is this tendency? We are going to use two measures of the dependence, covariance and the coefficient of linear correlation, to measure the relationship between two variables. We'll begin our discussion by examining a set of bivariate data and identifying some related facts as we prepare to define covariance.

Illustration 13–1

Let's consider the following sample of six pieces of bivariate data: (2, 1), (3, 5), (6, 3), (8, 2), (11, 6), (12, 1); see figure 13–1. The mean of the six x values (2, 3, 6, 8, 11, 12) is $\bar{x} = 7$. The mean of the six y values (1, 5, 3, 2, 6, 1) is $\bar{y} = 3$.

The point $(\bar{x}, \bar{y})$, which is (7, 3), is located as shown on the graph of the sample points in figure 13–2. The point $(\bar{x}, \bar{y})$ is called the **centroid** of the data. If a vertical and a horizontal line are drawn through the centroid, the graph is divided into four sections, as shown in figure 13–2. Each point (x, y) lies a certain distance from each of these two lines. $(x - \bar{x})$ is the horizontal distance from (x, y) to the vertical line passing through the centroid. $(y - \bar{y})$ is the vertical distance from (x, y) to the horizontal line passing through the centroid. Both the horizontal and the vertical distances of each data point from the centroid can be measured, as shown in figure 13–3. The

FIGURE 13–1

Graph of Data for Illustration 13–1

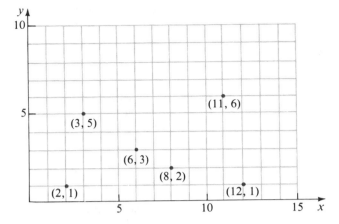

FIGURE 13–2

The Point (7, 3) Is the Centroid

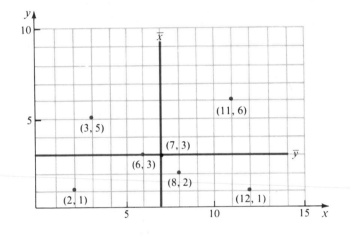

FIGURE 13–3

Measuring the Distance of Each Data
Point from the Centroid

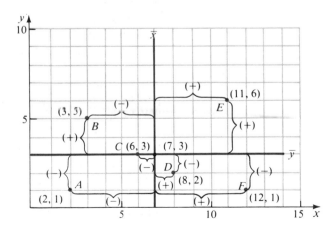

distances may be positive, negative, or zero, depending on the position of the point (x, y) in reference to $(\bar{x}, \bar{y})$. [$(x - \bar{x})$ and $(y - \bar{y})$ are represented by means of braces, with positive or negative signs, as shown in figure 13–3.]

☐

covariance   One measure of linear dependency is the covariance. The **covariance of x and y** is defined as the sum of the products of the distances of all values of x and y from the centroid, $\sum [(x - \bar{x})(y - \bar{y})]$, divided by $n - 1$:

$$\text{covar}(x, y) = \frac{\sum_{i=1}^{n} (x_i - \bar{x})(y_i - \bar{y})}{n - 1} \tag{13-1}$$

The covariance for the data given in illustration 13–1 is calculated in table 13–1. The covariance, written as covar (x, y), of the data is $+\frac{3}{5} = 0.6$.

TABLE 13–1

Calculations for Finding covar (x, y) for the Data of Illustration 13–1

Points	$x - \bar{x}$	$y - \bar{y}$	$(x - \bar{x})(y - \bar{y})$
A (2, 1)	−5	−2	10
B (3, 5)	−4	2	−8
C (6, 3)	−1	0	0
D (8, 2)	1	−1	−1
E (11, 6)	4	3	12
F (12, 1)	5	−2	−10
Total	0	0	3

NOTE: $\sum (x - \bar{x}) = 0$ and $\sum (y - \bar{y}) = 0$. This will always happen. Why? (See page 49.)

The covariance is positive if the graph is dominated by points to the upper right and to the lower left of the centroid. The products of $(x - \bar{x})$ and $(y - \bar{y})$ are positive in these two sections. If the majority of the points are in the upper left and lower right sections relative to the centroid, the sum of the products is negative. Figure 13–4 shows data that represent a positive dependency (a), a negative dependency (b), and little or no dependency (c). The covariances for these three situations would definitely be positive in part (a), negative in (b), and near zero in (c). (The sign of the covariance is always the same as the sign of the slope of the regression line.)

The biggest disadvantage of covariance as a measure of linear dependency is that it does not have a standardized unit of measure. One reason for this is that the spread of the data is a strong factor in the size of the covariance. For example, if we were to multiply each data point in illustration 13–1 by 10, we would have (20, 10), (30, 50), (60, 30), (80, 20), (110, 60), and (120, 10). The relationship of the points to each other would be changed

FIGURE 13-4
Data and Covariance

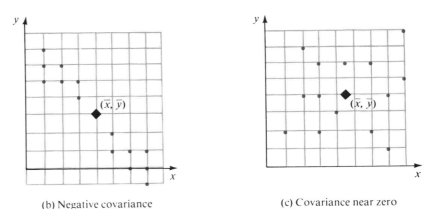

(a) Positive covariance   (b) Negative covariance   (c) Covariance near zero

only in that they would be much more spread out. However, the covariance for this new set of data is 60. (This calculation is assigned in exercise 13–3.) Does this mean that the amount of dependency between the x and y variables is stronger than in the original case? No, it does not; the relationship is the same, even though each data point was multiplied by 10. This is the trouble with covariance as a measure. We must find some way to eliminate the effect of the spread of the data when we measure dependency.

If we standardize x and y by dividing the distance of each from the respective mean by the respective standard deviation,

$$x' = \frac{x - \bar{x}}{s_x} \quad \text{and} \quad y' = \frac{y - \bar{y}}{s_y}$$

and then compute the covariance of x' and y', we will have a covariance that is *not* affected by the spread of the data. This is exactly what is accomplished by the linear correlation coefficient. It divides the covariance of x and y by a measure of the spread of x and by a measure of the spread of y (the standard deviations of x and of y are used as measures of spread). Therefore, by definition, the **coefficient of linear correlation** is

$$r = \text{covar}(x', y') = \frac{\text{covar}(x, y)}{s_x \cdot s_y} \tag{13-2}$$

The coefficient of linear correlation standardizes the measure of dependency and allows us to compare the relative strengths of dependency of different sets of data. [Formula (13–2) for linear correlation is also commonly referred to as **Pearson's product moment**, r.]

Pearson's product moment

The value of r, the coefficient of linear correlation, for the data in illustration 13–1 can be found by calculating the two standard deviations and then dividing:

$$s_x = 4.099 \quad \text{and} \quad s_y = 2.098$$

$$r = \frac{0.6}{(4.099)(2.098)} = 0.07$$

Finding the correlation coefficient by using formula (13–2) can be a very tedious arithmetic process. The formula can be written in a more workable form, as it was given in chapter 3:

$$r = \frac{\text{covar}(x, y)}{s_x \cdot s_y} = \frac{\sum [(x - \bar{x}) \cdot (y - \bar{y})]/(n - 1)}{s_x \cdot s_y}$$

$$= \frac{\text{SS}(xy)}{\sqrt{\text{SS}(x) \cdot \text{SS}(y)}} \tag{13–3}$$

Formula (13–3) avoids the separate calculations of $\bar{x}$, $\bar{y}$, s_x, and s_y, and, more importantly, the calculations of the deviations from the means. Therefore, formula (13–3) is much easier to use. (Refer to chapter 3 for an illustration of the use of this formula.)

Exercises

13–1 Explain why $\sum (x - \bar{x}) = 0$ and $\sum (y - \bar{y}) = 0$.

13–2 (a) Construct a scatter diagram of the accompanying bivariate data.

					Point					
	A	B	C	D	E	F	G	H	I	J
x	1	1	3	3	5	5	7	7	9	9
y	1	2	2	3	3	4	4	5	5	6

(b) Calculate the covariance.

(c) Calculate s_x and s_y.

(d) Calculate r using formula (13–2).

(e) Calculate r using formula (13–3).

13–3 (a) Calculate the covariance of the set of data (20, 10), (30, 50), (60, 30), (80, 20), (110, 60), and (120, 10).

(b) Calculate the standard deviation of the six x values and the standard deviation of the six y values.

(c) Calculate r, the coefficient of linear correlation, for the data in part (a).

(d) Compare these results to those found in the text for illustration 13–1.

13-4 Consider the accompanying bivariate data.

Point

	A	B	C	D	E	F	G	H	I	J
x	0	1	1	2	3	4	5	6	6	7
y	6	6	7	4	5	2	3	0	1	1

(a) Draw a scatter diagram for the data.

(b) Calculate the covariance.

(c) Calculate s_x and s_y.

(d) Calculate r by formula (13–2).

(e) Calculate r by formula (13–3).

Section 13–2   Inferences about the Linear Correlation Coefficient

After the linear correlation coefficient r has been calculated for the sample data, it seems necessary to ask this question: Does the value of r indicate that there is a dependency between the two variables in the population from which the sample was drawn? To answer this question we can perform a hypothesis test. The null hypothesis is "the two variables are linearly unrelated" ($\rho = 0$), where ρ (the Greek lowercase letter rho) is the **linear correlation coefficient for the population**. The alternative hypothesis may be either one-tailed or two-tailed. Most frequently it is two-tailed. However, when we suspect that there is only a positive or only a negative correlation, we should use a one-tailed test. The alternative hypothesis of a one-tailed test is $\rho > 0$ or $\rho < 0$.

The critical region for the test is on the right when a positive correlation is expected and on the left when a negative correlation is expected. The test statistic used to test the null hypothesis is the calculated value of r from the sample. Critical values for r are found in table 9 of appendix D at the intersection of the column identified by the appropriate value of α and the row identified by the degrees of freedom. The number of degrees of freedom for the r statistic is 2 less than the sample size, $df = n - 2$.

The rejection of the null hypothesis means that there is evidence of a linear dependency between the two variables in the population.

Caution:

The sample evidence may only say that the pattern of behavior of the two variables is related in that one can be used effectively to predict the other. **This does not mean that you have established a cause-and-effect relationship.**

Failure to reject the null hypothesis is interpreted as meaning that linear dependency between the two variables in the population has not been shown.

Now let's look at such a hypothesis test.

Illustration 13–2

For illustration 13–1, where $n = 6$, we found $r = 0.07$. Is this significantly different from zero, at the 0.02 level of significance?

Solution

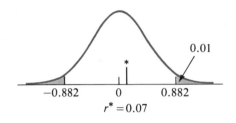

−0.882   0   0.882

$r^* = 0.07$

0.01

STEP 1 $H_0: \rho = 0$.

STEP 2 $H_a: \rho \neq 0$.

STEP 3 $\alpha = 0.02$, df $= n - 2 = 6 - 2 = 4$. See the figure.

The critical values (-0.882 and 0.882) were obtained from table 9 of appendix D.

STEP 4   The calculated value of r, r^*, was found earlier. It is $r^* = 0.07$.

STEP 5   *Decision:* Fail to reject H_0.

Conclusion: We have failed to show that x and y are correlated.   □

As in other problems, sometimes a confidence interval estimate of the population correlation coefficient is required. It is possible to estimate the value of ρ, the linear correlation coefficient of the population. Usually this is accomplished by using a table showing **confidence belts**. Table 10 in appendix D gives confidence belts for 95% confidence interval estimates. This table is a bit tricky to read, so be extra careful when you use it. The next illustration demonstrates the procedure for estimating ρ.

confidence belts

Illustration 13–3

A sample of 15 ordered pairs of data have a calculated r value of 0.35. Find the 95% confidence interval estimate for ρ, the population linear correlation coefficient.

Solution   Find $r = 0.35$ at the bottom of table 10; see the arrow on figure 13–5. Visualize a vertical line through that point. Find the two points where the belts marked for the correct sample size cross the vertical line. The sample size is 15. These two points are circled in figure 13–5. Now look horizontally from the two circled points to the vertical scale on the left and read the confidence interval. The values are 0.72 and -0.20. Thus the 95% confidence interval estimate for ρ, the population coefficient of linear correlation, is -0.20 to 0.72 □

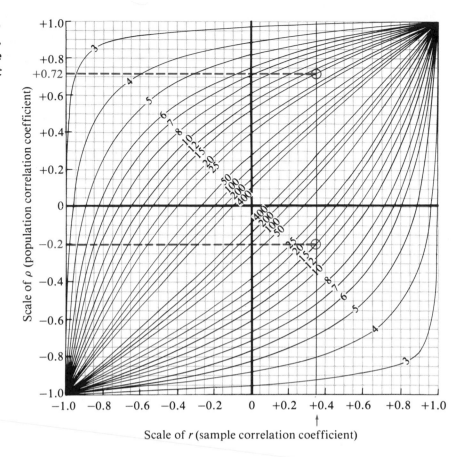

FIGURE 13–5
*Using Table 10 of Appendix D,
Confidence Belts for the
Correlation Coefficient*

Scale of ρ (population correlation coefficient)

Scale of r (sample correlation coefficient)

Exercises

13–5 Using graphs to illustrate, explain the meaning of a correlation coefficient whose value is, (a) -1, (b) 0, (c) $+1$.

13–6 A sample of 20 pieces of bivariate data has a linear correlation coefficient of $r = 0.43$. Does this provide sufficient evidence to reject the null hypothesis that $\rho = 0$ in favor of a two-sided alternative? Use $\alpha = 0.10$.

13–7 If a sample of size 18 has a linear correlation coefficient of -0.50, do we have significant reason to conclude that the linear correlation coefficient of the population is negative? Use $\alpha = 0.01$.

13–8 A sample of size 10 produced $r = -0.67$. Is this sufficient evidence to conclude that ρ is different from zero, at the 0.05 level of significance?

13–9 Is a value of $r = +0.24$ significant in trying to show that ρ is greater than zero for a sample of 62 data, at the 0.05 level of significance?

13-10 Use table 10 of appendix D to determine a 0.95 confidence interval estimate for the true population linear correlation coefficient based on these sample statistics:

(a) $n = 8, r = 0.20$ (b) $n = 100, r = -0.40$

(c) $n = 25, r = +0.65$ (d) $n = 15, r = -0.23$

13-11 An analysis of the nature of state taxes was concerned with the percentage of personal income spent for state taxes and the amount of tax collected per person. Data for 9 randomly selected states are listed here:

State	Percentage of Personal Income	Tax Collections per Person
California	12.14%	$1,060
Georgia	11.36	752
Iowa	11.31	894
Maryland	12.57	1,050
Missouri	9.87	726
New Jersey	12.19	1,070
Ohio	9.76	768
South Carolina	10.93	680
Vermont	14.04	905

(a) Calculate the correlation coefficient for this data.

(b) Test, using $\alpha = 0.05$, to see if the correlation coefficient is significantly different from zero.

(c) If the correlation coefficient is different from zero, estimate the true value of the correlation coefficient using a 0.95 confidence interval.

Section 13–3 Linear Regression Analysis

Recall that the line of best fit results from an analysis of a situation in which we have two or more related variables. (We will restrict our work to two variables. However, on occasion more than two will be mentioned to clarify the analysis.) When two variables are studied jointly, we often would like to control one variable by means of controlling the other. Or we might want to predict the value of a variable based on knowledge about another variable. In both cases we want to find the line of best fit, provided one exists, that will best predict the value of the dependent, or output, variable. Recall that the variable we know or can control is called the **independent, or input, variable**; the variable resulting from the use of the equation of the line of best fit is called the **dependent, or predicted, variable**.

When the line of best fit is plotted, it does more than just show us a pictorial representation. It tells us two things: (1) there really is a functional (equational) relationship between the two variables, and (2) the line expresses the quantitative relationship between the two variables. Recall that the line of best fit is of no use when a change in the input variable does not seem to have a definite effect on the output variable. When there is no relationship between the variables, a horizontal line of best fit is plotted. A horizontal line has a slope of zero, which implies that the value of the input variable has no effect on the output variable. This idea about the slope being zero will be amplified later in this chapter.

The result of regression analysis is the mathematical equation that is the equation of the line of best fit. We will, as mentioned before, restrict our work to the simple linear case, that is, one input variable and one output variable, where the line of best fit is straight. However, you should be aware of the fact that not all relationships are of this nature. If the scatter diagram suggests something other than a straight line, we have **curvilinear regression**. In cases of this type we must introduce terms to higher powers, x^2, x^3, and so on, or other functions, e^x, $\log x$, and so on; or we would introduce other input variables. Maybe two or three input variables would improve the usefulness of our regression equation. These possibilities are examples of curvilinear regression and multiple regression.

curvilinear regression

The **linear model** used to explain the behavior of linear bivariate data **in the population** is

$$y = \beta_0 + \beta_1 x + \varepsilon \qquad (13\text{--}4)$$

This equation represents the linear relationship between the two variables in a population. β_0 **is the y-intercept and** β_1 **is the slope.** ε (Greek lowercase letter epsilon) **is the random experimental error** in the observed value of y at a given value of x.

The regression line from the sample data gives us b_0, which is **our estimate of β_0,** and b_1, which is **our estimate of β_1.** The **error ε is approximated by $e = y - \hat{y}$,** the difference between the observed value of y and the predicted value of y, $\hat{y}$, at a given value of x.

experimental error (ε or e)

$$e = y - \hat{y} \qquad (13\text{--}5)$$

The random variable e is positive when the observed value of y is larger than the predicted value $\hat{y}$; e is negative when y is smaller than $\hat{y}$. The sum of the errors for the different values of y for a given value of x is exactly zero. (This is part of the least squares criteria.) Thus the mean value of the experimental error is zero; its variance is σ_ε^2. Our next goal is to estimate this variance of the experimental error.

Before we estimate the variance of ε, let's try to understand exactly what the error represents. ε is the amount of error in our observed value of y. That is, it is the difference between the observed value of y and the mean value of y at that particular value of x. Since we do not know the mean value of y, we will use the regression equation and estimate it with $\hat{y}$, the predicted value of y at this same value of x. Thus the best estimate that we have for ε is $e = (y - \hat{y})$, as shown in figure 13–6.

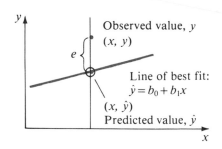

FIGURE 13–6
The Error e Is $y - \hat{y}$

Observed value, y
(x, y)
e
Line of best fit:
$\hat{y} = b_0 + b_1 x$
$(x, \hat{y})$
Predicted value, $\hat{y}$

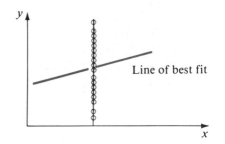

FIGURE 13–7
Sample of y Values at a Given x

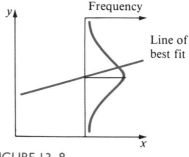

FIGURE 13–8
*Theoretical Distribution of y
Values for a Given x*

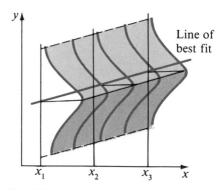

FIGURE 13–9
*Standard Deviation of the Distribution
of y Values for All x Is the Same*

NOTE: e is the observed error in measuring y at a specified value of x.

If we were to observe several values of y at a given value of x, we could plot a distribution of y values about the line of best fit (about $\hat{y}$, in particular). Figure 13–7 shows a sample of bivariate values for which the value of x is the same. Figure 13–8 shows the theoretical distribution of all possible y values at a given x value. A similar distribution occurs at each different value of x. The mean of the observed y's at a given value of x varies, but it can be estimated by $\hat{y}$.

Before we can make any inferences about a regression line, we must assume that the distribution of y's is approximately normal and that the variances of the distributions of y at all values of x are the same. That is, the standard deviation of the distribution of y about $\hat{y}$ is the same for all values of x, as shown in figure 13–9.

Before looking at the variance of e, let's review the definition of sample variance. The sample variance s^2 is defined as $\sum (x - \bar{x})^2/(n - 1)$, the sum of the squares of each deviation divided by the number of degrees of freedom, $n - 1$, associated with a sample of size n. The variance of y involves an additional complication: there is a different mean for y at each value of x. (Notice the many distributions in figure 13–9.) However, each of these "means" is actually the predicted value, $\hat{y}$, that corresponds to the x that fixes the distribution. So the **variance of the error e** is estimated by the formula

$$s_e^2 = \frac{\sum (y - \hat{y})^2}{n - 2} \qquad (13\text{-}6)$$

where $n - 2$ is the number of degrees of freedom.

NOTE: The variance of y about the line of best fit is the same as the variance of the error e. Recall that $e = y - \hat{y}$.

Formula (13–6) can be rewritten by substituting $b_0 + b_1 x$ for $\hat{y}$, since $\hat{y} = b_0 + b_1 x$, then s_e^2 becomes

$$s_e^2 = \frac{\sum (y - b_0 - b_1 x)^2}{n - 2} \qquad (13\text{-}7)$$

With some algebra and some patience, this formula can be rewritten once again into a more workable form. The form we will use is

$$s_e^2 = \frac{(\sum y^2) - (b_0)(\sum y) - (b_1)(\sum xy)}{n - 2} \qquad (13\text{-}8)$$

For ease of discussion let's agree to call the numerator of formulas (13–6), (13–7), and (13–8) the **sum of squares for error (SSE)**.

Now let's see how all this information can be used.

Illustration 13-4

Suppose you move into a new city and take a job. One of your concerns will be the problems you will face commuting to and from work. For example, you would like to know how long (in minutes) it will take you to drive to work each morning. Let's use one-way distance to work as a measure of where you live. You live x miles away from work and want to know how long it will take you to commute each day. Your new employer, foreseeing this question, has already collected a random sample of data to be used in answering your question. Fifteen of your coworkers were asked to give their one-way travel time and distance to work. The resulting data are shown in table 13-2. (For convenience the data have been arranged so that the x values are in numerical order.) Find the line of best fit and the variance of y about the line of best fit, s_e^2.

TABLE 13-2
Data for Illustration 13-4

Coworker	Miles, x	Minutes, y	x^2	xy	y^2
1	3	7	9	21	49
2	5	20	25	100	400
3	7	20	49	140	400
4	8	15	64	120	225
5	10	25	100	250	625
6	11	17	121	187	289
7	12	20	144	240	400
8	12	35	144	420	1,225
9	13	26	169	338	676
10	15	25	225	375	625
11	15	35	225	525	1,225
12	16	32	256	512	1,024
13	18	44	324	792	1,936
14	19	37	361	703	1,369
15	20	45	400	900	2,025
Total	184	403	2,616	5,623	12,493

Solution The extensions and summations needed for this problem are shown in table 13-2. The line of best fit, using formulas (2-9), (3-4), (3 6) and (3-7), can now be calculated.

Using formula (2-9):

$$SS(x) = 2{,}616 - \frac{(184)^2}{15} = 358.9333$$

Using formula (3-4):

$$SS(xy) = 5{,}623 - \frac{(184)(403)}{15} = 679.5333$$

Using formula (3-7):

$$b_1 = \frac{679.5333}{358.9333} = 1.893202 = 1.89$$

Using formula (3-6):

$$b_0 = \frac{1}{15}[403 - (1.893202)(184)] = 3.643387 = \mathbf{3.64}$$

Therefore, the equation for the line of best fit is:

$$\hat{y} = \mathbf{3.64 + 1.89}x$$

The variance of y about the regression line is calculated by using formula (13-8).

$$s_e^2 = \frac{SSE}{n-2} = \frac{(\sum y^2) - (b_0)(\sum y) - (b_1)(\sum xy)}{n-2}$$

$$= \frac{12{,}493 - (3.643387)(403) - (1.893202)(5{,}623)}{15-2}$$

$$= \frac{379.2402}{13} = 29.1723 = \mathbf{29.17}$$

NOTE: Extra decimal places are often needed for this type of calculation. Notice that b_1 (1.893202) was multiplied by 5,623. If 1.89 had been used instead, that one product would have changed the numerator by approximately 18. That, in turn, would have changed the final answer by almost 1.5, and that is a sizable round-off error.

$s_e^2 = 29.17$ is the variance of the 15 e's. In figure 13–10 the 15 e's are shown as vertical line segments. □

FIGURE 13–10
*The 15 Random Errors as
Line Segments*

The variance of e will be used in the following sections of this chapter in much the same way as the variance of x (as calculated in chapter 2) was used in chapters 8, 9, and 10 to complete the statistical inferences studied there.

Exercises

13-12 (a) Find the equation of the line of best fit, $\hat{y} = b_0 + b_1 x$, and graph it on the scatter diagram of the 10 points shown in the table below.

Point

	A	B	C	D	E	F	G	H	I	J
x	1	1	3	3	5	5	7	7	9	9
y	1	2	2	3	3	4	4	5	5	6

(b) Find the ordinates, $\hat{y}$, for the points on the line of best fit whose abscissas are $x = 1, 3, 5, 7$, and 9.

(c) Find the value of e for each of the points in the given data

$$(e = y - \hat{y})$$

(d) Find the variance s_e^2 of those points about the line of best fit by using formula (13-6).

(e) Find the variance s_e^2 by using formula (13-8). [Answers to parts (d) and (e) should be the same.]

13-13 Consider the accompanying bivariate data.

Point

	A	B	C	D	E	F	G	H	I	J
x	0	1	1	2	3	4	5	6	6	7
y	6	6	7	4	5	2	3	0	1	1

Answer the questions in exercise 13-12. In part (b) use $x = 0, 1, 2, \ldots, 7$.

13-14 The accompanying data show the number, x, of hours studied for an exam and the grade, y, received on the exam. (y is measured in 10s; that is, $y = 8$ means that the grade, rounded to the nearest 10 points, is 80.)

x	2	3	3	4	4	5	5	6	6	6	7	7	7	8	8
y	5	5	7	5	7	7	8	6	9	8	7	9	10	8	9

(a) Draw a scatter diagram of the data.

(b) Find the equation of the line of best fit and graph it on the scatter diagram.

(c) Find the ordinates $\hat{y}$ that correspond to $x = 2, 3, 4, 5, 6, 7$, and 8.

(d) Find the five values of e that are associated with the points where $x = 3$ and $x = 6$.

(e) Find the variance s_e^2 of all the points about the line of best fit.

Retain these answers for use in exercise 13–19.

⚇ 13–15   In a regression problem where x = the number of days on a diet, and y = the number of pounds lost, we obtain the following:

$$n = 4 \qquad \sum xy = 132$$
$$\sum x = 20 \qquad \sum x^2 = 110$$
$$\sum y = 24 \qquad \sum y^2 = 162$$

(a) Find the equation of the line of best fit.

(b) How much is the average weight loss in pounds for each additional day on the diet?

Section 13–4   Inferences Concerning the Slope of the Regression Line

Now that the equation of the line of best fit has been determined, and the linear model has been verified (by inspection of the scatter diagram), we are ready to determine whether we can use the equation to predict y. We will test the null hypothesis "the equation of the line of best fit is of no value in predicting y given x." That is, the null hypothesis to be tested is β_1 (the slope of the relationship in the population) is 0. If $\beta_1 = 0$, then the linear equation will be of no real use in predicting y. To test this hypothesis we will use a t test.

Before we look at the hypothesis test, let's discuss the sampling distribution of the slope. If random samples of size n are repeatedly taken from a bivariate population, the calculated slopes, the b_1's, would form a sampling distribution that is approximately normally distributed with a mean of β_1, the population value of the slope, and with a variance of $\sigma_{b_1}^2$, where

$$\sigma_{b_1}^2 = \frac{\sigma_\varepsilon^2}{\sum (x - \bar{x})^2} \qquad (13\text{–}9)$$

provided there is no lack of fit. An appropriate estimator for $\sigma_{b_1}^2$ is obtained by replacing σ_ε^2 by s_e^2, the estimate of the variance of the error about the regression line:

$$s_{b_1}^2 = \frac{s_e^2}{\sum (x - \bar{x})^2} \qquad (13\text{–}10)$$

This formula may be rewritten in the following, more manageable form:

$$s_{b_1}^2 = \frac{s_e^2}{SS(x)} = \frac{s_e^2}{\sum x^2 - [(\sum x)^2/n]} \qquad (13\text{–}11)$$

NOTE: The "standard error of —" is the standard deviation of the sampling distribution of —. Therefore, the standard error of regression (slope) is σ_{b_1} and is estimated by s_{b_1}.

We are now ready to **test the hypothesis** $\beta_1 = 0$. Let's use the line of best fit determined in illustration 13–4, $\hat{y} = 3.64 + 1.89x$. That is, we want to determine if this equation is of any use in predicting travel time y. In this type of hypothesis test, the null hypothesis is always $H_0: \beta_1 = 0$.

STEP 1 $H_0: \beta_1 = 0$. (This implies that x is of no use in predicting y; that is, that $\hat{y} = \bar{y}$ would be as effective.)

The alternative hypothesis can be either one-tailed or two-tailed. If we suspect that the slope is positive, as in illustration 13–4 [we would expect travel time (y) to increase as the distance (x) increased], a one-tailed test is appropriate.

STEP 2 $H_a: \beta_1 > 0$.

STEP 3   The test statistic is t. The number of degrees of freedom for this test is $n - 2$, **df** $= n - 2$. Thus for our example, df $= 15 - 2 = 13$. If we use $\alpha = 0.05$, the critical value of t is $t(13, 0.05) = 1.77$, as found in table 6 of appendix D; see the accompanying figure.

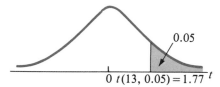

0.05

$0 \quad t(13, 0.05) = 1.77 \quad t$

The formula used to calculate the value of the **test statistic** t for inferences about the slope is

$$t = \frac{b_1 - \beta_1}{s_{b_1}} \qquad (13-12)$$

STEP 4   In our illustration of travel times and distances, the variance among the b_1's is estimated by use of formula (13–11):

$$s_{b_1}^2 = \frac{29.1723}{358.9333} = 0.081275 = 0.0813$$

Using formula (13–12), the observed value of t becomes

$$t = \frac{b_1 - \beta_1}{s_{b_1}} = \frac{1.89 - 0}{\sqrt{0.0813}} = 6.629$$

$$t^* = 6.63$$

STEP 5   *Decision*: Reject H_0 (t^* is in the critical region; see the figure).

Conclusion: The slope of the line of best fit in the population is greater than zero. The evidence indicates that there is a linear relationship and that the travel time to work (y) can be predicted based on the one-way distance (x).

The slope β_1 of the regression line of the population can be estimated by means of a confidence interval. The **confidence interval** is given by

$$b_1 \pm t(n - 2, \alpha/2) \cdot s_{b_1} \qquad (13-13)$$

The 95% confidence interval for the estimate of the population's slope, β_1, for illustration 13–4 is

$$1.89 \pm (2.16)(\sqrt{0.0813})$$

$$1.89 \pm 0.6159$$

$$1.89 \pm 0.62$$

$$\mathbf{1.27} \quad \text{to} \quad \mathbf{2.51}$$

Thus 1.27 to 2.51 is the 0.95 confidence interval for β_1. That is, we can say that the slope of the line of best fit of the population from which the sample was drawn is between 1.27 and 2.51, with 95% confidence.

Exercises

13–16 A sample of 10 students was asked for the distance and the time required to commute to college yesterday. The data collected are shown in the accompanying table.

x, Distance	1	3	5	5	7	7	8	10	10	12
y, Time	5	10	15	20	15	25	20	25	35	35

(a) Draw a scatter diagram of these data.

(b) Find the equation that describes the regression line for these data.

(c) Does the value of b_1 show sufficient strength to conclude that β_1 is greater than zero, at the $\alpha = 0.05$ level?

(d) Find the 0.98 confidence interval for the estimation of β_1.

Retain these answers for use in exercise 13–20.

13–17 It is said that when the going gets tough then the tough get going. A record was kept for some basketball stars who have reputations for scoring points. This analysis uses the points scored in the first quarter and the points scored in the fourth quarter of games where the final margin of victory was less than five points.

Player	1	2	3	4	5	6	7	8	9
First Quarter	8	14	11	4	15	16	9	13	18
Fourth Quarter	13	4	10	3	6	8	6	9	10

(a) Determine the line of best fit.

(b) Estimate the mean points scored in the fourth quarter when 16 points are scored in the first quarter.

(c) Is the relationship significant at the 0.05 level?

(d) Does the evidence suggest that the high-scoring players perform better in the fourth quarter, when the going gets tough, than in the first quarter? (There may be other factors that affect fourth quarter performances.)

13–18 Interest rates are alleged to have an effect on the level of employment. The data below show by quarters, the bank interest rates on short-term loans and the unemployment rate.

Interest Rate	12.27	12.34	12.31	15.81	15.67	17.75	11.56	15.71	19.91	19.99	21.11
Unemployment Rate	5.9	5.6	5.9	5.9	6.2	7.6	7.5	7.3	7.6	7.2	8.3

(a) Calculate the equation of the line of best fit.

(b) Does this sample present sufficient evidence to reject the null hypothesis (slope is zero) in favor of the alternative of the null hypothesis that the slope is positive, at the 0.05 level of significance?

Section 13–5 Confidence Interval Estimates for Regression

Once the equation for the time of best fit has been obtained and determined usable, we are ready to use the equation to make predictions. There are two different quantities that we can estimate: (1) the mean of the population y values at a given value of x, written as $\mu_{y|x_0}$, and (2) the individual y value selected at random that will occur at a given value of x, written as y_{x_0}. The best point estimate or prediction for both $\mu_{y|x_0}$ and y_{x_0} is $\hat{y}$. This is the y value obtained when an x value is substituted into the equation of the line of best fit. Like other point estimates it is seldom right. The actual values for both $\mu_{y|x_0}$ and y_{x_0} will vary above and below the calculated value of $\hat{y}$.

predicted value of y ($\hat{y}$)

Before developing confidence intervals for $\mu_{y|x_0}$ and y_{x_0}, recall the development of confidence intervals for the population mean μ in chapter 8 when the variance was known and in chapter 9 when the variance was estimated. The sample mean $\bar{x}$ was the best point estimate of μ. We used the fact that $\bar{x}$ is normally distributed with a standard deviation of $\sigma/\sqrt{n}$ to construct formula (8–2) for the confidence interval for μ. When σ had to be estimated, we used formula (9–2) for the confidence interval.

The confidence intervals for $\mu_{y|x_0}$ and y_{x_0} are constructed in a similar fashion. $\hat{y}$ replaces $\bar{x}$ as our point estimate. If we were to take random samples from the population, construct the line of best fit for each sample, calculate $\hat{y}$ for a given x using each regression line, and plot the various $\hat{y}$ values (they would vary since each sample would yield a slightly different regression line), we would find that the $\hat{y}$ values form a normal distribution. That is, the sampling distribution of $\hat{y}$ is normal, just as the sampling distribution of $\bar{x}$ is normal. What about the appropriate standard deviation of $\hat{y}$? The

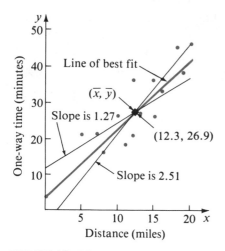

FIGURE 13–11

Lines Representing the Confidence Interval for Slope

standard deviation in both cases ($\mu_{y|x_0}$ and y_{x_0}) is calculated by multiplying the square root of the variance of the error by an appropriate correction factor. Recall that the variance of the error, s_e^2, is calculated by means of formula (13–8).

Before looking at the correction factors for the two cases, let's see why they are necessary. Recall that the line of best fit passes through the point $(\bar{x}, \bar{y})$, the centroid. Earlier in this chapter we formed a confidence interval estimate for the slope β_1 (see illustration 13–4) by using formula (13–13). If we draw lines with slopes equal to the extremes of that confidence interval, 1.27 to 2.51, through the point $(\bar{x}, \bar{y})$ [which is (12.3, 26.9)] on the scatter diagram, we will see that the value for $\hat{y}$ fluctuates considerably for different values of x (figure 13–11). Therefore, we should suspect a need for a wider confidence interval as we select values of x that are further away from $\bar{x}$. Hence we need a correction factor to adjust for the distance between x_0 and $\bar{x}$. This factor must also adjust for the variation of the y values about $\hat{y}$.

First, let's estimate the **mean value of y** at a given value of x, $\mu_{y|x_0}$. The confidence interval estimate formula is

$$\hat{y} \pm t(n-2, \alpha/2) \cdot s_e \cdot \sqrt{\frac{1}{n} + \frac{(x_0 - \bar{x})^2}{\sum (x - \bar{x})^2}} \qquad (13\text{–}14)$$

NOTE: The numerator of the second term under the radical sign is the square of the distance of x_0 from $\bar{x}$. The denominator is closely related to the variance of x and has a "standardizing" effect on this term.

Formula (13–14) can be modified to avoid having $\bar{x}$ in the denominator. The new form is

$$\hat{y} \pm t(n-2, \alpha/2) \cdot s_e \cdot \sqrt{\frac{1}{n} + \frac{(x_0 - \bar{x})^2}{SS(x)}} \qquad (13\text{–}15)$$

Let's compare formula (13–14) with formula (9–2). $\hat{y}$ replaces $\bar{x}$, and

$$s_e \sqrt{\frac{1}{n} + \frac{(x_0 - \bar{x})^2}{\sum (x - \bar{x})^2}} \qquad \text{(the standard error of } \hat{y})$$

the estimated standard deviation of $\hat{y}$ in predicting $\mu_{y|x_0}$, replaces $s/\sqrt{n}$, the standard deviation of $\bar{x}$. The degrees of freedom are now $n - 2$ instead of $n - 1$ as before.

These ideas are explored in the next illustration.

Illustration 13–5

Construct a 95% confidence interval estimate for the mean travel time for all coworkers who travel 7 miles (refer to illustration 13–4).

Solution

STEP 1 Find $\hat{y}_{x_0}$ when $x_0 = 7$:

$$\hat{y} = 3.64 + 1.89x = 3.64 + (1.89)(7) = \mathbf{16.87}$$

STEP 2 Find s_e:

$$s_e^2 = 29.17 \qquad \text{(found previously)}$$

$$s_e = \sqrt{29.17} = \mathbf{5.40}$$

STEP 3 Find $t(13, 0.025) = 2.16$ (from table 6 in appendix D).

STEP 4 Use formula (13–15):

$$16.87 \pm (2.16) \cdot (5.40) \cdot \sqrt{\frac{1}{15} + \frac{(7 - 12.27)^2}{358.9333}}$$

$$16.87 \pm (2.16)(5.40)\sqrt{0.06667 + 0.07738}$$

$$16.87 \pm (2.16)(5.40)\sqrt{0.14405}$$

$$16.87 \pm (2.16)(5.40)(0.38)$$

$$16.87 \pm 4.43$$

$$\mathbf{12.44} \quad \text{to} \quad \mathbf{21.30}$$

is the 0.95 confidence interval for $\mu_{y|x_0 = 7}$. This confidence interval estimate is shown in figure 13–12 by the heavy vertical line. The **confidence interval belt** showing the upper and lower boundaries of all interval estimates at 95% confidence is shown also. Notice that the boundary lines for x values far away from $\bar{x}$ become close to the two lines that represent the equations having slopes equal to the extreme values of the 95% confidence interval estimate for the slope (see figure 13–11). ☐

Often when making a prediction, we want to predict the value of an individual y. For example, you live 7 miles from this place of business. You are interested in an estimate of how long it will take you to get to work. You are somewhat less interested in the average time for all those who live 7 miles away. The formula for the **confidence interval** estimate for the value of a **single randomly selected** y is

$$\hat{y} \pm t(n - 2, \alpha/2) \cdot s_e \cdot \sqrt{1 + \frac{1}{n} + \frac{(x_0 - \bar{x})^2}{SS(x)}} \qquad (13\text{–}16)$$

Illustration 13–6

What is the 95% confidence interval estimate for the time it will take you to commute to work if we know that you live 7 miles away?

FIGURE 13–12

Confidence Belts for μ_{yx_0}

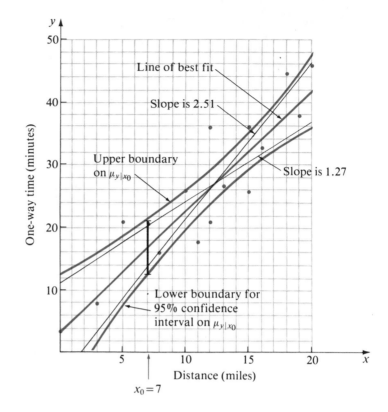

Solution

STEP 1 $\hat{y}$ at $x_0 = 7$ is 16.87, as found previously.

STEP 2 $s_e = 5.40$, as found previously.

STEP 3 $t(13, 0.025) = 2.16$.

STEP 4   Use formula (13–16). (Notice that some of the calculations needed were found in illustration 13–5.)

$$16.87 \pm (2.16)(5.40)\sqrt{1 + 0.06667 + 0.07738}$$

$$16.87 \pm (2.16)(5.40)(1.0696)$$

$$16.87 \pm 12.48$$

$$\mathbf{4.39} \quad \text{to} \quad \mathbf{29.35}$$

is the 0.95 confidence interval for $y_{x_0 = 7}$.

The confidence interval is shown in figure 13–13 as the vertical line segment at $x_0 = 7$. Notice that it is much longer than the confidence interval for $\mu_{y|x_0 = 7}$. The dashed lines represent the upper and lower boundaries of the confidence intervals for individual y values for all given x values.

FIGURE 13–13
Confidence Belts for y_{x_0}

FIGURE 13–14
*Confidence Belts for Individual y's and
for the Mean Value of y*

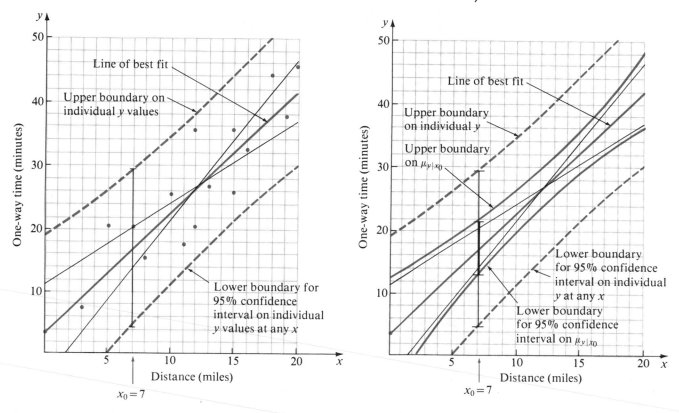

Can you justify the fact that the confidence interval for individual values of y is wider than the confidence interval for the mean values? Think about "individual values" and "mean values," and study figure 13–14. □

Computer Solution
Minitab Printout for parts of Illustrations 13–4, 13–5 and 13–6

Information given to computer

```
--READ DATA, X-VALUES INTO C1, Y-VALUES INTO C2
     COLUMN  C1              C2
     COUNT   15              15
ROW
     1          3.              7.
     2          5.             20.
```

(continued)

3	7.	20.
4	8.	15.
5	10.	25.
6	11.	17.
7	12.	20.
8	12.	35.
9	13.	26.
10	15.	25.
11	15.	35.
12	16.	32.
13	18.	44.
14	19.	37.
15	20.	45.

Scatter diagram of given data

```
--PLOT Y-VALUES IN C2 VS X-VALUES IN C1
  C2
50.0+
    -
    -                                                              *
    -                                                          *
    -
40.0+
    -                                                      *
    -                                  *            *
    -                                          *
    -                                               *
30.0+
    -                            *        *        *
    -
20.0+                *        *              *
    -                                    *
    -                            *
    -
10.0+
    -        *
    -
    -
    -
 0.0+
    +-------+-------+-------+-------+---- ----+C1
   0.0     4.0     8.0    12.0    16.0    20.0
        4.0          8.0    12.0       16.0    20.0
```

(continued)

Calculated correlation coefficient, $r = +0.879$

--CORRELATION COEFFICIENT BETWEEN DATA IN C1 AND C2
CORRELATION OF C1 AND C2 = 0.879

Equation of line of best fit, $\hat{y} = 3.64 + 1.89x$, see page 447, solution of illustration 13–4

--REGRESS Y–VARIABLE IN C2 ON 1 X–VARIABLE IN C1
THE REGRESSION EQUATION IS
Y = 3.64 + (1.89) × 1

Given data

Values of $\hat{y}$ for each given x-value using $\hat{y} = 3.6434 + 1.8932x$

ROW	X IN COL. 1	Y IN COL. 2	PRED Y VALUES	STD DEV PRED Y	RES.	STD. RES.
1	3.000	7.000	9.323	2.987	−2.323	−0.516
2	5.000	20.000	13.109	2.497	6.891	1.439
3	7.000	20.000	16.896	2.049	3.104	0.621
4	8.000	15.000	18.789	1.851	−3.789	−0.747
5	10.000	25.000	22.575	1.537	2.425	0.468
6	11.000	17.000	24.469	1.441	−7.469	−1.435
7	12.000	20.000	26.362	1.397	−6.362	−1.219
8	12.000	35.000	26.362	1.397	8.638	1.656
9	13.000	26.000	28.255	1.410	−2.255	−0.433
10	15.000	25.000	32.041	1.598	−7.041	−1.365
11	15.000	35.000	32.041	1.598	2.959	0.573
12	16.000	32.000	33.935	1.754	−1.935	−0.379
13	18.000	44.000	37.721	2.149	6.279	1.267
14	19.000	37.000	39.614	2.373	−2.614	−0.539
15	20.000	45.000	41.507	2.609	3.493	0.738

Calculated values of b_0 and b_1

X	COLUMN	ESTIMATES COEFFICIENTS	S.D. OF COEFF.	RATIO
0	**	3.6434	3.76	0.97
1	1	1.8932	0.285	6.64

ESTIMATES FROM LEAST SQUARES FIT

Calculated value of S_{b_1},
$S_{b_1} = 0.285 (\sqrt{0.0813})$; compare to $S_{b_1}^2 = 0.0813$ as found in step 4 on page 451

THE ESTIMATED ST. DEV. OF Y ABOUT THE REGRESSION LINE IS 5.4011 WITH (15 − 2) = 13 DEGREES OF FREEDOM

Calculated value of S_e,
$S_e = 5.4011 (\sqrt{29.1723})$; compare to $S_e^2 = 29.1723$ as found in solution of illustration 13–4, page 448

R–SQUARED = 77.2 PERCENT
R–SQUARED = 75.5 PERCENT (ADJUSTED FOR D.F.)

ANALYSIS OF VARIANCE

	DF	SS	MS = SS/DF
TOTAL	15	0.1249D 05	
DUE TO MEAN	1	0.1083D 05	
TOTAL DUE TO REGRESSION	1	1286.	1286.
RESIDUAL	13	379.2	29.17

(*continued*)

```
X TRANSPOSE X INVERSE IN LOWER TRIANGLE
CORRELATIONS OF ESTIMATES OF COEFFICIENTS IN UPPER
TRIANGLE
                          ***              1
  ***               0.4858841           −.929
   1              −0.341753D−01      0.278603D−02


 −−STOP
```

The techniques of coding (for large numbers) and the techniques of frequency distributions (for large sets of numbers) are adaptable to the calculations presented in this chapter. Information about these techniques and about models other than linear models may be found in many other textbooks. It is worth noting here that if the data suggest a logarithmic or an exponential functional relationship, a simple change of variable will allow for the use of linear analysis.

There are three basic precautions that you need to be aware of as you work with regression analysis.

1. Remember that the regression equation is meaningful only in the domain of the x variable studied. Prediction outside this domain is extremely dangerous; it requires that we know or assume that the relationship between x and y remains the same outside the domain of the sample data. For example, Joe says that he lives 75 miles from work, and he wants to know how long it will take him to commute. We certainly can use $x = 75$ in all the formulas, but we do not expect the answers to carry the confidence or validity of the values of x between 3 and 20, which were in the sample. The 75 miles may represent a distance to the heart of a nearby major city. Do you think the estimated times, which were based on local distances of 3 to 20 miles, would be good predictors in this situation? Also, at $x = 0$ the equation has no real meaning. However, although projections outside the interval may be somewhat dangerous, they may be the best predictors available.

2. Don't get caught by the common fallacy of applying the regression results inappropriately. For example, this fallacy would include applying the results of illustration 13–4 to another company. But suppose the second company had a city location, whereas the first company had a rural location, or vice versa. Do you think the results for a rural location would also be valid for a city location? Basically, the results of one sample should not be used to make inferences about a population other than the one from which the sample was drawn.

3. Perhaps the most common fallacy is to jump to the conclusion that the results of the regression prove that x causes y to change. Regressions only measure movement between x and y; they **never prove**

causation. A judgment of causation can be made only when it is based on theory or knowledge of the relationship separate from the regression results. The most common difficulty in this regard occurs because of what is called the missing variable, or third-variable, effect. That is, we observe a relationship between x and y because a third variable, one that is not in the regression, affects both x and y.

Exercises

13-19 Use the results obtained in answering exercises 13-14 to make the following estimates:

(a) Give a point estimate for the mean test score one might expect after studying for 6 hours.

(b) Find the 0.95 confidence interval estimate for the mean test score when $x = 6$.

(c) Give a point estimate for an individual's test score if he had studied for 6 hours.

(d) Give a 0.95 confidence interval estimate for the expected exam score after 6 hours of study.

13-20 Use the data and the answers found in exercise 13-16 to make the following estimates:

(a) Give a point estimate for the mean time required to commute 4 miles.

(b) Give a 0.90 confidence interval estimate for the mean travel time required to commute 4 miles.

(c) Give a 0.90 confidence interval estimate for the travel time required for one person to commute the 4 miles.

(d) Answer parts (a), (b), and (c) for $x = 9$.

13-21 Explain why a 95% confidence interval estimate for the mean value of y at a particular x is much narrower than a 95% confidence interval for an individual y value at the same value of x.

13-22 When $x_0 = \bar{x}$, is the formula for the standard error of $\hat{y}_{x_0}$ what you might have expected it to be, $s \cdot \sqrt{1/n}$?

Section 13-6 Understanding the Relationship between Correlation and Regression

Now that we have taken a closer look at both correlation and regression analysis, it is necessary to decide when to use them. Do you see any duplication of work?

The primary use of the liner correlation coefficient is in answering the question: Are these two variables linearly related? (There are other words that may be used to ask this basic question.) For example: Is there a linear correlation between the annual consumption of alcoholic beverages and the salary paid to firemen?

The linear correlation coefficient can be used to indicate the usefulness of x as a predictor of y in the case where the linear model is appropriate. The test concerning the slope of the regression line ($H_0: \beta_1 = 0$) also tests this same basic concept. Either one of the two is sufficient to determine the answer to this query.

Although the "lack-of-fit" test can be statistically determined, it is beyond the scope of this text. However, we are very likely to carry this test out at a subjective level when we view the scatter diagram. If the scatter diagram is carefully constructed, this subjective decision will determine the mathematical model for the regression line that you believe will fit the data.

The two concepts, linear correlation and regression, are quite different, because each measures different characteristics. It is possible to have data that yield a strong linear correlation coefficient and have the wrong model. For example, the straight line can be used to approximate almost any curved line if the interval of domain is restricted sufficiently. In such a case the linear correlation coefficient can become quite high but the curve will still not be a straight line. Figure 13–15 suggests one such interval where r could be significant, but the scatter diagram does not suggest a straight line.

Regression analysis should be used to answer questions about the relationship between two variables. Such questions as "What is the relationship?" "How are two variables related?" and so on, require this regression analysis.

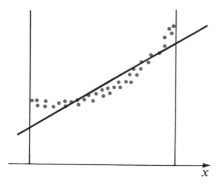

FIGURE 13–15
The Value of r Is High but the Relationship Is Not Linear

In Retrospect

In this chapter we have made a more thorough inspection of the linear relationship between two variables. Although the curvilinear and multiple regression situations were only mentioned in passing, the basic techniques and concepts have been explored. We would only have to modify our mathematical model and our formulas if we wanted to deal with these other relationships.

In this chapter we have drawn scatter diagrams and constructed confidence belts for estimations. But both were done for one population at a time. The news article at the beginning of the chapter shows a unique graphic presentation. The graph in the article shows a set of confidence belts for one population (the weights of Lubchenco et al.) and then plots a scatter diagram of the very low-birth-weight infants on the same graph. One definitely gets the feeling that there is a difference between the weights of the infants from these two different populations.

Although it was not directly emphasized, we have applied many of the topics of earlier chapters in this chapter. The ideas of hypothesis testing and confidence interval estimates were applied to the regression problem. Reference was made to the sampling distribution of the sample slope b_1. This allowed us to make inferences about β_1, the slope of the population from which the sample was drawn. We estimated the mean value of y at a fixed value of x by pooling the variance for the slope with the variance of the y's. This was allowable since they are independent. You should recall that in chapter 10 we presented formulas for combining the variances of independent samples. The idea here is much the same. Finally, we added a measure of variance for individual values of y and made estimates for these individual values of y at fixed values of x.

As this chapter ends you should be aware of the basic concepts of regression analysis and correlation analysis. You should now be able to collect the data for and do a complete analysis on any two-variable linear relationship.

Chapter Exercises

13-23   Answer the following as "Sometimes," "Always," or "Never." Explain each "Never" and "Sometimes" response.

(a) The correlation coefficient has the same sign as the slope of the least squares line fitted to the same data.

(b) A correlation coefficient of 0.99 indicates a strong causal relationship between the variables under consideration.

(c) An r value greater than zero indicates that ordered pairs with high x-values will have low y-values.

(d) The two coefficients for the line of best fit have the same sign.

(e) If x and y are independent, then the population correlation coefficient equals zero.

13-24   Four social-service agency caseworkers in a large urban area were randomly selected from all social-service agency caseworkers in the area. Their salaries (in thousands of dollars) and ages follow:

Age	37	41	62	54
Salary	20	18	27	24

(a) Calculate the correlation coefficient r and test its significance at the $\alpha = 0.05$ level.

(b) Find the equation of the line of best fit and use it to predict the salary of a 52-year-old caseworker.

13-25   One would usually expect that the number of hours studied, x, in preparation for a particular examination would have direct (positive) correlation with the grade, y, attained on that exam. The hours studied and the grades obtained by ten students selected at random from a large class are shown in the accompanying table.

	Student									
	1	*2*	*3*	*4*	*5*	*6*	*7*	*8*	*9*	*10*
Hours Studied, x	10	6	15	11	7	19	17	3	13	17
Exam Grade, y	51	36	67	63	44	89	80	26	50	85

(a) Draw the scatter diagram and estimate r.

(b) Calculate r. (How close was your estimate?)

(c) Is there significant positive correlation shown by this data? Test at $\alpha = 0.01$.

(d) Find the 95% confidence interval estimate for the true population value of ρ.

13-26 When buying almost any item, it is often advantageous to buy in as large a quantity as possible. The unit price is usually less for the larger quantities. The accompanying data were obtained to test this theory.

Number of Units, x	1	3	5	10	12	15	24
Cost per Unit, y	55	52	48	36	32	30	25

(a) Calculate r.

(b) Does the number of units ordered have an effect on the unit cost? Test at $\alpha = 0.05$.

(c) Find the 95% confidence interval estimate for the true value of ρ.

13-27 When two dice are rolled simultaneously, it is expected that the result on each roll will be independent of the other. To test this, roll a pair of dice 12 times. Identify one die as x and the other as y. Record the number observed on each of the two dice.

(a) Draw the scatter diagram of x versus y.

(b) If the two do behave independently, what value of r can be expected?

(c) Calculate r.

(d) Test for independence of these dice at $\alpha = 0.10$.

13-28 In economics one frequently hears about the consumption function, the relationship between Gross National Product (GNP) and Personal Consumption Expenditures (C). The data below are for the years of the great depression. The data are expressed in billions of 1926 dollars (constant dollars). Find the relationship between C and GNP. Use GNP as the input variable for these years.

Year	GNP	C	Year	GNP	C
1929	104.4	79.0	1935	72.5	56.3
1930	91.1	71.0	1936	82.7	62.6
1931	76.3	61.3	1937	90.8	67.3
1932	58.5	49.3	1938	85.2	64.6
1933	56.0	46.4	1939	91.1	67.6
1934	65.0	51.9	1940	100.6	81.9

(a) Determine the line of best fit.

(b) Is there sufficient reason to conclude that there is evidence of positive slope at the 0.05 level of significance?

(c) Based on the sample information and given that the 1941 GNP was 125.8, find the 0.95 confidence interval estimate for the 1941 mean value of C.

13–29 The accompanying data resulted from an experiment performed for the purpose of regression analysis. The input variable x was set at five different levels, and observations were made at each level.

x	0.5	1.0	2.0	3.0	4.0
y	3.8	3.2	2.9	2.4	2.3
	3.5	3.4	2.6	2.5	2.2
	3.8	3.3	2.7	2.7	2.3
		3.6	3.2	2.3	

(a) Draw a scatter diagram of the data.

(b) Draw the regression line by eye.

(c) Place an asterisk, *, at each level, approximately where the mean of the observed y values is located. Does your regression line look like the line of best fit for these five mean values?

(d) Calculate the equation of the regression line.

(e) Find the standard deviation of y about the regression line.

(f) Construct a 95% confidence interval estimate for the true value of β_1.

(g) Construct a 95% confidence interval estimate for the mean value of y at $x = 3.0$. At $x = 3.5$.

(h) Construct a 95% confidence interval estimate for an individual value of y at $x = 3.0$. At $x = 3.5$.

P 13-30   The accompanying set of 25 scores was randomly selected from a teacher's class list. Let x be the prefinal average and y the final examination score. (The final examination had a maximum of 75 points.)

Student	x	y	Student	x	y
1	75	64	14	73	62
2	86	65	15	78	66
3	68	57	16	71	62
4	83	59	17	86	71
5	57	63	18	71	55
6	66	61	19	96	72
7	55	48	20	96	75
8	84	67	21	59	49
9	61	59	22	81	71
10	68	56	23	58	58
11	64	52	24	90	67
12	76	63	25	92	75
13	71	66			

(a) Draw a scatter diagram for these data.

(b) Draw the regression line (by eye) and estimate its equation.

(c) Estimate the value of the coefficient of linear correlation.

(d) Calculate the equation of the line of best fit.

(e) Draw this line on your graph. How does it compare with your estimate?

(f) Calculate the linear correlation coefficient. How does it compare with your estimate?

(g) Test the significance of r at $\alpha = 0.10$.

(h) Find the 95% confidence interval estimate for the true value of ρ.

(i) Find the standard deviation of the y values about the regression line.

(j) Calculate a 95% confidence interval estimate for the true value of the slope, β_1.

(k) Test the significance of the slope at $\alpha = 0.05$.

(l) Predict the mean final exam grade that all students with an 85 prefinal average will obtain (95% confidence interval).

(m) Predict, using the 95% confidence interval, the grade that John Henry will receive on his final, knowing that his prefinal average is 78.

& 13-31   Twenty-one mature flowers of a particular species were dissected and the number of stamens and carpels present in each flower were counted. See the accompanying table.

(a) Is there sufficient evidence to claim a linear relationship between these two variables, at $\alpha = 0.05$?

(b) What is the relationship between the number of stamens and carpels in this variety of flower?

(c) Is the slope of the regression line significant, at $\alpha = 0.05$?

Stamens, x	Carpels, y	Stamens, x	Carpels, y
52	20	74	29
68	31	38	28
70	28	35	25
38	20	45	27
61	19	72	21
51	29	59	35
56	30	60	27
65	30	73	33
43	19	76	35
37	25	68	34
36	22		

(d) Give the 0.95 confidence interval estimate for the number of carpels that one would expect to find in a mature flower of this variety if the number of stamens were 64.

13–32 It is believed that the amount of nitrogen fertilizer used per acre has a direct effect on the amount of wheat produced. The accompanying data show the amount of nitrogen fertilizer used per test plot and the amount of wheat harvested per test plot.

Pounds of Fertilizer, x	100 Pounds of Wheat, y	Pounds of Fertilizer, y	100 Pounds of Wheat, y
30	5	70	19
30	9	70	23
30	14	70	31
40	6	80	24
40	14	80	32
40	18	80	35
50	12	90	27
50	14	90	32
50	23	90	38
60	18	100	34
60	24	100	35
60	28	100	39

(a) Is there sufficient reason to conclude that the use of more fertilizer results in a higher yield? Use $\alpha = 0.05$.

(b) Estimate, with a 0.98 confidence interval, the mean yield that could be expected if 50 pounds of fertilizer is used per plot.

(c) Estimate, with a 0.98 confidence interval, the mean yield that could be expected if 75 pounds of fertilizer is used per plot.

Vocabulary List

Be able to define each term. In addition, describe in your own words and give an example of each term. Your examples should not be ones given in class or in the textbook.

The terms followed by bracketed numbers were defined in earlier chapters (the bracketed numbers are those of the earlier chapters) but we ask you to define them again to show your increased understanding of their meaning.

bivariate data [3]
coefficient of linear
  correlation [3]
confidence belts
confidence interval
  estimate [8, 9, 10]
covariance
curvilinear regression
experimental error (ε or e)
intercept (b_0 or β_0) [3]
linear regression [3]
line of best fit [3]

multiple regression
Pearson's product moment r
predicted value of y ($\hat{y}$)
regression line [3]
rho (ρ)
sampling distribution [7, 10]
scatter diagram [3]
slope (b_1 or β_1) [3]
standard error [7, 8, 9]
sum of squares for error [1, 2]
variance (s^2 or σ^2)
  [2, 8, 9, 10, 12]

Quiz

Answer "True" if the statement is always true. If the statement is not always true, replace the words shown in color with words that make the statement always true.

13–1   The error **must be** normally distributed if inferences are to be performed.

13–2 x and y **must both be** normally distributed.

13–3   A high correlation between x and y **proves** that x causes y.

13–4   The values of the input variable **must be** randomly selected to achieve valid results.

13–5   The output variable must be **normally distributed** about the regression line for each value of x.

13–6   **Covariance** measures the strength of the linear relationship and is a standardized measure.

13–7   The **sum of squares for error** is the name given to the numerator portion of the formula for the calculation of the variance of y about the line of regression.

13–8   **Correlation** analysis attempts to find the equation of the line of best fit for two variables.

13–9   There are $n - 3$ degrees of freedom involved with the inferences about the regression line.

13–10 $\hat{y}$ serves as the **point estimate** for both $\mu_{y|x_0}$ and y_{x_0}.

14 Elements of Nonparametric Statistics

Study Finds That Money Can't Buy Job Satisfaction

When it comes to getting workers to produce—do their level best—money is less than everything. Feeling appreciated—having a sense of being recognized—is more important.

That is the thesis of Kenneth A. Kovach, of George Mason University....

Kovach reproduces a consequential survey of 30 years ago in which workers and supervisors gave their opinions of "what workers want from their jobs and what management thinks they want." Here it is:

Managements today are more "behavioral" in their approach to managing. Nevertheless, Kovach contends that a wide gap still exists, a gap that is suggested by what workers rank as No. 1 in importance to them ("appreciation") and what supervisors consider No. 1 to workers ("good wages")....

	Worker Ranking	Boss Ranking
Full appreciation of work done	1	8
Feeling of being in on things	2	10
Sympathetic help on personal problems	3	9
Job security	4	2
Good wages	5	1
Interesting work	6	5
Promotion and growth in the organization	7	3
Personal loyalty to employees	8	6
Good working conditions	9	4
Tactful disciplining	10	7

Reprinted by permission of the *Philadelphia Inquirer*, 29 December 1976.

Chapter Objectives

Nonparametric methods of statistics dominate the success story of statistics in recent years. Unlike their parametric counterparts, many of the best-known nonparametric tests, also known as distribution-free tests, are founded on a basis of elementary probability theory. The derivation of most of these tests is well within the grasp of the student who is competent in high school algebra and understands binomial probability. Thus the nonmathematical statistics user is much more at ease with the nonparametric techniques.

This chapter is intended to give you a feeling for basic concepts involved in nonparametric techniques and to show you that nonparametric methods are extremely versatile and easy to use, once a table of critical values is developed for a particular application. The selection of the nonparametric methods presented here includes only a few of the common tests and applications. You will learn about the sign test, the Mann-Whitney U test, the runs test, and Spearman's rank correlation test. These will be used to make inferences corresponding to both one- and two-sample situations.

Section 14-1 Nonparametric Statistics

parametric method

Most of the statistical procedures that we have studied in this book are known as **parametric methods**. For a statistical procedure to be parametric, we either assume that the parent population is at least approximately normally distributed or we rely on the central limit theorem to give us a normal approximation. This is particularly true of the statistical methods studied in chapters 8, 9, and 10.

non-parametric or distribution-free methods

The **nonparametric methods**, or **distribution-free methods**, as they are also known, do not depend on the distribution of the population being sampled. The nonparametric statistics are usually subject to much less confining restrictions than are their parametric counterparts. Some, for example, require only that the parent population be continuous.

The recent popularity of nonparametric statistics can be attributed to the following characteristics:

1. Nonparametric methods require few assumptions about the parent population.

2. Nonparametric methods are generally easier to apply than their parametric counterparts.

3. Nonparametric methods are relatively easy to understand.

4. Nonparametric methods can be used in situations where the normality assumptions cannot be made.

5. Nonparametric methods appear to be wasteful of information in that they sacrifice the value of the variable for only a sign or a rank number. However, nonparametric statistics are generally only slightly less efficient than their parametric counterparts.

Section 14–2 The Sign Test

sign test The ordinary **sign test** is a very versatile and exceptionally easy-to-apply nonparametric method that **uses only plus and minus signs**. (There are several specific techniques.) The sign test is useful in two situations: (1) a hypothesis test concerning the value of the median for one population and (2) a hypothesis test concerning the median difference (paired difference) for two dependent samples. Both tests are carried out using the same basic procedures and are nonparametric alternatives to the t tests used with one mean (section 9–1) and the difference between two dependent means (section 10–5).

Single-Sample Hypothesis Test Procedure

The sign test can be used when a random sample is drawn from a population with an unknown **median**, M, and the population is assumed to be continuous in the vicinity of M. The null hypothesis to be tested concerns the value of the population median M. The test may be either one- or two-sided. We present the test procedure in the next illustration.

Illustration 14–1

A random sample of 75 students was selected and each was asked to carefully measure the amount of time required to commute from their front door to the college parking lot. The data collected were used to test the hypothesis "the median time required for students to commute is 15 minutes" against the alternative that the median is unequal to 15 minutes. The 75 pieces of data were summarized as follows:

Under 15:   18

15:               15

Over 15:   42

Use the sign test to test the null hypothesis against the alternative hypothesis.

Solution   The data are converted to $(+)$ and $(-)$ signs. A plus sign will be assigned to each piece of data larger than 15, a minus sign to each piece of data smaller than 15, and a zero to those data equal to 15. The sign test uses only the plus and minus signs; therefore, the zeros are discarded and the usable sample size becomes 60. That is, $n(+) = 42$, $n(-) = 18$, and $n = n(+) + n(-) = 42 + 18 = 60$.

STEP 1 $H_0: M = 15$

STEP 2 $H_a: M \neq 15$

STEP 3 $\alpha = 0.05$ for a two-sided test. The **test statistic** that will be used is the **number of the less frequent sign**, the smaller of $n(+)$ and $n(-)$, which is $n(-)$ for our illustration. We will want to reject the null hypothesis whenever the number of the less frequent sign is extremely small. Table 11 of appendix D gives the maximum allowable number of the less frequent sign, k, that will allow us to reject the null hypothesis. That is, if the number of the less frequent sign is less than or equal to the critical value in the table, we will reject H_0. If the observed value of the less frequent sign is larger than the table value, we will fail to reject H_0. n in the table is the total number of signs, not including zeros.

For our illustration $n = 60$ and the critical value from the table is 21. See the diagram.

Reject H_0	Fail to reject H_0
0 21	22

Number of less frequent sign

STEP 4 The observed statistic is $x = n(-) = 18$ and it falls in the critical region.

STEP 5 *Decision*: Reject H_0.

Conclusion: The sample shows sufficient evidence at the 0.05 level to reject the claim that the median is 15 minutes. ☐

Two-Sample Hypothesis Test Procedure

The sign test may also be applied to a hypothesis test dealing with the **median difference** between paired data that result from **two dependent samples**. A familiar application is the use of before-and-after testing to determine the effectiveness of some activity. In a test of this nature the signs of the differences are used to carry out the test. Again zeros are disregarded. The following illustration shows the procedure.

Illustration 14–2

A new no-exercise, no-starve weight-reducing plan has been developed and advertised. To test the claim that "you will lose weight within 2 weeks or...," a local statistician obtained the before-and-after weights of 18 people who had used this plan. Table 14–1 lists the people, their weights, and a — for those who lost weight during the 2 weeks, a 0 for those who remained the same, and a + for those who actually gained weight.

The claim being tested is that people are able to lose weight. The null hypothesis that will be tested is that there is no weight loss (or the median weight loss is zero), meaning that only a rejection of the null hypothesis will allow us to conclude in favor of the advertised claim. Actually we will be testing to see if there are significantly more minus signs than plus signs. If the weight-reducing plan is of absolutely no value, we would expect to find an equal number of plus and minus signs. If it works, there should be significantly more minus signs than plus signs. Thus the test performed here will be a one-tailed test. (We will want to reject the null hypothesis in favor of the advertised claim if there are "many" minus signs.)

Solution

STEP 1 H_0: $M = 0$ (no weight loss).

STEP 2 H_a: $M < 0$ (weight loss).

TABLE 14-1

Sample Results for Illustration 14-2

Person	Weight Before	Weight After	Sign of Difference, Before to After
Mrs. Smith	146	142	−
Mr. Brown	175	178	+
Mrs. White	150	147	−
Mr. Collins	190	187	−
Mr. Gray	220	212	−
Miss Collins	157	160	+
Mrs. Allen	136	135	−
Mrs. Noss	146	138	−
Miss Wagner	128	132	+
Mr. Carroll	187	187	0
Mrs. Black	172	171	−
Mrs. McDonald	138	135	−
Miss Henry	150	151	+
Miss Greene	124	126	+
Mr. Tyler	210	208	−
Mrs. Williams	148	148	0
Mrs. Moore	141	138	−
Mrs. Sweeney	164	159	−

Reject H_0					Fail to reject H_0
0	1	2	3	4	5

Number of less frequent sign

STEP 3   Use $\alpha = 0.05$. $n(+) = 5$, $n(-) = 11$, and $n = 16$. The critical value from table 11 shows $k = 4$ as the maximum allowable number. (You must use the $\alpha = 0.10$ column, as the table is set up for two-tailed tests.)

STEP 4 $x = n(+) = 5$

STEP 5   *Decision*: Fail to reject H_0 (we have too many plus signs).

Conclusion: The evidence observed is not sufficient to allow us to reject the no-weight-loss null hypothesis. □

The sign test may be carried out by means of a **normal approximation** using the standard normal variable z. The normal approximation will be used if table 11 does not show the particular critical value desired or if n is large. z will be calculated by using the formula

$$z = \frac{x' - n/2}{(\frac{1}{2})\sqrt{n}}$$ (14-1)

(See note 3 with regard to x'.)

NOTES:

1. x may be the number of the less frequent sign or the most frequent sign. You will have to determine this in such a way that your test is consistent with the interpretation of the situation.

2. x is really a binomial random variable, where $p = \frac{1}{2}$. The sign test statistic satisfies the properties of a binomial experiment (see page 184). Each sign is the result of an independent trial. There are n trials, and each trial has two possible outcomes ($+$ or $-$). Since the median is used, the probabilities for each outcome are both $\frac{1}{2}$. Therefore, the mean μ_x is equal to

$$n/2(\mu = np = n \cdot \tfrac{1}{2} = n/2)$$

and the standard deviation σ_x is equal to

$$(\tfrac{1}{2})\sqrt{n}\,(\sigma = \sqrt{npq} = \sqrt{n \cdot \tfrac{1}{2} \cdot \tfrac{1}{2}} = \tfrac{1}{2}\sqrt{n})$$

3. x is a discrete variable. But recall that the normal distribution must be used only with continuous variables. However, although the binomial random variable is discrete, it does become approximately normally distributed for large n. Nevertheless, when using the normal distribution for testing, we should make an adjustment in the variable so that the approximation is more accurate. (See section 6–5 on the normal approximation.) This adjustment is illustrated in figure 14–1 and is called a **continuity correction**. For this discrete variable the area that represents the probability is a rectangular bar. Its width is 1 unit wide, from $\frac{1}{2}$ unit below to $\frac{1}{2}$ unit above the value of interest. Therefore, when z is to be used, we will need to make a $\frac{1}{2}$-unit adjustment before calculating the observed value of z. x' will be the adjusted value for x. If x is larger than $n/2$, $x' = x - \frac{1}{2}$. If x is smaller than $n/2$, $x' = x + \frac{1}{2}$. The test is then completed by the usual procedure.

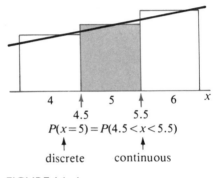

FIGURE 14–1
Continuity Correction

Illustration 14–3

Use the sign test to test the hypothesis that the median number of hours, M, worked by students of our college is at least 15 hours per week. A survey of 120 students was taken; a plus sign was recorded if the student responded with a number of hours worked last week that was equal to or larger than 15 and a minus sign was recorded if the response was less than 15 hours. Totals showed 80 minus signs and 40 plus signs.

Solution

STEP 1 $H_0: M = 15\,(\geq)$ (at least as many plus signs as minus signs).

STEP 2 $H_a: M < 15$ (fewer plus signs than minus signs).

STEP 3 $\alpha = 0.05$; x is the number of plus signs. The critical value is shown in the figure.

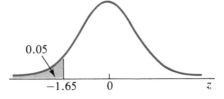

STEP 4

$$z = \frac{x' - n/2}{(\frac{1}{2})\sqrt{n}} = \frac{40.5 - 60}{(\frac{1}{2})\sqrt{120}} = \frac{-19.5}{(\frac{1}{2})(10.95)}$$

$$= \frac{-19.5}{5.475} = -3.562$$

$$z^* = -3.56$$

STEP 5   *Decision*: Reject H_0.

Conclusion: There are significantly more minus signs than plus signs, thereby implying that the median is less than the claimed 15 hours.   □

Confidence Interval Procedure

The sign test techniques can be applied to obtain a confidence interval estimate for the unknown population median M. To accomplish this we will need to arrange the data in ascending order (smallest to largest). The data are identified as x_1 (smallest), x_2, x_3, ..., x_n (largest). The critical value k for the maximum allowable number of signs, obtained from table 11, tells us the number of positions to be dropped from each end of the ordered data. The remaining extreme values become the bounds of the $1 - \alpha$ confidence interval. That is, the lower boundary for the confidence interval is x_{k+1}, the $(k + 1)$th piece of data; the upper boundary is x_{n-k}, the $(n - k)$th piece of data. The following illustration will clarify this procedure.

Illustration 14–4

Suppose we have 12 pieces of data in ascending order $(x_1, x_2, x_3, ..., x_{12})$ and we wish to form a 0.95 confidence interval for the population median. Table 11 shows a critical value of 2 ($k = 2$) for $n = 12$ and $\alpha = 0.05$ for a hypothesis test. This means that the critical region for a two-tailed hypothesis test would contain the last two values on each end (x_1 and x_2 on the left; x_{11} and x_{12} on the right). The noncritical region is then from x_3 to x_{10}, inclusive. The 0.95 confidence interval is then x_3 to x_{10}, expressed as

$$x_3 \quad \text{to} \quad x_{10}, \qquad 0.95 \text{ confidence interval for } M \qquad □$$

In general, the two data that bound the confidence interval will occupy positions $k + 1$ and $n - k$, where k is the table value read from table 11. Thus

$$x_{k+1} \quad \text{to} \quad x_{n-k}, \qquad 1 - \alpha \text{ confidence interval for } M$$

If the normal approximation is to be used (including the continuity correction), the position numbers become

$$(\tfrac{1}{2})n \pm (\tfrac{1}{2} + \tfrac{1}{2} \cdot z(\alpha/2) \cdot \sqrt{n}) \qquad (14–2)$$

The interval is

$$x_L \quad \text{to} \quad x_U, \qquad 1 - \alpha \text{ confidence interval for } M$$

where

$$L = \frac{n}{2} - \frac{1}{2} - \frac{z(\alpha/2)}{2} \cdot \sqrt{n}$$

$$U = \frac{n}{2} + \frac{1}{2} + \frac{z(\alpha/2)}{2} \cdot \sqrt{n}$$

(L should be rounded down and U should be rounded up to be sure that the level of confidence is at least $1 - \alpha$.)

Illustration 14-5

Estimate the population median with a 0.95 confidence interval given a set of 60 pieces of data, $x_1, x_2, x_3, \ldots, x_{59}, x_{60}$.

Solution   When we use formula (14-2), the position numbers L and U are

$$(\tfrac{1}{2})(60) \pm [\tfrac{1}{2} + \tfrac{1}{2}(1.96)\sqrt{60}]$$

$$30 \pm 0.50 + 7.59$$

$$30 \pm 8.09$$

Thus

$$L = 30 - 8.09 = 21.91 \quad \text{(21st piece of data)}$$

$$U = 30 + 8.09 = 38.09 \quad \text{(39th piece of data)}$$

Therefore,

$$x_{21} \text{ to } x_{39}, \qquad 0.95 \text{ confidence interval for } M \qquad \square$$

Exercises

14-1   Use the sign test to test the hypothesis that the median daily high temperature in the city of Rochester, New York, during the month of December is 48 degrees. The following daily highs were recorded on 20 randomly selected days during December:

47 46 40 40 46 35 34 59 54 33
65 39 48 47 46 46 42 36 45 38

(a) State the null hypothesis under test.

(b) State specifically what it is that you are actually testing when using the sign test.

(c) Complete the test for $\alpha = 0.05$, and carefully state your findings.

14-2 The coach of our school's swimming team has kept close records on the time that it takes his freestyle swimmer to swim 50 yards. The swimmer claims that his median time must be no more than 24.8 seconds. Use the sample drawn from the coach's records and test the swimmer's claim by using the sign test (use $\alpha = 0.05$).

24.7 24.7 24.6 25.5 25.7 25.8 26.5 24.5 25.3
26.2 25.5 26.3 24.2 25.3 24.3 24.2 24.2

(a) Perform the sign test. Use both techniques shown in the text:
   (1) the number of the least frequent sign and table 11
   (2) the number of signs and the standard normal z

(b) Perform the hypothesis test by using the parametric method, the t test.

(c) Compare the results of the three methods.

14-3 Determine the 0.95 confidence interval for the median daily high temperature in Rochester during December based on the sample given in exercise 14-1.

14-4 Find the 0.75 confidence interval estimate for the median swim time for the swimmer in exercise 14-2.

14-5 Sixteen students were given elementary statistics tests on a hot day. Eight randomly selected students took the test in a room that was not air-conditioned and then after a short break they completed a similar test in an air-conditioned room. The procedure was reversed for the other eight students.

Student	1	2	3	4	5	6	7	8	9	10	11	12	13	14	15	16
Not air-conditioned	52	90	63	74	87	77	92	72	77	94	67	86	78	84	57	55
Air-conditioned	49	94	60	78	93	77	93	74	78	93	78	89	92	83	49	68

Does the sample provide sufficient reason to conclude that the use of air conditioning on a hot day has an effect on test grades? Use $\alpha = 0.05$.

Section 14-3 The Mann-Whitney U Test

Mann-Whitney U test   The **Mann-Whitney U test** is a nonparametric alternative for the t test for the **difference between two independent means**. It can be applied when we have two independent random samples (independent within each sample as well as between samples) in which the random variable is continuous. This test is often applied in situations in which the two samples are drawn from the same population but different "treatments" are used on each set. We will demonstrate the procedure in the following illustration.

Illustration 14–6

In a large lecture class, when the instructor gives an hour exam, she makes two "equivalent" examinations. Students in even-numbered seats take form A and those in the odd-numbered seats take form B. It is very reasonable to ask: "Are these two different exam forms equivalent?" Assuming that the odd- or even-numbered seats had no effect, we would want to test the hypothesis "the test forms yielded scores that had identical distributions."

To test this hypothesis the following two random samples were taken:

A	52	78	56	90	65	86	64	90	49	78
B	72	62	91	88	90	74	98	80	81	71

The size of the individual samples will be called n_a and n_b; actually it makes no difference which way these are assigned. In our illustration they both have the value 10.

The first thing that must be done with the entire sample (all $n_a + n_b$ pieces of data) is to order it into one sample, smallest to largest:

49 52 56 62 64 65 71 72 74 78

78 80 81 86 88 90 90 90 91 98

rank

Each piece of data is then assigned a **rank number**. The smallest (49) is assigned rank 1, the next smallest (52) is assigned rank 2, and so on up to the largest, which is assigned rank $n_a + n_b$ (20). Ties are handled by assigning to each of the tied observations the mean rank of those rank positions that they occupy. For example, in our illustration there are two 78s; they are the 10th and 11th pieces of data. The mean rank for each is then $(10 + 11)/2 = 10.5$. In the case of the three 90s, the 16th, 17th, and 18th pieces of data, each is assigned 17, since $(16 + 17 + 18)/3 = 17$. The rankings are shown in table 14–2.

TABLE 14–2
Ranked Data for Illustration 14–6

Ranked Data	Rank	Source	Ranked Data	Rank	Source
49	1	A	78	10.5	A
52	2	A	80	12	B
56	3	A	81	13	B
62	4	B	86	14	A
64	5	A	88	15	B
65	6	A	90	17	A
71	7	B	90	17	A
72	8	B	90	17	B
74	9	B	91	19	B
78	10.5	A	98	20	B

The calculation of the test statistic U is a two-step procedure. We first determine the **sum of the ranks** for each of the two samples. Then using the two sum of ranks, we calculate a U **score** for each sample. The **smaller U score is the test statistic**.

The sum of ranks R_a for sample A is computed as

$$R_a = 1 + 2 + 3 + 5 + 6 + 10.5 + 10.5 + 14 + 17 + 17 = 86$$

The sum of ranks R_b for sample B is

$$R_b = 4 + 7 + 8 + 9 + 12 + 13 + 15 + 17 + 19 + 20 = 124$$

The U score for each sample is obtained by use of the following pair of formulas:

$$U_a = n_a \cdot n_b + \frac{(n_b)(n_b + 1)}{2} - R_b \qquad (14\text{--}3)$$

$$U_b = n_a \cdot n_b + \frac{(n_a)(n_a + 1)}{2} - R_a \qquad (14\text{--}4)$$

and U, the test statistic, will be the smaller of U_a and U_b.

For our illustration we obtain

$$U_a = (10)(10) + \frac{(10)(10 + 1)}{2} - 124 = 31$$

$$U_b = (10)(10) + \frac{(10)(10 + 1)}{2} - 86 = 69$$

Therefore,

$$U = 31$$

Before we carry out the test for the illustration, let's try to understand some of the underlying possibilities. Recall that the null hypothesis is that the distributions are the same and that we will most likely want to conclude from this that the means are approximately equal. Suppose for a moment that they are indeed quite different—say all of one sample comes before the smallest piece of data in the second sample when they are ranked together. This would certainly mean that we want to reject the null hypothesis. What kind of a value can we expect for U in this case? Suppose in illustration 14–6 that the 10 A values had ranks 1 through 10 and the 10 B values ranks 11 through 20. Then we would obtain

$$R_a = 55 \qquad R_b = 155$$

$$U_a = (10)(10) + \frac{(10)(10 + 1)}{2} - 155 = 0$$

$$U_b = (10)(10) + \frac{(10)(10 + 1)}{2} - 55 = 100$$

Therefore,

$$U = 0$$

Suppose, on the other hand, that both samples were perfectly matched, that is, a score in each set identical to one in the other. Now what would happen?

54	54	62	62	71	71	72	72	$\cdots$
A	B	A	B	A	B	A	B	$\cdots$
1.5	1.5	3.5	3.5	5.5	5.5	7.5	7.5	$\cdots$

$$R_a = R_b = 105$$

$$U_a = U_b = (10)(10) + \frac{(10)(10 + 1)}{2} - 105 = 50$$

Therefore,

$$U = 50$$

If this case were to exist, we certainly would want to reach the decision "fail to reject the null hypothesis."

It should be noted that the sum of the two U's ($U_a + U_b$) will always be equal to the product of the two sample sizes ($n_a \cdot n_b$). For this reason we need to concern ourselves with only one of them, the smaller one.

Now let's return to the solution of illustration 14–6.

In order to complete our hypothesis test, we need to be able to determine a critical value for U. Table 12 in appendix D gives us the critical value for some of the more common testing situations, as long as both sample sizes are reasonably small. Table 12 shows only the critical region in the left-hand tail, and the null hypothesis will be rejected if the observed value for U is less than or equal to the value read from the table. For our example $n_a = n_b = 10$; at $\alpha = 0.05$ in a two-tailed test, the critical value is 23. We observed a value of 31 and therefore we make the decision "fail to reject the null hypothesis." This means that we do not have sufficient evidence to reject the "equivalent" hypothesis. □

If the **samples are larger than size 20**, we may make the test decision with the aid of the standard normal variable z. This is possible since the distribution of U is approximately normal with a mean

$$\mu_U = \frac{n_a \cdot n_b}{2} \tag{14-5}$$

and a standard deviation

$$\sigma_U = \sqrt{\frac{n_a n_b (n_a + b_b + 1)}{12}} \tag{14-6}$$

The null hypothesis is then tested by use of

$$z = \frac{U - \mu_U}{\sigma_U} \tag{14-7}$$

in the typical fashion. The test statistic z may be used whenever n_a and n_b are both greater than 10.

Illustration 14-7

A dog obedience trainer is training 27 dogs to obey a certain command. The trainer is using two different training techniques, the reward-and-encouragement method (I) and the no-reward method (II). The accompanying table shows the number of obedience sessions that were necessary before the dog would obey the command. Does the trainer have sufficient evidence to claim that the reward method will on the average require less training time? ($\alpha = 0.05$).

I	29	27	32	25	27	28	23	31	37	28	22	24	28	31	34
II	40	44	33	26	31	29	34	31	38	33	42	35			

Solution

STEP 1 H_0: The average amount of training time required is the same for both methods.

STEP 2 H_a: The reward method requires less time on the average.

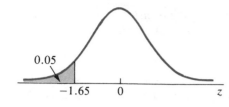

STEP 3 $\alpha = 0.05$; the critical value is shown in the figure.

STEP 4   The two sets of data are ranked jointly and ranks are assigned, as shown in table 14–3. Then the sums are

$$R_\mathrm{I} = 1 + 2 + 3 + 4 + 6.5 + \cdots + 20.5 + 23 = 151.0$$

$$R_\mathrm{II} = 5 + 11.5 + \cdots + 26 + 27 = 227.0$$

TABLE 14–3
Rankings for Training Methods

Number of Sessions	Group	Rank		Number of Sessions	Group	Rank	
22	I	1		31	II	15	14.5
23	I	2		31	II	16	14.5
24	I	3		32	I	17	
25	I	4		33	II	18	18.5
26	II	5		33	II	19	18.5
27	I	6	6.5	34	I	20	20.5
27	I	7	6.5	34	II	21	20.5
28	I	8	9	35	II	22	
28	I	9	9	37	I	23	
28	I	10	9	38	II	24	
29	I	11	11.5	40	II	25	
29	II	12	11.5	42	II	26	
31	I	13	14.5	44	II	27	
31	I	14	14.5				

Now by using formulas (14–3) and (14–4), we can obtain the value of the U scores:

$$U_\mathrm{I} = (15)(12) + \frac{(12)(12 + 1)}{2} - 227 = 180 + 78 - 227 = 31$$

$$U_\mathrm{II} = (15)(12) + \frac{(15)(15 + 1)}{2} - 151 = 180 + 120 - 151 = 149$$

Therefore,

$$U = 31$$

Now we use formulas (14–5), (14–6), and (14–7) to determine the z statistic.

$$\mu_U = \frac{12 \cdot 15}{2} = 90$$

$$\sigma_U = \sqrt{\frac{(12)(15)(12 + 15 + 1)}{12}} = \sqrt{\frac{(180)(28)}{12}} = \sqrt{420} = 20.49$$

$$z = \frac{31 - 90}{20.49} = \frac{-59}{20.49} = -2.879$$

$$z^* = -2.88$$

STEP 5   *Decision*: Reject H_0.

Conclusion: The data show sufficient evidence to conclude that the reward method does on the average require less training time.   □

Exercises

14-6   Use the Mann-Whitney U test to test the claim that the amount of soft drink dispensed by each of two coin-operated vending machines is the same. ($\alpha = 0.05$.)

Machine	Amount Dispensed (Ounces)									
C	6.5	6.8	3.0	6.0	5.8	6.1	3.3	3.4	5.9	6.2
P	7.6	6.0	6.1	6.0	7.4	7.3	6.0	6.2	6.3	6.1

14-7   To determine if there is a difference in the breaking strength of lightweight monofilament fishing line produced by two companies, the accompanying test data (pounds of force required to break the line) were obtained. Use the Mann-Whitney U test and $\alpha = 0.05$ to determine if there is a difference.

A	12.4	11.9	11.8	13.5	11.6	12.0	12.9	11.3	13.8	12.3	
B	12.7	10.7	13.2	11.8	11.8	12.1	12.5	11.9	12.8	11.2	11.5	11.3

14-8   The accompanying set of data represents the ages of drivers involved in automobile accidents. Do these data present sufficient evidence to conclude that there is a difference in the average of men and women drivers involved in accidents? Use a two-tailed test at $\alpha = 0.05$.

Men	70	60	77	39	36	28	19	40
	23	23	63	31	36	55	24	76

Women	62	46	43	28	21	22	27	42
	21	46	33	29	44	29	56	70

(a) State the null hypothesis that is being tested.
(b) Complete the test.

14-9   In a heart study the systolic blood pressure was measured for 24 men aged 25 and for 30 men aged 40. (For your convenience, the data are ranked.) Do

25-year-olds	95	100	100	105	106	108	110	110		
	115	118	120	122	124	125	130	130		
	130	132	136	138	140	148	150	156		

40-year-olds	108	110	110	114	114	116	118	120	122	124
	126	126	128	130	130	132	136	136	136	140
	142	142	146	148	150	152	154	160	164	176

these data show sufficient evidence to conclude that the older men have a higher systolic blood pressure, at the 0.02 level of significance?

Section 14–4   The Runs Test

runs test
run

The **runs test** is most frequently used to test a question involving the **randomness of data** (or lack of randomness). A **run** is a sequence of data that possesses a common property. One run ends and another starts when the piece of data does not display the property in question. The random variable that will be used in this test is V, the number of runs that are observed.

Illustration 14–8

To illustrate the idea of runs, let's draw a sample of 10 single-digit numbers from the telephone book, using the next-to-last digit from each of the selected telephone numbers.

Sample: 2,   3,   1,   1,   4,   2,   6,   6,   6,   7

Let's consider the property of "odd" (o) or "even" (e). The sample, as it was drawn, becomes $e, o, o, o, e, e, e, e, e, o$, which displays four runs.

e o o o e e e e e o

So $V = 4$.   □

In illustration 14–8, if the sample contained no randomness, there would be only two runs—all the evens, then all the odds, or the other way around. We would also not expect to see them alternate: odd, even, odd, even. The maximum number of possible runs would be $n_1 + n_2$, or less (provided n_1 and n_2 are not equal), where n_1 and n_2 are the number of data that possess each of the two properties being identified.

We will often want to interpret the maximum number of runs as a rejection of a null hypothesis of randomness, since we often want to test randomness of the data in reference to how they were obtained. For example, if the data alternated all the way down the line, we might suspect that the data had been tampered with. There are many aspects to the concept of randomness. The occurrence of odd and even as discussed in illustration 14–8 is one. Another aspect of randomness that we might wish to check is the ordering of fluctuations of the data above (a) or below (b) the mean or median of the sample.

Illustration 14–9

Consider the sequence that results from determining whether each of the data points in the sample of illustration 14–8 is above or below the median value. Test the null hypothesis that this sequence is random. Use $\alpha = 0.05$.

Solution

STEP 1 H_0: The numbers in the sample form a random sequence with respect to the two properties "above" and "below" the median value.

STEP 2 H_a: The sequence is not random.

$$\text{Sample:} \quad 2, \quad 3, \quad 1, \quad 1, \quad 4, \quad 2, \quad 6, \quad 6, \quad 6, \quad 7$$

First we must rank the data and find the median. The ranked data are 1, 1, 2, 2, 3, 4, 6, 6, 6, 7. Since there are 10 pieces of data, the median is at the $i = 5.5$ position. Thus $\tilde{x} = (3 + 4)/2 = 3.5$. By comparing each number in the original sample to the value of the median, we obtain the following sequence of a's and b's:

$$b \quad b \quad b \quad b \quad a \quad b \quad a \quad a \quad a \quad a$$

We observe $n_a = 5$, $n_b = 5$, and 4 runs. So $V = 4$.

If n_1 and n_2 are both less than or equal to 20 and a two-tailed test at $\alpha = 0.05$ is desired, then table 13 of appendix D will give us the two critical values for the test. For our illustration with $n_a = 5$ and $n_b = 5$, table 13 shows critical values of 2 and 10. This means that if 2 or fewer, or 10 or more, runs are observed, the null hypothesis will be rejected. If between 3 and 9 runs are observed, we will fail to reject the null hypothesis.

Reject H_0	Fail to reject H_0	Reject H_0
0 2	3 9	10

V, number of runs

STEP 3 $\alpha = 0.05$ and a two-tailed test is used. The critical values for V are found in table 13 (see the diagram).

STEP 4   Four runs were observed; $V^* = 4$.

STEP 5   *Decision*: Fail to reject H_0.

Conclusion: We are unable to reject the hypothesis of randomness.   □

To complete the hypothesis test about randomness when n_1 or n_2 is larger than 20 or when α is other than 0.05, we will use z, the standard normal random variable. V is approximately normally distributed with a mean of μ_V and a standard deviation of σ_V. The formulas are as follows:

$$\mu_V = \frac{2n_1 \cdot n_2}{n_1 + n_2} + 1 \qquad (14\text{--}8)$$

$$\sigma_V = \sqrt{\frac{(2n_1 \cdot n_2)(2n_1 \cdot n_2 - n_1 - n_2)}{(n_1 + n_2)^2(n_1 + n_2 - 1)}} \qquad (14\text{--}9)$$

$$z = \frac{V - \mu_V}{\sigma_V} \qquad (14\text{--}10)$$

Illustration 14–10

Test the null hypothesis that the sequence that results from classifying the sample data in illustration 14–8 as "odd" or "even" is a random sequence. Use $\alpha = 0.10$.

Solution

STEP 1 H_0: The sequence of odd and even occurrences is a random sequence.

STEP 2 H_a: The sequence is not random.

STEP 3 $\alpha = 0.10$ and a two-tailed test is to be used. The test criteria are shown in the figure.

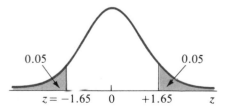

0.05   0.05

$z = -1.65$ 0 $+1.65$ z

STEP 4   The sample and the sequence of odd and even properties are shown in illustration 14–8. $n_1 = n(\text{even}) = 6$ and $n_2 = n(\text{odd}) = 4$. There were 4 runs, so $V = 4$.

$$\mu_V = \frac{2n_1 n_2}{n_1 + n_2} + 1 = \frac{2 \cdot 6 \cdot 4}{6 + 4} + 1 = \frac{48}{10} + 1 = 5.8$$

$$\sigma_V = \sqrt{\frac{(2n_1 n_2)(2n_1 n_2 - n_1 - n_2)}{(n_1 + n_2)^2 (n_1 + n_2 - 1)}}$$

$$= \sqrt{\frac{(2 \cdot 6 \cdot 4)(2 \cdot 6 \cdot 4 - 6 - 4)}{(6 + 4)^2 (6 + 4 - 1)}}$$

$$= \sqrt{\frac{(48)(38)}{(10)^2 (9)}} = \sqrt{\frac{1824}{900}} = \sqrt{2.027} = 1.42$$

$$z = \frac{V - \mu_V}{\sigma_V} = \frac{4.0 - 5.8}{1.42} = \frac{-1.8}{1.42} = -1.268$$

$$z^* = -1.27$$

STEP 5   *Decision*: Fail to reject H_0.

Conclusion: We are unable to reject the hypothesis of randomness.   ☐

Exercises

14–10   A student was asked to perform an experiment that involved tossing a coin 25 times. After each toss the student recorded the results. The following data were reported (H = heads, T = tails):

H, T, H, T, H, T, H, T, H, H, T, T, H,
H, T, T, H, T, H, T, H, T, H, T, H

Use the runs test at a 5% level of significance to test the student's claim that the results reported are random.

14–11   In an attempt to answer the question, "Does the husband (h) or wife (w) do the family banking?" the results of a sample of 28 married customers doing the family banking show arrivals at the bank in the following sequence:

w, w, w, w, h, w, h, h, h, h, w, w, w, w,
w, h, h, w, w, w, h, h, h, h, w, h, h, w

Do these data show lack of randomness with regard to whether it is the husband or wife who is doing the family banking? Use $\alpha = 0.05$.

14-12 The following data were collected in an attempt to show that the lateness (in number of minutes) of the city bus I commute to work on is steadily growing larger, The data are in order of occurrence.

$$6, \quad 1, \quad 3, \quad 9, \quad 10, \quad 10, \quad 2, \quad 5, \quad 5, \quad 6,$$
$$12, \quad 3, \quad 7, \quad 8, \quad 9, \quad 4, \quad 5, \quad 8, \quad 11, \quad 14$$

At $\alpha = 0.05$, do these data show sufficient lack of randomness to support my claim?

14-13 The number of absences recorded at a large 8 A.M. lecture that met on Monday and Thursday last semester were (in order of occurrence)

$$5, \quad 16, \quad 6, \quad 9, \quad 18, \quad 11, \quad 16, \quad 21, \quad 14, \quad 17, \quad 12, \quad 14, \quad 10,$$
$$6, \quad 8, \quad 12, \quad 13, \quad 4, \quad 5, \quad 5, \quad 6, \quad 1, \quad 7, \quad 18, \quad 26, \quad 6$$

Do these data show a randomness about the median value, at $\alpha = 0.05$? Complete this test by use of (1) critical values from table 13, and (2) the standard normal distribution.

Section 14-5 Rank Correlation

The rank correlation coefficient was developed by C. Spearman in the early 1900s. It is a nonparametric alternative to the linear correlation coefficient that was discussed in chapters 3 and 13. Only rankings are used in the calculation of this coefficient. If the data are quantitative, each of the two variables must be ranked separately.

Spearman's rank correlation coefficient

The **Spearman rank correlation coefficient**, r_s, is found by using the formula

$$r_s = 1 - \frac{6 \sum (d_i)^2}{n(n^2 - 1)} \tag{14-11}$$

where d_i is the difference in the rankings and n is the number of pairs of data. The value of r_s will range from -1 to $+1$ and will be used in much the same manner as the linear correlation coefficient was used previously.

The null hypothesis that we will be testing is "there is no correlation between the two rankings." The alternative hypothesis may be either two-tailed, "there is correlation," or one-tailed, if we anticipate the existence of either positive or negative correlation. The critical region will be on the side(s) corresponding to the specific alternative that is expected. For example, if we suspect negative correlation, the critical region will be in the left-hand tail.

Illustration 14-11

Let's consider a hypothetical situation in which four judges rank five contestants in a contest. Let's identify the judges as *A*, *B*, *C*, and *D* and the contestants as *a*, *b*, *c*, *d*, and *e*. Table 14-4 lists the awarded rankings.

TABLE 14–4

Rankings for Five Contestants

Contestant	Judge			
	A	B	C	D
a	1	5	1	5
b	2	4	2	2
c	3	3	3	1
d	4	2	4	4
e	5	1	5	3

When we compare judges A and B, we see that they ranked the contestants in exactly the opposite order, perfect disagreement (see table 14–5). From our previous work with correlation, we expect the calculated value for r_s to be exactly -1 for these data.

$$r_s = 1 - \frac{6[\sum (d_i)^2]}{n(n^2 - 1)} = 1 - \frac{(6)(40)}{(5)(5^2 - 1)} = 1 - \frac{240}{120} = -1$$

TABLE 14–5

Rankings of A and B

Contestant	A	B	$d_i = A - B$	$(d_i)^2$
a	1	5	-4	16
b	2	4	-2	4
c	3	3	0	0
d	4	2	2	4
e	5	1	4	16
				40

When judges A and C are compared, we see that their rankings of the contestants are identical (see table 14–6). We would expect to find a calculated correlation coefficient of $+1$ for these data.

$$r_s = 1 - \frac{(6)(0)}{(5)(5^2 - 1)} = 1 - 0 = 1$$

TABLE 14–6

Rankings of A and C

Contestant	A	C	$d_i = A - C$	$(d_i)^2$
a	1	1	0	0
b	2	2	0	0
c	3	3	0	0
d	4	4	0	0
e	5	5	0	0
				0

By comparing judge A with judge B and then with judge C, we have seen the extremes, total agreement and total disagreement. Now let's compare judge A with judge D (see table 14–7). There seems to be no real agreement or disagreement here. Let's compute r_s:

$$r_s = 1 - \frac{(6)(24)}{(5)(5^2 - 1)} = 1 - \frac{144}{120} = 1 - 1.2 = -0.2$$

TABLE 14–7

Rankings of A and D

Contestant	A	D	$d_i = A - D$	$(d_i)^2$
a	1	5	−4	16
b	2	2	0	0
c	3	1	2	4
d	4	4	0	0
e	5	3	2	4
				24

This is fairly close to zero, which is what we should have suspected since there was no real agreement or disagreement.

The test of significance will result in a failure to reject the null hypothesis when r_s is close to zero and will result in a rejection of the null hypothesis in cases where r_s is found to be close to +1 or −1. The critical values found in table 14 of appendix D are positive critical values only. Since the null hypothesis is "the population correlation coefficient is zero" (that is, $\rho = 0$), we have a symmetric test statistic. Hence we need only add a plus or minus sign to the value found in the table, as appropriate. This will be determined by the specific alternative that we have in mind.

In our illustration the critical values for a two-tailed test at $\alpha = 0.10$ are ± 0.900. (Remember that n represents the number of pairs.) If the calculated value for r_s is between 0.9 and 1.0 or between −0.9 and −1.0, we will reject the null hypothesis in favor of the alternative "there is a correlation." □

The Spearman rank coefficient is defined by formula (3–1) (Pearson's product moment r) with rankings used in place of the quantitative x and y values. If there are no ties in the rankings, formula (14–11) is equivalent to formula (3–1). Formula (14–1) provides us with a much easier to use procedure for calculating the r_s statistic. When there are only a few ties, it is common practice to use formula (14–11). The resulting value of r_s is not exactly the same; however, it is generally considered to be an acceptable estimate. Illustration 14–12 shows the procedure for handling ties and uses formula (14–11) for the calculation of r_s. For comparison exercise 14–15 asks you to calculate r_s using formula (3–1).

If ties occur in either set of the ordered pairs of rankings, assign each tied observation the mean of the ranks that would have been assigned had there been no ties, as was done in the Mann–Whitney U test.

Illustration 14–12

Students who finish exams quicker than the rest of the class are often thought to be the smarter students. The accompanying set of data shows the score and order of finish for 12 students on a recent hour exam. At the 0.05 level, do these data support the alternative hypothesis that the first students to complete an exam have higher grades?

Order of Finish	1	2	3	4	5	6	7	8	9	10	11	12
Exam Score	90	74	76	60	68	86	92	60	78	70	78	64

Solution

STEP 1 H_0: Order of finish has no relationship to exam score.

STEP 2 H_a: First to finish tend to have higher grades.

STEP 3 $\alpha = 0.05$ and $n = 12$; the critical region is shown in the diagram.

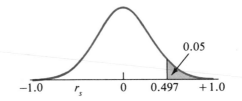

STEP 4 Rank the scores from highest to lowest, assigning the highest score the rank number 1, as shown.

92	90	86	78	78	76	74	70	68	64	60	60
1	2	3	4	5	6	7	8	9	10	11	12
			4.5	4.5						11.5	11.5

The rankings and the preliminary calculations are shown in table 14–8.
Using formula (14–11), we obtain

$$r_s = 1 - \frac{(6)(235.0)}{(12)(143)} = 1 - \frac{1410}{1716} = 1 - 0.822$$

$$= 0.178$$

Test Score Rank	Order of Finish	Difference, d_i	$(d_i)^2$
TABLE 14–8 *Rankings for Test Scores and Differences*			
1	7	-6	36.00
2	1	1	1.00
3	6	-3	9.00
4.5	9	-4.5	20.25
4.5	11	-6.5	42.25
6	3	3	9.00
7	2	5	25.00
8	10	-2	4.00
9	5	4	16.00
10	12	-2	4.00
11.5	4	7.5	56.25
11.5	8	3.5	12.25
			235.00

STEP 5 *Decision*: Fail to reject H_0.

Conclusion: There is not sufficient evidence presented by these sample data to enable us to conclude that the first students to finish have the higher grades. ☐

Exercises

14–14 The accompanying table displays the final class averages of 12 students who completed both elementary statistics and psychology I last semester. Is there sufficient evidence provided by this sample to reject the null hypothesis that the final class averages for the two courses are independent? Use $\alpha = 0.02$.

						Student						
Course	1	2	3	4	5	6	7	8	9	10	11	12
Elementary statistics	73	60	92	91	80	70	81	83	97	76	68	88
Psychology I	90	70	81	97	85	66	75	82	89	79	80	91

14–15 Calculate the Spearman rank correlation coefficient for the data in illustration 14–12 using formula (3–2). Recall that formula (3–2) is equivalent to the definition formula (3–1) and that rank numbers must be used with this formula in order for the resulting statistic to be Spearman's r_s.

14–16 The accompanying sets of air pollution readings were taken from our local newspaper on 10 separate days.

Pollution	Day									
	1	2	3	4	5	6	7	8	9	10
Sulfur dioxide	0.07	0.01	0.01	0.02	0.07	0.02	0.00	0.01	0.05	0.04
Carbon monoxide	0.9	2.0	0.8	0.8	0.9	1.3	3.8	2.0	1.3	2.5
Soiling or dust	0.3	0.5	0.3	0.3	0.3	0.5	0.3	0.5	0.3	0.5

(a) Draw a scatter diagram showing the sulfur dioxide and carbon monoxide readings as paired data. Use Spearman's rank correlation coefficient to test the correlation between the sulfur dioxide and the carbon monoxide readings. Complete a two-tailed test at $\alpha = 0.10$.

(1) State the null hypothesis being tested.

(2) Complete the test and carefully state the conclusion reached.

(b) Repeat part (a) using the sulfur dioxide and the soiling readings. ($\alpha = 0.05$.)

(c) Repeat part (a) using the carbon monoxide and the soil readings. ($\alpha = 0.01$.)

14-17 Life expectancy at birth for the states listed is ranked by two different standards. Is there sufficient evidence to show a correlation between the two ranking systems? Use $\alpha = 0.10$.

State	1940 Standard	1970 Standard
Alabama	12	12
Alaska	11	8
Arkansas	6	5
California	3	3
Hawaii	1	1
Maine	8	10
Montana	7	7
Michigan	9	9
New York	10	11
Texas	5	4
Virginia	4	6
Wisconsin	2	2

14-18 Use the bivariate data shown in the accompanying table.

x	-2	-1	1	2
y	4	1	1	4

(a) Construct a scatter diagram.

(b) Calculate Spearman's correlation coefficient r_s [formula (14-11)].

(c) Calculate Pearson's correlation coefficient r [formula (3–2)].

(d) Compare the two results [(b) and (c)]. Do the two measures of correlation measure the same thing?

Section 14–6 Comparing Statistical Tests

Only a few nonparametric tests have been presented in this chapter. Many more nonparametric tests can be found in other books. In addition, most of these other nonparametric tests can be used in place of certain parametric tests. The question is, then, which test do we use, the parametric or the nonparametric? Furthermore, when there is more than one nonparametric test, which one do we use?

The decision about which test to use must be based on the answer to the question, which test will do the job best? First, let's agree that we are dealing with two or more tests that are equally qualified to be used. That is, each test has a set of assumptions that must be satisfied before they can be applied. From this starting point we will attempt to define "best" to mean the test that is best able to control the risks of error and at the same time keep the size of the sample to a number that is reasonable to work with. (Sample size means cost, cost to you or your employer.)

Let's look first at the ability to **control the risk of error**. The risk that is associated with a type I error is controlled directly by the level of significance α. Recall that P (type I error) $= \alpha$ and P (type II error) $= \beta$. Therefore, it is β that we must control. Statisticians like to talk about power (as do **power of a statistical test** others) and the **power of a statistical test** is defined to be $1 - \beta$. Thus the power of a test, $1 - \beta$, is the probability that we reject the null hypothesis when we should have rejected it. If two tests with the same α are equal candidates for use, the one with the greater power is the one you would want to choose.

The other factor is the **sample size required** to do a job. Suppose that you set the levels of risk that you can tolerate, α and β, and then are able to determine the sample size that it would take to meet your specified challenge. The test that required the smaller sample size would then seem to have the edge. Statisticians usually use the term "efficiency" to talk about this con- **efficiency** cept. **Efficiency** is defined to be the ratio of the sample size of the best parametric test to the sample size of the best nonparametric test, when compared under a fixed set of risk values. For example, the efficiency rating for the sign test is approximately 0.63. This means that a sample of size 63 with a parametric test will do the same job as a sample of size 100 will do with the sign test.

The power and the efficiency of a test cannot be used alone to determine the choice of test. Sometimes you will be forced to use a certain test because of the data you are given. When there is a decision to be made, the final decision rests in a trade-off of three factors: (1) the power of the test, (2) the efficiency of the test, and (3) the data (and the number of data) available. Table 14–9 shows how the nonparametric tests discussed in this chapter compare with the parametric tests covered in previous chapters.

TABLE 14–9
Comparison of Parametric and Nonparametric Tests

Test Situation	Parametric Test	Nonparametric Test	Efficiency of Nonparametric Test
One mean	*t* test (page 304)	Sign test (page 473)	0.63
Two independent means	*t* test (page 350)	*U* test (page 479)	0.95
Two dependent means	*t* test (page 358)	Sign test (page 474)	0.63
Correlation	Pearson's (page 441)	Spearman test (page 490)	0.91
Randomness		Runs test (page 486)	Not meaningful; there is no parametric test for comparison

In Retrospect

You have become acquainted in this chapter with some of the basic concepts of nonparametric statistics. While learning about the use of nonparametric methods and specific nonparametric tests of significance, you should have also come to realize and understand some of the basic assumptions that are needed when the parametric techniques of the earlier chapters are encountered. You now have seen a variety of tests, many of which somewhat duplicate the job done by others. What you must keep in mind is that you should use the best test for your particular needs. The power of the test and the cost of sampling, as related to size and availability of the desired response variable, will play important roles in determining the specific test to be used.

The news article at the beginning of this chapter illustrates only one of many situations in which a nonparametric test can be used. Spearman's rank correlation test may be used to statistically compare the two rankings. They appear to be quite different, but are they significantly different? The answer to this is left for you to determine in exercise 14–24.

Chapter Exercises

14–19   Nonparametric tests are also called "distribution-free" tests. However, the normal and the chi-square distributions are used in the inference-making procedures.

(a) To what does the "distribution-free" term apply? (The population? The sample? The sampling distribution?) Explain.

(b) What is it that has the normal or the chi-square distribution? Explain.

14-20 Track coaches, runners, and fans talk a lot about the "speed of the track." The surface of the track is believed to have a direct effect on the amount of time that it takes a runner to cover the required distance. To test this effect, 10 trackmen were asked to run a 220-yard sprint on each of two tracks. Track A is a cinder track and track B is made of a new synthetic material. The times for the runners are given in the accompanying table. Test the claim that the surface on track B is conducive to faster running times.

					Runner					
Track	1	2	3	4	5	6	7	8	9	10
A	27.7	26.8	27.0	25.5	26.6	27.4	27.2	27.4	25.8	25.1
B	27.0	26.7	25.3	26.0	26.1	25.3	26.7	27.1	24.8	27.1

(a) State the null and alternative hypotheses being tested. Complete the test by using the sign test with $\alpha = 0.05$.

(b) State your conclusions.

14-21 Can today's high temperature be effectively predicted using yesterday's high? Pairs of yesterday's and today's high temperatures were randomly selected; the results are shown in the accompanying table. Do the data present sufficient evidence to justify the statement, "Today's high temperature tends to correlate with yesterday's high temperature"? Use $\alpha = 0.05$.

										Pair of Days								
Reading	1	2	3	4	5	6	7	8	9	10	11	12	13	14	15	16	17	18
Yesterday's	40	58	46	33	40	51	55	81	85	83	89	64	73	63	46	58	28	69
Today's	40	56	34	59	46	51	74	77	83	84	85	68	65	60	54	62	34	66

14-22 In her last article, the sports editor of our student newspaper made the statement that the basketball team's leading scorer was averaging at least 25 points per game. The following data are the number of points scored by this player in the games for which records could be found:

$$14, \quad 24, \quad 23, \quad 20, \quad 13, \quad 33, \quad 17, \quad 20, \quad 20,$$
$$17, \quad 27, \quad 21, \quad 21, \quad 16, \quad 17, \quad 19, \quad 27, \quad 28$$

Use the sign test and check the sports editor's claim.

(a) State the null hypothesis that you wish to test.

(b) Complete the test and state a conclusion. Use $\alpha = 0.05$.

(c) Complete the test and state the conclusion for $\alpha = 0.10$.

(d) Explain the difference between the results in parts (c) and (d).

14-23 A candy company has developed two new chocolate-covered candy bars. Six randomly selected people all preferred candy bar I. Is this statistical evidence, at $\alpha = 0.05$, that the general public will prefer candy bar I?

14-24 The news article at the beginning of this chapter shows the rankings assigned to ten components of job satisfaction. Do the rankings assigned by the workers and the boss show a significant difference in what each thinks is important? Test by using $\alpha = 0.05$.

14-25 A patient, given two different types of vitamin pills, one containing iron and one iron-free, was instructed to take the pills on alternate days. To free himself from remembering which pill he needed to take, he mixed all of the pills together in a larger bottle. Each morning he took the first pill that came out of the bottle. To see if this was a random process, for 25 days he recorded an "I" each morning that he took a vitamin with iron and an "N" for no iron.

Day	1	2	3	4	5	6	7	8	9	10	11	12	13
Type	I	I	N	I	I	N	N	I	N	N	N	N	N

Day	14	15	16	17	18	19	20	21	22	23	24	25
Type	I	I	I	N	I	I	I	I	N	I	I	N

Is there sufficient reason to reject the null hypothesis, that the vitamins were taken in a random order, at the 0.05 level of significance?

14-26 In a study to see if spouses are consistent in their preferences for television programs, a market research firm asked several husbands and wives to rank a list of 12 programs (1 represents the highest score; 12 represents the lowest). The average ranks for the programs, rounded to the nearest integer, were

	Program											
Rank	1	2	3	4	5	6	7	8	9	10	11	12
Husband	12	2	6	10	3	11	7	1	9	5	8	4
Wife	5	4	1	9	3	12	2	8	6	10	7	11

Is there significant evidence of negative correlation at the 0.01 level of significance?

14-27 Mrs. Brown attended a special reading course that claimed to improve reading speed and comprehension. She took a pretest (on which she scored 106) and a test at the end of each of 15 sessions. Her test scores are shown in the accompanying table. Do these data support the claim of improvement, at the $\alpha = 0.025$ level?

Session	1	2	3	4	5	6	7	8	9	10	11	12	13	14	15
Score	109	108	110	112	108	111	112	113	110	112	115	114	116	116	118

14–28 While trying to decide on the best time to harvest his crop, a commercial apple farmer recorded the day on which the first apple on the top half and the first apple on the bottom half of 20 randomly selected trees were ripe. The variable x was assigned a value of 1 on the first day that the first ripe apple appeared on 1 of the 20 trees. The days were then numbered sequentially. The observed data are shown in the accompanying table. Do these data provide convincing evidence that the apples on the top of the trees start to ripen before the apples on the bottom half? Use $\alpha = 0.05$.

										Tree										
Position	1	2	3	4	5	6	7	8	9	10	11	12	13	14	15	16	17	18	19	20
Top	5	6	1	4	5	3	6	7	8	5	8	6	4	7	8	10	3	2	9	7
Bottom	6	5	5	7	3	6	6	8	9	4	10	7	5	11	6	11	5	6	9	8

14–29 A firm is currently testing two different procedures for adjusting the cutting machines used in the production of greeting cards. The results of two samples show the following recorded adjustment times.

Method 1	17	15	14	18	16	15	17	18	15	14	14	16	15
Method 2	14	14	13	13	15	12	16	14	16	13	14	13	12	15	17	13

Is there sufficient reason to conclude that method 2 requires less time (on the average) than method 1 at the 0.05 level of significance?

14–30 Two table tennis ball manufacturers have agreed that the quality of their products can be measured by the height to which the balls rebound. A test is arranged, the balls are dropped from a constant height, and the rebound heights are measured, with the accompanying results (in inches). Manufacturer A claims, "The results show my product to be superior." Manufacturer B replies, "I know of no statistical test that supports this claim." Can you find a test that supports A's claim?

A	14.0	12.5	11.5	12.2	12.4	12.3	11.8	11.9	13.7	13.2
B	12.0	12.5	11.6	13.3	13.0	13.0	12.1	12.8	12.2	12.6

(a) Does the appropriate parametric test show that A's product is superior? [What parametric test (or tests) is appropriate and what exactly does it show?]

(b) Does the appropriate nonparametric test show that A's product is superior?

14–31   A sample of 32 students received the following grades on an exam:

41	42	48	46	50	54	51	42
51	50	45	42	32	45	43	56
55	47	45	51	60	44	57	57
47	28	41	42	54	48	47	32

(a) Does this sample show that the median score for the exam differs from 50? Use $\alpha = 0.05$?

(b) Does this sample show that the median score for the exam is less than 50? Use $\alpha = 0.05$.

Vocabulary List

Be able to define each term. In addition, describe in your own words and give an example of each term. Your examples should not be ones given in class or in the textbook.

The terms followed by bracketed numbers were defined in earlier chapters (the bracketed numbers are those of the earlier chapters) but we ask you to define them again to show your increased understanding of their meaning.

binomial random variable [5, 9]
continuity correction [6]
correlation [3, 13]
distribution [2, 5, 6]
distribution-free test
efficiency
independent sample [10]
Mann-Whitney U test
median [2]
nonparametric test
normal approximation [6]

paired data [3, 10, 13]
parametric test
power
randomness [2]
rank
run
runs test
sign test
Spearman's rank
   correlation coefficient

Quiz

Answer "True" if the statement is always true. If the statement is not always true, replace the words shown in color with words that make the statement always true.

14–1   One of the advantages that the nonparametric tests have is the necessity for **less restrictive** assumptions.

14–2   The sign test is a possible replacement for the **F test**.

14–3   The **sign test** can be used to test the randomness of a set of data.

14–4   If a tie occurs in a set of ranked data, the data that form the tie are **removed from the set**.

14–5   Two dependent **means** can be compared nonparametrically by use of the sign test.

14–6   The sign test is a possible alternative to the Student's t test for **one mean value**.

14–7   The **runs test** is a nonparametric alternative to the difference between two independent means.

14–8   The **confidence level** of a statistical hypothesis test is measured by $1 - \beta$.

14–9   Spearman's rank correlation coefficient is an alternative to the use of the **linear correlation coefficient**.

14–10   The **efficiency** of a nonparametric test is the probability that a false null hypothesis is rejected.

Working with Your Own Data

Many variables in everyday life can be treated as bivariate. Often two such variables have a mathematical relationship that can be approximated by means of a straight line. The following illustration demonstrates such a situation.

A The Age and Value of Peggy's Car

Peggy would like to sell her 1981 Rabbit. She needs to know what price to ask for it in order to write an advertisement to be used in the newspaper. The car is in average condition and Peggy expects to ask and receive an average price for it. She must answer the question, "What is an average price to be asking for a Rabbit of this age?"

Inspection of many classified sections of newspapers turned up only four advertisements for 1981 Rabbits. The prices listed varied a great deal. Peggy finally decided, in order to determine an accurate selling price, she would define two variables and collect several pairs of values based on the following definitions.

Population: Used Volkswagon Rabbits advertised for sale by individual owners, dealers not included.

Independent variable, x: The age of the car as measured in years and defined by

$$x = \text{(present calendar year)} - \text{(year of manufacture)} + 1.$$

Example: During 1983 Peggy's 1981 Rabbit is considered to be 3 years old.

$$x = (1983 - 1981) + 1 = 2 + 1 = 3.$$

Dependent variable, y: The advertised asking price.

The accompanying table lists the data collected in February 1983.

Year of Manufacture:			1981	1981	1981	1981	1980	1982
Asking Price:			5500	4500	3600	4800	4300	6200

Year:	1978	1980	1976	1977	1977	1978	1978	1979
Price:	2250	3300	2000	1400	2900	2350	2800	3500

Year:	1980	1977	1976	1978	1979	1980	1982	1976
Price:	4100	3200	1950	3300	3300	5000	5400	2450

Year:	1976	1977	1978	1979	1976	1978	1979	1982
Price:	1500	2000	2100	2700	1000	1500	3900	5000

Year:	1979	1976
Price:	3100	1850

1. Construct and label a scatter diagram of Peggy's data.

2. Determine the equation for the line of best fit.

3. Draw the line of best fit on the scatter diagram.

4. Test the equation of the line of best fit to see if the linear model is appropriate for the data. Use $\alpha = 0.05$.

5. Construct a 0.95 confidence interval estimate for the mean advertised price for 1981 Rabbits.

6. Draw a line segment on the scatter diagram that represents the interval estimate found for question 5.

7. What does the value of the slope, b_1, represent? Explain.

8. What does the value of the y-intercept, b_0, represent? Explain.

B   Your Own Investigation

Identify a situation of interest to you that can be investigated statistically using bivariate data. (Consult your instructor for specific guidance.)

1. Define the population, the independent variable, the dependent variable, and the purpose for studying these two variables as a regression analysis.

2. Collect 15 to 20 ordered pairs of data.

3. Construct and label a scatter diagram of your data.

4. Determine the equation for the line of best fit.

5. Draw the line of best fit on the scatter diagram.

6. Test the equation of the line of best fit to see if the linear model is appropriate for the data. Use $\alpha = 0.05$.

7. Construct a 0.95 confidence interval estimate for the mean value of the dependent variable at the following value of x:

Let x be equal to one third the sum of the lowest value of x in your sample and twice the largest value. That is:

$$x = \frac{L + 2H}{3}$$

8. Draw a line segment on the scatter diagram that represents the interval estimate found for question 7.

9. What does the value of the slope, b_1, represent? Explain.

10. What does the value of the y-intercept, b_0, represent? Explain.

Appendixes

A Summation Notation

The capital Greek letter sigma (Σ) is used in mathematics to indicate the summation of a set of addends. Each of these addends must be of the form of the variable following Σ. For example:

1. $\sum x$ means sum the variable x.
2. $\sum (x - 5)$ means sum the set of addends that are each 5 less than the values of each x.

When large quantities of data are collected, it is usually convenient to index the response variable so that at a future time its source will be known. This indexing is shown on the notation by using i (or j or k) and affixing the index of the first and last addend at the bottom and top of the Σ. For example,

$$\sum_{i=1}^{3} x_i$$

means to add all the consecutive values of x's starting with source number 1 and proceeding to source number 3.

Illustration A–1

Consider the inventory in the accompanying table about the number of defective stereo tapes per lot of 100.

Lot Number, i	1	2	3	4	5	6	7	8	9	10
Number of Defective Tapes per Lot, x	2	3	2	4	5	6	4	3	3	2

(a) Find $\sum_{i=1}^{10} x_i$.

(b) Find $\sum_{i=4}^{8} x_i$.

Solution

(a) $\sum_{i=1}^{10} x_i = x_1 + x_2 + x_3 + x_4 + \cdots + x_{10}$

$= 2 + 3 + 2 + 4 + 5 + 6 + 4 + 3 + 3 + 2 = 34$

(b) $\sum_{i=4}^{8} x_i = x_4 + x_5 + x_6 + x_7 + x_8 = 4 + 5 + 6 + 4 + 3 = 22$ ☐

This index system must be used whenever only part of the available information is to be used. In statistics, however, we will usually use all the available information, and to simplify the formulas we will make an adjustment. This adjustment is actually an agreement that allows us to do away with the index system in situations where all values are used. Thus in our previous illustration, $\sum_{i=1}^{10} x_i$ could have been written simply as $\sum x$.

NOTE: The lack of the index indicates that all data are being used.

Illustration A–2

Given the following six values for x, 1, 3, 7, 2, 4, 5, find $\sum x$.

Solution

$$\sum x = 1 + 3 + 7 + 2 + 4 + 5 = 22 \qquad \square$$

Throughout the study and use of statistics you will find many formulas that use the $\sum$ symbol. Care must be taken so that the formulas are not misread. Symbols like $\sum x^2$ and $(\sum x)^2$ are quite different. $\sum x^2$ means "square each x value and then add up the squares," while $(\sum x)^2$ means "sum the x values and then square the sum."

Illustration A–3

Find (a) $\sum x^2$ and (b) $(\sum x)^2$ for the sample in illustration A–2.

Solution

(a)

x	1	3	7	2	4	5
x^2	1	9	49	4	16	25

$$\sum x^2 = 1 + 9 + 49 + 4 + 16 + 25 = 104$$

(b) $\sum x = 22$, as found in illustration A–2. Thus

$$(\sum x)^2 = (22)^2 = 484$$

As you can see, there is quite a difference between $\sum x^2$ and $(\sum x)^2$. $\qquad \square$

Likewise, $\sum xy$ and $\sum x \sum y$ are different. These forms will appear only when there are paired data, as shown in the following illustration.

Illustration A–4

Given the five pairs of data shown in the accompanying table, find (a) $\sum xy$ and (b) $\sum x \sum y$.

x	1	6	9	3	4
y	7	8	2	5	10

Solution

(a) $\sum xy$ means to sum the products of the corresponding x and y values. Therefore, we have

x	1	6	9	3	4
y	7	8	2	5	10
xy	7	48	18	15	40

$$\sum xy = 7 + 48 + 18 + 15 + 40 = 128$$

(b) $\sum x \sum y$ means the product of the two summations, $\sum x$ and $\sum y$. Therefore, we have

$$\sum x = 1 + 6 + 9 + 3 + 4 = 23$$
$$\sum y = 7 + 8 + 2 + 5 + 10 = 32$$
$$\sum x \sum y = (23)(32) = 736$$

There are three basic rules for algebraic manipulation of the $\sum$ notation.

NOTE: c represents any constant value.

Rule I

$$\sum_{i=1}^{n} c = nc$$

To prove this rule, we need only write down the meaning of $\sum_{i=1}^{n} c$:

$$\sum_{i=1}^{n} c = \underbrace{c + c + c + \cdots + c}_{n \text{ addends}}$$

Therefore,

$$\sum_{i=1}^{n} c = n \cdot c$$

Illustration A–5

Show that $\sum_{i=1}^{5} 4 = (5)(4) = 20$.

Solution

$$\sum_{i=1}^{5} 4 = \underbrace{4_{(\text{when } i=1)} + 4_{(\text{when } i=2)} + 4_{(i=3)} + 4_{(i=4)} + 4_{(i=5)}}_{\text{five 4s added together}}$$

$$= (5)(4) = 20$$

Rule 2

$$\sum_{i=1}^{n} cx_i = c \cdot \sum_{i=1}^{n} x_i$$

To demonstrate the truth of rule 2, we will need to expand the term $\sum_{i=1}^{n} cx_i$ and then factor out the common term c.

$$\sum_{i=1}^{n} cx_i = cx_1 + cx_2 + cx_3 + \cdots + cx_n$$

$$= c(x_1 + x_2 + x_3 + \cdots + x_n)$$

Therefore,

$$\sum_{i=1}^{n} cx_i = c \cdot \sum_{i=1}^{n} x_i$$

Rule 3

$$\sum_{i=1}^{n} (x_i + y_i) = \sum_{i=1}^{n} x_i + \sum_{i=1}^{n} y_i$$

The expansion and regrouping of $\sum_{i=1}^{n} (x_i + y_i)$ is all that is needed to show this rule.

$$\sum_{i=1}^{n} (x_i + y_i) = (x_1 + y_1) + (x_2 + y_2) + \cdots + (x_n + y_n)$$

$$= (x_1 + x_2 + \cdots + x_n) + (y_1 + y_2 + \cdots + y_n)$$

Therefore,

$$\sum_{i=1}^{n} (x_i + y_i) = \sum_{i=1}^{n} x_i + \sum_{i=1}^{n} y_i$$

Illustration A–6

Show that $\sum_{i=1}^{3} (2x_i + 6) = 2 \cdot \sum_{i=1}^{3} x_i + 18$.

Solution

$$\sum_{i=1}^{3} (2x_i + 6) = (2x_1 + 6) + (2x_2 + 6) + (2x_3 + 6)$$

$$= (2x_1 + 2x_2 + 2x_3) + (6 + 6 + 6)$$

$$= (2)(x_1 + x_2 + x_3) + (3)(6)$$

$$= 2 \sum_{i=1}^{3} x_i + 18 \qquad \square$$

Illustration A–7

Let $x_1 = 2$, $x_2 = 4$, $x_3 = 6$, $f_1 = 3$, $f_2 = 4$, and $f_3 = 2$. Find $\sum\limits_{i=1}^{3} x_i \cdot \sum\limits_{i=1}^{3} f_i$.

Solution

$$\sum_{i=1}^{3} x_i \cdot \sum_{i=1}^{3} f_i = (x_1 + x_2 + x_3) \cdot (f_1 + f_2 + f_3)$$

$$= (2 + 4 + 6) \cdot (3 + 4 + 2)$$

$$= (12)(9) = 108 \qquad \square$$

Illustration A–8

Using the same values for the x's and f's as in illustration A–7, find $\sum (xf)$.

Solution   Recall that the use of no index numbers means "use all data."

$$\sum (xf) = \sum_{i=1}^{3} (x_i f_i) = (x_1 f_1) + (x_2 f_2) + (x_3 f_3)$$

$$= (2 \cdot 3) + (4 \cdot 4) + (6 \cdot 2) = 6 + 16 + 12 = 34 \qquad \square$$

Exercises

A–1   Write each of the following in expanded form (without the summation sign):

(a) $\sum\limits_{i=1}^{4} x_i$ (b) $\sum\limits_{i=1}^{3} (x_i)^2$ (c) $\sum\limits_{i=1}^{5} (x_i + y_i)$

(d) $\sum\limits_{i=1}^{5} (x_i + 4)$ (e) $\sum\limits_{i=1}^{8} x_i y_i$ (f) $\sum\limits_{i=1}^{4} x_i^2 f_i$

A–2   Write each of the following expressions as summations, showing the subscripts and the limits of summation:

(a) $x_1 + x_2 + x_3 + x_4 + x_5 + x_6$
(b) $x_1 y_1 + x_2 y_2 + x_3 y_3 + \cdots + x_7 y_7$
(c) $x_1^2 + x_2^2 + \cdots + x_9^2$
(d) $(x_1 - 3) + (x_2 - 3) + \cdots + (x_n - 3)$

A–3   Show each of the following to be true:

(a) $\sum\limits_{i=1}^{4} (5x_i + 6) = 5 \cdot \sum\limits_{i=1}^{4} x_i + 24$

(b) $\sum\limits_{i=1}^{n} (x_i - y_i) = \sum\limits_{i=1}^{n} x_i - \sum\limits_{i=1}^{n} y_i$

A–4   Given $x_1 = 2$, $x_2 = 7$, $x_3 = -3$, $x_4 = 2$, $x_5 = -1$, and $x_6 = 1$, find each of the following:

(a) $\sum\limits_{i=1}^{6} x_i$ (b) $\sum\limits_{i=1}^{6} x_i^2$ (c) $\left(\sum\limits_{i=1}^{6} x_i\right)^2$

A-5   Given $x_1 = 4$, $x_2 = -1$, $x_3 = 5$, $f_1 = 4$, $f_2 = 6$, $f_3 = 2$, $y_1 = -3$, $y_2 = 5$, and $y_3 = 2$, find each of the following:

(a) $\sum x$ (b) $\sum y$ (c) $\sum f$ (d) $\sum (x - y)$

(e) $\sum x^2$ (f) $(\sum x)^2$ (g) $\sum xy$ (h) $\sum x \cdot \sum y$

(i) $\sum xf$ (j) $\sum x^2 f$ (k) $(\sum xf)^2$

A-6   Suppose you take out a \$12,000 small-business loan. The terms of the loan are that each month for 10 years (120 months) you will pay back \$100 plus accrued interest. The accrued interest is calculated by multiplying 0.005 (6%/12) times the amount of the loan still outstanding. That is, the first month you pay \$12,000 × 0.005 in accrued interest, the second month (\$12,000 − 100) × 0.005 in interest, the third month [\$12,000 − (2)(100)] × 0.005, and so forth. Express the total amount of interest paid over the life of the loan by using summation notation.

The answers to these exercises are on page 549.

B Using the Random Number Table

The random number table is a collection of "random" digits. The term *random* means that each of the ten digits (0, 1, 2, 3, ..., 9) has an equal chance of occurrence. The digits in table 1 of appendix D can be thought of as single-digit numbers (0–9), as two-digit numbers (00–99), as three-digit numbers (000–999), or as numbers of any desired size. The digits presented in table 1 are arranged in pairs and grouped into blocks of five rows and five columns. This format is used for convenience. Tables in other books may be arranged differently.

Random numbers are used primarily for one of two reasons: (1) to identify the source element of a population (the source of data) or (2) to simulate an experiment.

Illustration B–1

A simple random sample of 10 people is to be drawn from a population of 7564 people. Each person will be assigned a number, using the numbers from 0001 to 7564. We will view table 1 as a collection of four-digit numbers (two columns used together), where the numbers 0001, 0002, 0003, ..., 7564 identify the 7564 people. The numbers 0000, 7565, 7566, ..., 9999 represent no one in our population; that is, they will be discarded if selected.

Now we are ready to select our 10 people. Turn to table 1 (page 516). We need to select a starting point and a "path" to be followed. Perhaps the most common way to locate a starting point is to look away and arbitrarily point to a starting point. The number we located this way was 3909. (It is located in the upper left corner of the block that is in the fourth large block from the left and the ninth large block down.) From here we will proceed down the column, then go to the top of the next set of columns, if necessary. The person identified by number 3909 is the first source of data selected. Proceeding down the column, we find 8869 next. This number is discarded. The number 2501 is next. Therefore, the person identified by 2501 is the second source of data to be selected. Continuing down this column, our sample will be obtained from those people identified by the numbers 3909, 2501, 7485, 0545, 6104, 3347, 6743, 1168, 3398, 3852. (The number 6104 is the second four-digit number at the top of the next set of columns. The numbers 9091 and 9074 were discarded.) □

Illustration B–2

Let's use the random number table and simulate 100 tosses of a coin. The simulation is accomplished by assigning numbers to each of the possible outcomes of a particular experiment. The assignment must be done in such a way as to preserve the probabilities.

Perhaps the simplest way to make the assignment for the coin toss is to let the even digits (0, 2, 4, 6, 8) represent heads and the odd digits (1, 3, 5, 7, 9) represent tails. The correct probabilities are maintained: $P(H) = P(0, 2, 4, 6, 8) = \frac{5}{10} = 0.5$ and $P(T) = P(1, 3, 5, 7, 9) = \frac{5}{10} = 0.5$. Once this assignment is complete, we are ready to obtain our sample.

Since the question asked for 100 tosses and there are 50 digits to a "block" in table 1, let's select two blocks as our sample of random one-digit numbers (instead of a column 100 lines long). Let's look away and point to one block on page 516 and then do the same to select one block from page 517. We picked the sixth block down in the first column of blocks on page 516 (24 even and 26 odd numbers) and the fifth block down in the third column of blocks on page 517 (23 even and 27 odd numbers). Thus we obtain a sample of 47 heads and 53 tails for our 100 simulated tosses. ☐

There are, of course, many ways to use the random number table. You must use your good sense in assigning the numbers to be used and in choosing the "path" to be followed through the table. One bit of advice is to make the assignments in as simple and easy a method as possible to avoid errors.

Exercises

B–1   A random sample of size 8 is to be selected from a population that contains 75 elements. Describe how the random sample of the 8 objects could be made with the aid of the random number table.

B–2   A coin-tossing experiment is to be simulated. Two coins are to be tossed simultaneously and the number of heads appearing is to be recorded for each toss. Ten such tosses are to be observed. Describe two ways to use the random number table to simulate this experiment.

B–3   Simulate five rolls of three dice by using the random number table.

The answers to these exerises are on page 549.

C Round-Off Procedure

When rounding off a number, we use the following procedure.

STEP 1   Identify the position where the round-off is to occur. This is shown by use of a vertical line that separates the part of the number to be kept from the part to be discarded. For example,

| 125.267 | to the nearest tenth is written as | 125.2|67 |
| 7.8890 | to the nearest hundredth is written as | 7.88|90 |

STEP 2   Step 1 has separated all numbers into one of four cases. (X's will be used as placeholders for number values in front of the vertical line. These X's can represent any number value.)

Case I: $XXXX|000...$

Case II: $XXXX|----$ (any value from 000 ... 1 to 499 ... 9)

Case III: $XXXX|5000...0$

Case IV: $XXXX|----$ (any value from 5000 ... 1 to 999 ... 9)

STEP 3   Perform the rounding off.

Case I requires no round-off. It is exactly XXXX.

Illustration C–1

Round 3.5000 to the nearest tenth.

$$3.5|000 \quad \text{becomes} \quad \textbf{3.5} \qquad \square$$

Case II requires rounding. We will round down for this case. That is, just drop the part of the number that is behind the vertical line.

Illustration C–2

Round 37.6124 to the nearest hundredth.

$$37.61|24 \quad \text{becomes} \quad \textbf{37.61} \qquad \square$$

Case III requires rounding. This is the case that requires special attention. **When a 5 (exactly a 5) is to be rounded off, round to the even digit.** In the long run, half of the time the 5 will be preceded by an even digit (0, 2, 4, 6, 8) and you will round down, while the other half of the time the 5 will be preceded by an odd digit (1, 3, 5, 7, 9) and you will round up.

Illustration C–3

Round 87.35 to the nearest tenth.

$$87.3|5 \quad \text{becomes} \quad \textbf{87.4}$$

Round 93.445 to the nearest hundredth.

$$93.44|5 \quad \text{becomes} \quad \textbf{93.44}$$

(Note: 87.35 is 87.35000 . . . and 93.445 is 93.445000. . . .) □

Case IV requires rounding. We will round up for this case. That is, we will drop the part of the number that is behind the vertical line and we will increase the last digit in front of the vertical line by one.

Illustration C–4

Round 7.889 to the nearest tenth.

$$7.8|89 \quad \text{becomes} \quad \textbf{7.9}$$ □

NOTE: **Cases I, II, and IV describe what is commonly done. Our guidelines for case III are the only ones that are different from typical procedure.**

If the typical round-off rule (0, 1, 2, 3, 4 are dropped; 5, 6, 7, 8, 9 are rounded up) is followed, then $(n + 1)/(2n + 1)$ of the situations are rounded up. (n is the number of different sequences of digits that fall into each of case II and case IV.) That is more than half. You (as many others have) may say, "So what?" In today's world that tiny, seemingly insignificant amount becomes very significant when applied repeatedly to large numbers.

Exercises

C–1   Round each of the following to the nearest integer:

(a) 12.94   (b) 8.762   (c) 9.05   (d) 156.49

(e) 45.5   (f) 42.5   (g) 102.51   (h) 16.5001

C–2   Round each of the following to the nearest tenth:

(a) 8.67   (b) 42.333   (c) 49.666   (d) 10.25   (e) 10.35

(f) 8.4501   (g) 27.35001   (h) 5.65   (i) 3.05   (j) $\frac{1}{4}$

C–3   Round each of the following to the nearest hundredth:

(a) 17.6666   (b) 4.444   (c) 54.5454   (d) 102.055   (e) 93.225

(f) 18.005   (g) 18.015   (h) 5.555   (i) 44.7450   (j) $\frac{2}{3}$

The answers to these exercises are on page 550.

D Tables

Table I   Random Numbers

```
10 09 73 25 33   76 52 01 35 86   34 67 35 48 76   80 95 90 91 17   39 29 27 49 45
37 54 20 48 05   64 89 47 42 96   24 80 52 40 37   20 63 61 04 02   00 82 29 16 65
08 42 26 89 53   19 64 50 93 03   23 20 90 25 60   15 95 33 47 64   35 08 03 36 06
99 01 90 25 29   09 37 67 07 15   38 31 13 11 65   88 67 67 43 97   04 43 62 76 59
12 80 79 99 70   80 15 73 61 47   64 03 23 66 53   98 95 11 68 77   12 17 17 68 33

66 06 57 47 17   34 07 27 68 50   36 69 73 61 70   65 81 33 98 85   11 19 92 91 70
31 06 01 08 05   45 57 18 24 06   35 30 34 26 14   86 79 90 74 39   23 40 30 97 32
85 26 97 76 02   02 05 16 56 92   68 66 57 48 18   73 05 38 52 47   18 62 38 85 79
63 57 33 21 35   05 32 54 70 48   90 55 35 75 48   28 46 82 87 09   83 49 12 56 24
73 79 64 57 53   03 52 96 47 78   35 80 83 42 82   60 93 52 03 44   35 27 38 84 35

98 52 01 77 67   14 90 56 86 07   22 10 94 05 58   60 97 09 34 33   50 50 07 39 98
11 80 50 54 31   39 80 82 77 32   50 72 56 82 48   29 40 52 42 01   52 77 56 78 51
83 45 29 96 34   06 28 89 80 83   13 74 67 00 78   18 47 54 06 10   68 71 17 78 17
88 68 54 02 00   86 50 75 84 01   36 76 66 79 51   90 36 47 64 93   29 60 91 10 62
99 59 46 73 48   87 51 76 49 69   91 82 60 89 28   93 78 56 13 68   23 47 83 41 13

65 48 11 76 74   17 46 85 09 50   58 04 77 69 74   73 03 95 71 86   40 21 81 65 44
80 12 43 56 35   17 72 70 80 15   45 31 82 23 74   21 11 57 82 53   14 38 55 37 63
74 35 09 98 17   77 40 27 72 14   43 23 60 02 10   45 52 16 42 37   96 28 60 26 55
69 91 62 68 03   66 25 22 91 48   36 93 68 72 03   76 62 11 39 90   94 40 05 64 18
09 89 32 05 05   14 22 56 85 14   46 42 75 67 88   96 29 77 88 22   54 38 21 45 98

91 49 91 45 23   68 47 92 76 86   46 16 28 35 54   94 75 08 99 23   37 08 92 00 48
80 33 69 45 98   26 94 03 68 58   70 29 73 41 35   53 14 03 33 40   42 05 08 23 41
44 10 48 19 49   85 15 74 79 54   32 97 92 65 75   57 60 04 08 81   22 22 20 64 13
12 55 07 37 42   11 10 00 20 40   12 86 07 46 97   96 64 48 94 39   28 70 72 58 15
63 60 64 93 29   16 50 53 44 84   40 21 95 25 63   43 65 17 70 82   07 20 73 17 90

61 19 69 04 46   26 45 74 77 74   51 92 43 37 29   65 39 45 95 93   42 58 26 05 27
15 47 44 52 66   95 27 07 99 53   59 36 78 38 48   82 39 61 01 18   33 21 15 94 66
94 55 72 85 73   67 89 75 43 87   54 62 24 44 31   91 19 04 25 92   92 92 74 59 73
42 48 11 62 13   97 34 40 87 21   16 86 84 87 67   03 07 11 20 59   25 70 14 66 70
23 52 37 83 17   73 20 88 98 37   68 93 59 14 16   26 25 22 96 63   05 52 28 25 62

04 49 35 24 94   75 24 63 38 24   45 86 25 10 25   61 96 27 93 35   65 33 71 24 72
00 54 99 76 54   64 05 18 81 59   96 11 96 38 96   54 69 28 23 91   23 28 72 95 29
35 96 31 53 07   26 89 80 93 54   33 35 13 54 62   77 97 45 00 24   90 10 33 93 33
59 80 80 83 91   45 42 72 68 42   83 60 94 97 00   13 02 12 48 92   78 56 52 01 06
46 05 88 52 36   01 39 09 22 86   77 28 14 40 77   93 91 08 36 47   70 61 74 29 41

32 17 90 05 97   87 37 92 52 41   05 56 70 70 07   86 74 31 71 57   85 39 41 18 38
69 23 46 14 06   20 11 74 52 04   15 95 66 00 00   18 74 39 24 23   97 11 89 63 38
19 56 54 14 30   01 75 87 53 79   40 41 92 15 85   66 67 43 68 06   84 96 28 52 07
45 15 51 49 38   19 47 60 72 46   43 66 79 45 43   59 04 79 00 33   20 82 66 95 41
94 86 43 19 94   36 16 81 08 51   34 88 88 15 53   01 54 03 54 56   05 01 45 11 76

98 08 62 48 26   45 24 02 84 04   44 99 90 88 96   39 09 47 34 07   35 44 13 18 80
33 18 51 62 32   41 94 15 09 49   89 43 54 85 81   88 69 54 19 94   37 54 87 30 43
80 95 10 04 06   96 38 27 07 74   20 15 12 33 87   25 01 62 52 98   94 62 46 11 71
79 75 24 91 40   71 96 12 82 96   69 86 10 25 91   74 85 22 05 39   00 38 75 95 79
18 63 33 25 37   98 14 50 65 71   31 01 02 46 74   05 45 56 14 27   77 93 89 19 36
```

* For specific details on the use of this table, see page 512.

74	02	94	39	02	77	55	73	22	70	97	79	01	71	19	52	52	75	80	21	80	81	45	17	48
54	17	84	56	11	80	99	33	71	43	05	33	51	29	69	56	12	71	92	55	36	04	09	03	24
11	66	44	98	83	52	07	98	48	27	59	38	17	15	39	09	97	33	34	40	88	46	12	33	56
48	32	47	79	28	31	24	96	47	10	02	29	53	68	70	32	30	75	75	46	15	02	00	99	94
69	07	49	41	38	87	63	79	19	76	35	58	40	44	01	10	51	82	16	15	01	84	87	69	38
09	18	82	00	97	32	82	53	95	27	04	22	08	63	04	83	38	98	73	74	64	27	85	80	44
90	04	58	54	97	51	98	15	06	54	94	93	88	19	97	91	87	07	61	50	68	47	66	46	59
73	18	95	02	07	47	67	72	62	69	62	29	06	44	64	27	12	46	70	18	41	36	18	27	60
75	76	87	64	90	20	97	18	17	49	90	42	91	22	72	95	37	50	58	71	93	82	34	31	78
54	01	64	40	56	66	28	13	10	03	00	68	22	73	98	20	71	45	32	95	07	70	61	78	13
08	35	86	99	10	78	54	24	27	85	13	66	15	88	73	04	61	89	75	53	31	22	30	84	20
28	30	60	32	64	81	33	31	05	91	40	51	00	78	93	32	60	46	04	75	94	11	90	18	40
53	84	08	62	33	81	59	41	36	28	51	21	59	02	90	28	46	66	87	95	77	76	22	07	91
91	75	75	37	41	61	61	36	22	69	50	26	39	02	12	55	78	17	65	14	83	48	34	70	55
89	41	59	26	94	00	39	75	83	91	12	60	71	76	46	48	94	97	23	06	94	54	13	74	08
77	51	30	38	20	86	83	42	99	01	68	41	48	27	74	51	90	81	39	80	72	89	35	55	07
19	50	23	71	74	69	97	92	02	88	55	21	02	97	73	74	28	77	52	51	65	34	46	74	15
21	81	85	93	13	93	27	88	17	57	05	68	67	31	56	07	08	28	50	46	31	85	33	84	52
51	47	46	64	99	68	10	72	36	21	94	04	99	13	45	42	83	60	91	91	08	00	74	54	49
99	55	96	83	31	62	53	52	41	70	69	77	71	28	30	74	81	97	81	42	43	86	07	28	34
33	71	34	80	07	93	58	47	28	69	51	92	66	47	21	58	30	32	98	22	93	17	49	39	72
85	27	48	68	93	11	30	32	92	70	28	83	43	41	37	73	51	59	04	00	71	14	84	36	43
84	13	38	96	40	44	03	55	21	66	73	85	27	00	91	61	22	26	05	61	62	32	71	84	23
56	73	21	62	34	17	39	59	61	31	10	12	39	16	22	85	49	65	75	60	81	60	41	88	80
65	13	85	68	06	87	64	88	52	61	34	31	36	58	61	45	87	52	10	69	85	64	44	72	77
38	00	10	21	76	81	71	91	17	11	71	60	29	29	37	74	21	96	40	49	65	58	44	96	98
37	40	29	63	97	01	30	47	75	86	56	27	11	00	86	47	32	46	26	05	40	03	03	74	38
97	12	54	03	48	87	08	33	14	17	21	81	53	92	50	75	23	76	20	47	15	50	12	95	78
21	82	64	11	34	47	14	33	40	72	64	63	88	59	02	49	13	90	64	41	03	85	65	45	52
73	13	54	27	42	95	71	90	90	35	85	79	47	42	96	08	78	98	81	56	64	69	11	92	02
07	63	87	79	29	03	06	11	80	72	96	20	74	41	56	23	82	19	95	38	04	71	36	69	94
60	52	88	34	41	07	95	41	98	14	59	17	52	06	95	05	53	35	21	39	61	21	20	64	55
83	59	63	56	55	06	95	89	29	83	05	12	80	97	19	77	43	35	37	83	92	30	15	04	98
10	85	06	27	46	99	59	91	05	07	13	49	90	63	19	53	07	57	18	39	06	41	01	93	62
39	82	09	89	52	43	62	26	31	47	64	42	18	08	14	43	80	00	93	51	31	02	47	31	67
59	58	00	64	78	75	56	97	88	00	88	83	55	44	86	23	76	80	61	56	04	11	10	84	08
38	50	80	73	41	23	79	34	87	63	90	82	29	70	22	17	71	90	42	07	95	95	44	99	53
30	69	27	06	68	94	68	81	61	27	56	19	68	00	91	82	06	76	34	00	05	46	26	92	00
65	44	39	56	59	18	28	82	74	37	49	63	22	40	41	08	33	76	56	76	96	29	99	08	36
27	26	75	02	64	13	19	27	22	94	07	47	74	46	06	17	98	54	89	11	97	34	13	03	58
91	30	70	69	91	19	07	22	42	10	36	69	95	37	28	28	82	53	57	93	28	97	66	62	52
68	43	49	46	88	84	47	31	36	22	62	12	69	84	08	12	84	38	25	90	09	81	59	31	46
48	90	81	58	77	54	74	52	45	91	35	70	00	47	54	83	82	45	26	92	54	13	05	51	60
06	91	34	51	97	42	67	27	86	01	11	88	30	95	28	63	01	19	89	01	14	97	44	03	44
10	45	51	60	19	14	21	03	37	12	91	34	23	78	21	88	32	58	08	51	43	66	77	08	83
12	88	39	73	43	65	02	76	11	84	04	28	50	13	92	17	97	41	50	77	90	71	22	67	69
21	77	83	09	76	38	80	73	69	61	31	64	94	20	96	63	28	10	20	23	08	81	64	74	49
19	52	35	95	15	65	12	25	96	59	86	28	36	82	58	69	57	21	37	98	16	43	59	15	29
67	24	55	26	70	35	58	31	65	63	79	24	68	66	86	76	46	33	42	22	26	65	59	08	02
60	58	44	73	77	07	50	03	79	92	45	13	42	65	29	26	76	08	36	37	41	32	64	43	44
53	85	34	13	77	36	06	69	48	50	58	83	87	38	59	49	36	47	33	31	96	24	04	36	42
24	63	73	87	36	74	38	48	93	42	52	62	30	79	92	12	36	91	86	01	03	74	28	38	73
83	08	01	24	51	38	99	22	28	15	07	75	95	17	77	97	37	72	75	85	51	97	23	78	67
16	44	42	43	34	36	15	19	90	73	27	49	37	09	39	85	13	03	25	52	54	84	65	47	59
60	79	01	81	57	57	17	86	57	62	11	16	17	85	76	45	81	95	29	79	65	13	00	48	60

From tables of the RAND Corporation. Reprinted from Wilfred J. Dixon and Frank J. Massey, Jr., *Introduction to Statistical Analysis*, 3rd ed. (New York: McGraw-Hill, 1969), pp. 446–447. Reprinted by permission of the RAND Corporation.

Table 2   Factorials

n	$n!$
0	1
1	1
2	2
3	6
4	24
5	120
6	720
7	5,040
8	40,320
9	362,880
10	3,628,800
11	39,916,800
12	479,001,600
13	6,227,020,800
14	87,178,291,200
15	1,307,674,368,000
16	20,922,789,888,000
17	355,687,428,096,000
18	6,402,373,705,728,000
19	121,645,100,408,832,000
20	2,432,902,008,176,640,000

* For specific details on the use of this table, see page 186.

Table 3   Binomial Coefficients

n	$\binom{n}{0}$	$\binom{n}{1}$	$\binom{n}{2}$	$\binom{n}{3}$	$\binom{n}{4}$	$\binom{n}{5}$	$\binom{n}{6}$	$\binom{n}{7}$	$\binom{n}{8}$	$\binom{n}{9}$	$\binom{n}{10}$
0	1										
1	1	1									
2	1	2	1								
3	1	3	3	1							
4	1	4	6	4	1						
5	1	5	10	10	5	1					
6	1	6	15	20	15	6	1				
7	1	7	21	35	35	21	7	1			
8	1	8	28	56	70	56	28	8	1		
9	1	9	36	84	126	126	84	36	9	1	
10	1	10	45	120	210	252	210	120	45	10	1
11	1	11	55	165	330	462	462	330	165	55	11
12	1	12	66	220	495	792	924	792	495	220	66
13	1	13	78	286	715	1287	1716	1716	1287	715	286
14	1	14	91	364	1001	2002	3003	3432	3003	2002	1001
15	1	15	105	455	1365	3003	5005	6435	6435	5005	3003
16	1	16	120	560	1820	4368	8008	11440	12870	11440	8008
17	1	17	136	680	2380	6188	12376	19448	24310	24310	19448
18	1	18	153	816	3060	8568	18564	31824	43758	48620	43758
19	1	19	171	969	3876	11628	27132	50388	75582	92378	92378
20	1	20	190	1140	4845	15504	38760	77520	125970	167960	184756

If necessary, use the identity $\binom{n}{k} = \binom{n}{n-k}$

From John E. Freund, *Statistics, A First Course*, Prentice-Hall, Inc., Englewood Cliffs, N.J., 1970, p. 313. Reprinted by permission.

* For specific details on the use of this table, see page 186.

Table 4   Binomial Probabilities $\left[\binom{n}{x} \cdot p^x q^{n-x}\right]$

n	x	0.01	0.05	0.10	0.20	0.30	0.40	0.50	0.60	0.70	0.80	0.90	0.95	0.99	x
2	0	980	902	810	640	490	360	250	160	090	040	010	002	0+	0
	1	020	095	180	320	420	480	500	480	420	320	180	095	020	1
	2	0+	002	010	040	090	160	250	360	490	640	810	902	980	2
3	0	970	857	729	512	343	216	125	064	027	008	001	0+	0+	0
	1	029	135	243	384	441	432	375	288	189	096	027	007	0+	1
	2	0+	007	027	096	189	288	375	432	441	384	243	135	029	2
	3	0+	0+	001	008	027	064	125	216	343	512	729	857	970	3
4	0	961	815	656	410	240	130	062	026	008	002	0+	0+	0+	0
	1	039	171	292	410	412	346	250	154	076	026	004	0+	0+	1
	2	001	014	049	154	265	346	375	346	265	154	049	014	001	2
	3	0+	0+	004	026	076	154	250	346	412	410	292	171	039	3
	4	0+	0+	0+	002	008	026	062	130	240	410	656	815	961	4
5	0	951	774	590	328	168	078	031	010	002	0+	0+	0+	0+	0
	1	048	204	328	410	360	259	156	077	028	006	0+	0+	0+	1
	2	001	021	073	205	309	346	312	230	132	051	008	001	0+	2
	3	0+	001	008	051	132	230	312	346	309	205	073	021	001	3
	4	0+	0+	0+	006	028	077	156	259	360	410	328	204	048	4
	5	0+	0+	0+	0+	002	010	031	078	168	328	590	774	951	5
6	0	941	735	531	262	118	047	016	004	001	0+	0+	0+	0+	0
	1	057	232	354	393	303	187	094	037	010	002	0+	0+	0+	1
	2	001	031	098	246	324	311	234	138	060	015	001	0+	0+	2
	3	0+	002	015	082	185	276	312	276	185	082	015	002	0+	3
	4	0+	0+	001	015	060	138	234	311	324	246	098	031	001	4
	5	0+	0+	0+	002	010	037	094	187	303	393	354	232	057	5
	6	0+	0+	0+	0+	001	004	016	047	118	262	531	735	941	6
7	0	932	698	478	210	082	028	008	002	0+	0+	0+	0+	0+	0
	1	066	257	372	367	247	131	055	017	004	0+	0+	0+	0+	1
	2	002	041	124	275	318	261	164	077	025	004	0+	0+	0+	2
	3	0+	004	023	115	227	290	273	194	097	029	003	0+	0+	3
	4	0+	0+	003	029	097	194	273	290	227	115	023	004	0+	4
	5	0+	0+	0+	004	025	077	164	261	318	275	124	041	002	5
	6	0+	0+	0+	0+	004	017	055	131	247	367	372	257	066	6
	7	0+	0+	0+	0+	0+	002	008	028	082	210	478	698	932	7
8	0	923	663	430	168	058	017	004	001	0+	0+	0+	0+	0+	0
	1	075	279	383	336	198	090	031	008	001	0+	0+	0+	0+	1
	2	003	051	149	294	296	209	109	041	010	001	0+	0+	0+	2
	3	0+	005	033	147	254	279	219	124	047	009	0+	0+	0+	3
	4	0+	0+	005	046	136	232	273	232	136	046	005	0+	0+	4
	5	0+	0+	0+	009	047	124	219	279	254	147	033	005	0+	5
	6	0+	0+	0+	001	010	041	109	209	296	294	149	051	003	6
	7	0+	0+	0+	0+	001	008	031	090	198	336	383	279	075	7
	8	0+	0+	0+	0+	0+	001	004	017	058	168	430	663	923	8
9	0	914	630	387	134	040	010	002	0+	0+	0+	0+	0+	0+	0
	1	083	299	387	302	156	060	018	004	0+	0+	0+	0+	0+	1
	2	003	063	172	302	267	161	070	021	004	0+	0+	0+	0+	2
	3	0+	008	045	176	267	251	164	074	021	003	0+	0+	0+	3
	4	0+	001	007	066	172	251	246	167	074	017	001	0+	0+	4

* For specific details on the use of this table, see page 189.

Table 4 (Continued)

n	x	0.01	0.05	0.10	0.20	0.30	0.40	p 0.50	0.60	0.70	0.80	0.90	0.95	0.99	x
9	5	0+	0+	001	017	074	167	246	251	172	066	007	001	0+	5
	6	0+	0+	0+	003	021	074	164	251	267	176	045	008	0+	6
	7	0+	0+	0+	0+	004	021	070	161	267	302	172	063	003	7
	8	0+	0+	0+	0+	0+	004	018	060	156	302	387	299	083	8
	9	0+	0+	0+	0+	0+	0+	002	010	040	134	387	630	914	9
10	0	904	599	349	107	028	006	001	0+	0+	0+	0+	0+	0+	0
	1	091	315	387	268	121	040	010	002	0+	0+	0+	0+	0+	1
	2	004	075	194	302	233	121	044	011	001	0+	0+	0+	0+	2
	3	0+	010	057	201	267	215	117	042	009	001	0+	0+	0+	3
	4	0+	001	011	088	200	251	205	111	037	006	0+	0+	0+	4
	5	0+	0+	001	026	103	201	246	201	103	026	001	0+	0+	5
	6	0+	0+	0+	006	037	111	205	251	200	088	011	001	0+	6
	7	0+	0+	0+	001	009	042	117	215	267	201	057	010	0+	7
	8	0+	0+	0+	0+	001	011	044	121	233	302	194	075	004	8
	9	0+	0+	0+	0+	0+	002	010	040	121	268	387	315	091	9
	10	0+	0+	0+	0+	0+	0+	001	006	028	107	349	599	904	10
11	0	895	569	314	086	020	004	0+	0+	0+	0+	0+	0+	0+	0
	1	099	329	384	236	093	027	005	001	0+	0+	0+	0+	0+	1
	2	005	087	213	295	200	089	027	005	001	0+	0+	0+	0+	2
	3	0+	014	071	221	257	177	081	023	004	0+	0+	0+	0+	3
	4	0+	001	016	111	220	236	161	070	017	002	0+	0+	0+	4
	5	0+	0+	002	039	132	221	226	147	057	010	0+	0+	0+	5
	6	0+	0+	0+	010	057	147	226	221	132	039	002	0+	0+	6
	7	0+	0+	0+	002	017	070	161	236	220	111	016	001	0+	7
	8	0+	0+	0+	0+	004	023	081	177	257	221	071	014	0+	8
	9	0+	0+	0+	0+	001	005	027	089	200	295	213	087	005	9
	10	0+	0+	0+	0+	0+	001	005	027	093	236	384	329	099	10
	11	0+	0+	0+	0+	0+	0+	0+	004	020	086	314	569	895	11
12	0	886	540	282	069	014	002	0+	0+	0+	0+	0+	0+	0+	0
	1	107	341	377	206	071	017	003	0+	0+	0+	0+	0+	0+	1
	2	006	099	230	283	168	064	016	002	0+	0+	0+	0+	0+	2
	3	0+	017	085	236	240	142	054	012	001	0+	0+	0+	0+	3
	4	0+	002	021	133	231	213	121	042	008	001	0+	0+	0+	4
	5	0+	0+	004	053	158	227	193	101	029	003	0+	0+	0+	5
	6	0+	0+	0+	016	079	177	226	177	079	016	0+	0+	0+	6
	7	0+	0+	0+	003	029	101	193	227	158	053	004	0+	0+	7
	8	0+	0+	0+	001	008	042	121	213	231	133	021	002	0+	8
	9	0+	0+	0+	0+	001	012	054	142	240	236	085	017	0+	9
	10	0+	0+	0+	0+	0+	002	016	064	168	283	230	099	006	10
	11	0+	0+	0+	0+	0+	0+	003	017	071	206	377	341	107	11
	12	0+	0+	0+	0+	0+	0+	0+	002	014	069	282	540	886	12
13	0	878	513	254	055	010	001	0+	0+	0+	0+	0+	0+	0+	0
	1	115	351	367	179	054	011	002	0+	0+	0+	0+	0+	0+	1
	2	007	111	245	268	139	045	010	001	0+	0+	0+	0+	0+	2
	3	0+	021	100	246	218	111	035	006	001	0+	0+	0+	0+	3
	4	0+	003	028	154	234	184	087	024	003	0+	0+	0+	0+	4

Table 4 (Continued)

n	x	0.01	0.05	0.10	0.20	0.30	0.40	p 0.50	0.60	0.70	0.80	0.90	0.95	0.99	x
13	5	0+	0+	006	069	180	221	157	066	014	001	0+	0+	0+	5
	6	0+	0--	001	023	103	197	209	131	044	006	0+	0+	0+	6
	7	0+	0+	0+	006	044	131	209	197	103	023	001	0+	0+	7
	8	0+	0+	0+	001	014	066	157	221	180	069	006	0+	0+	8
	9	0+	0+	0+	0+	003	024	087	184	234	154	028	003	0+	9
	10	0+	0+	0+	0+	001	006	035	111	218	246	100	021	0+	10
	11	0+	0+	0+	0+	0+	001	010	045	139	268	245	111	007	11
	12	0+	0+	0+	0+	0+	0+	002	011	054	179	367	351	115	12
	13	0+	0+	0+	0+	0+	0+	0+	001	010	055	254	513	878	13
14	0	869	488	229	044	007	001	0+	0+	0+	0+	0+	0+	0+	0
	1	123	359	356	154	041	007	001	0+	0+	0+	0+	0+	0+	1
	2	008	123	257	250	113	032	006	001	0+	0+	0+	0+	0+	2
	3	0+	026	114	250	194	085	022	003	0+	0+	0+	0+	0+	3
	4	0+	004	035	172	229	155	061	014	001	0+	0+	0+	0+	4
	5	0+	0+	008	086	196	207	122	041	007	0+	0+	0+	0+	5
	6	0+	0+	001	032	126	207	183	092	023	002	0+	0+	0+	6
	7	0+	0+	0+	009	062	157	209	157	062	009	0+	0+	0+	7
	8	0+	0+	0+	002	023	092	183	207	126	032	001	0+	0+	8
	9	0+	0+	0+	0+	007	041	122	207	196	086	008	0+	0+	9
	10	0+	0+	0+	0+	001	014	061	155	229	172	035	004	0+	10
	11	0+	0+	0+	0+	0+	003	022	085	194	250	114	026	0+	11
	12	0+	0+	0+	0+	0+	001	006	032	113	250	257	123	008	12
	13	0+	0+	0+	0+	0+	0+	001	007	041	154	356	359	123	13
	14	0+	0+	0+	0+	0+	0+	0+	001	007	044	229	488	869	14
15	0	860	463	206	035	005	0+	0+	0+	0+	0+	0+	0+	0+	0
	1	130	366	343	132	031	005	0+	0+	0+	0+	0+	0+	0+	1
	2	009	135	267	231	092	022	003	0+	0+	0+	0+	0+	0+	2
	3	0+	031	129	250	170	063	014	002	0+	0+	0+	0+	0+	3
	4	0+	005	043	188	219	127	042	007	001	0+	0+	0+	0+	4
	5	0+	001	010	103	206	186	092	024	003	0+	0+	0+	0+	5
	6	0+	0+	002	043	147	207	153	061	012	001	0+	0+	0+	6
	7	0+	0+	0+	014	081	177	196	118	035	003	0+	0+	0+	7
	8	0+	0+	0+	003	035	118	196	177	081	014	0+	0+	0+	8
	9	0+	0+	0+	001	012	061	153	207	147	043	002	0+	0+	9

From Frederick Mosteller, Robert E. K. Rourke, and George B. Thomas, Jr., *Probability with Statistical Applications*, 2nd ed., © 1970, Addison-Wesley Publishing Company, Reading, Mass., pp. 475–477. Reprinted with permission.

Table 5   Areas of the Standard Normal Distribution

The entries in this table are the probabilities that a random variable having the standard normal distribution assumes a value between 0 and z; the probability is represented by the area under the curve shaded in the figure. Areas for negative values of z are obtained by symmetry.

z	0.00	0.01	0.02	0.03	0.04	0.05	0.06	0.07	0.08	0.09
					Second decimal place in z					
0.0	0.0000	0.0040	0.0080	0.0120	0.0160	0.0199	0.0239	0.0279	0.0319	0.0359
0.1	0.0398	0.0438	0.0478	0.0517	0.0557	0.0596	0.0636	0.0675	0.0714	0.0753
0.2	0.0793	0.0832	0.0871	0.0910	0.0948	0.0987	0.1026	0.1064	0.1103	0.1141
0.3	0.1179	0.1217	0.1255	0.1293	0.1331	0.1368	0.1406	0.1443	0.1480	0.1517
0.4	0.1554	0.1591	0.1628	0.1664	0.1700	0.1736	0.1772	0.1808	0.1844	0.1879
0.5	0.1915	0.1950	0.1985	0.2019	0.2054	0.2088	0.2123	0.2157	0.2190	0.2224
0.6	0.2257	0.2291	0.2324	0.2357	0.2389	0.2422	0.2454	0.2486	0.2517	0.2549
0.7	0.2580	0.2611	0.2642	0.2673	0.2704	0.2734	0.2764	0.2794	0.2823	0.2852
0.8	0.2881	0.2910	0.2939	0.2967	0.2995	0.3023	0.3051	0.3078	0.3106	0.3133
0.9	0.3159	0.3186	0.3212	0.3238	0.3264	0.3289	0.3315	0.3340	0.3365	0.3389
1.0	0.3413	0.3438	0.3461	0.3485	0.3508	0.3531	0.3554	0.3577	0.3599	0.3621
1.1	0.3643	0.3665	0.3686	0.3708	0.3729	0.3749	0.3770	0.3790	0.3810	0.3830
1.2	0.3849	0.3869	0.3888	0.3907	0.3925	0.3944	0.3962	0.3980	0.3997	0.4015
1.3	0.4032	0.4049	0.4066	0.4082	0.4099	0.4115	0.4131	0.4147	0.4162	0.4177
1.4	0.4192	0.4207	0.4222	0.4236	0.4251	0.4265	0.4279	0.4292	0.4306	0.4319
1.5	0.4332	0.4345	0.4357	0.4370	0.4382	0.4394	0.4406	0.4418	0.4429	0.4441
1.6	0.4452	0.4463	0.4474	0.4484	0.4495	0.4505	0.4515	0.4525	0.4535	0.4545
1.7	0.4554	0.4564	0.4573	0.4582	0.4591	0.4599	0.4608	0.4616	0.4625	0.4633
1.8	0.4641	0.4649	0.4656	0.4664	0.4671	0.4678	0.4686	0.4693	0.4699	0.4706
1.9	0.4713	0.4719	0.4726	0.4732	0.4738	0.4744	0.4750	0.4756	0.4761	0.4767
2.0	0.4772	0.4778	0.4783	0.4788	0.4793	0.4798	0.4803	0.4808	0.4812	0.4817
2.1	0.4821	0.4826	0.4830	0.4834	0.4838	0.4842	0.4846	0.4850	0.4854	0.4857
2.2	0.4861	0.4864	0.4868	0.4871	0.4875	0.4878	0.4881	0.4884	0.4887	0.4890
2.3	0.4893	0.4896	0.4898	0.4901	0.4904	0.4906	0.4909	0.4911	0.4913	0.4916
2.4	0.4918	0.4920	0.4922	0.4925	0.4927	0.4929	0.4931	0.4932	0.4934	0.4936
2.5	0.4938	0.4940	0.4941	0.4943	0.4945	0.4946	0.4948	0.4949	0.4951	0.4952
2.6	0.4953	0.4955	0.4956	0.4957	0.4959	0.4960	0.4961	0.4962	0.4963	0.4964
2.7	0.4965	0.4966	0.4967	0.4968	0.4969	0.4970	0.4971	0.4972	0.4973	0.4974
2.8	0.4974	0.4975	0.4976	0.4977	0.4977	0.4978	0.4979	0.4979	0.4980	0.4981
2.9	0.4981	0.4982	0.4982	0.4983	0.4984	0.4984	0.4985	0.4985	0.4986	0.4986
3.0	0.4987	0.4987	0.4987	0.4988	0.4988	0.4989	0.4989	0.4989	0.4990	0.4990
3.1	0.4990	0.4991	0.4991	0.4991	0.4992	0.4992	0.4992	0.4992	0.4993	0.4993
3.2	0.4993	0.4993	0.4994	0.4994	0.4994	0.4994	0.4994	0.4995	0.4995	0.4995
3.3	0.4995	0.4995	0.4995	0.4996	0.4996	0.4996	0.4996	0.4996	0.4996	0.4997
3.4	0.4997	0.4997	0.4997	0.4997	0.4997	0.4997	0.4997	0.4997	0.4997	0.4998
3.5	0.4998									
4.0	0.49997									
4.5	0.499997									
5.0	0.4999997									

Reprinted with permission from *Standard Mathematical Tables*, 15th ed. Copyright The Chemical Rubber, Co., CRC Press, Inc.

* For specific details on the use of this table, see page 206.

Table 6   Critical Values of Student's *t* Distribution

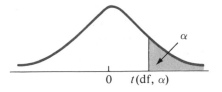

0   *t*(df, α)

The entries in this table are the critical values for Student's *t* for an area of α in the right-hand tail. Critical values for the left-hand tail are found by symmetry.

df	Amount of α in one-tail					
	0.25	0.10	0.05	0.025	0.01	0.005
1	1.000	3.08	6.31	12.7	31.8	63.7
2	0.816	1.89	2.92	4.30	6.97	9.92
3	0.765	1.64	2.35	3.18	4.54	5.84
4	0.741	1.53	2.13	2.78	3.75	4.60
5	0.727	1.48	2.02	2.57	3.37	4.03
6	0.718	1.44	1.94	2.45	3.14	3.71
7	0.711	1.42	1.89	2.36	3.00	3.50
8	0.706	1.40	1.86	2.31	2.90	3.36
9	0.703	1.38	1.83	2.26	2.82	3.25
10	0.700	1.37	1.81	2.23	2.76	3.17
11	0.697	1.36	1.80	2.20	2.72	3.11
12	0.695	1.36	1.78	2.18	2.68	3.05
13	0.694	1.35	1.77	2.16	2.65	3.01
14	0.692	1.35	1.76	2.14	2.62	2.98
15	0.691	1.34	1.75	2.13	2.60	2.95
16	0.690	1.34	1.75	2.12	2.58	2.92
17	0.689	1.33	1.74	2.11	2.57	2.90
18	0.688	1.33	1.73	2.10	2.55	2.88
19	0.688	1.33	1.73	2.09	2.54	2.86
20	0.687	1.33	1.72	2.09	2.53	2.85
21	0.686	1.32	1.72	2.08	2.52	2.83
22	0.686	1.32	1.72	2.07	2.51	2.82
23	0.685	1.32	1.71	2.07	2.50	2.81
24	0.685	1.32	1.71	2.06	2.49	2.80
25	0.684	1.32	1.71	2.06	2.49	2.79
26	0.684	1.32	1.71	2.06	2.48	2.78
27	0.684	1.31	1.70	2.05	2.47	2.77
28	0.683	1.31	1.70	2.05	2.47	2.76
29	0.683	1.31	1.70	2.05	2.46	2.76
z	0.674	1.28	1.65	1.96	2.33	2.58

NOTE: For df ≥ 30, the critical value $t(\text{df}, \alpha)$ is approximated by $z(\alpha)$, given in the bottom row of table.

Adapted from E. S. Pearson and H. O. Hartley, *Biometrika Tables for Statisticians*, vol. I (1966), p. 146. Reprinted by permission of the Biometrika Trustees. The two columns headed "0.10" and "0.01" are taken from Table III (adapted) on p. 46 of Fisher and Yates, *Statistical Tables for Biological, Agricultural and Medical Research*, 6th ed., published by Longman Group Ltd., London, 1974 (previously published by Oliver and Boyd, Edinburgh), and by permission of the authors and publishers.

* For specific details on the use of this table, see page 304.

Table 7   Critical Values of the χ^2 Distribution

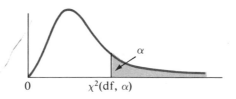

$$0 \qquad \chi^2(df, \alpha)$$

The entries in this table are the critical values for chi square for which the area to the right under the curve is equal to α.

df	Amount of α in Right-hand Tail									
	0.995	0.990	0.975	0.950	0.900	0.100	0.050	0.025	0.010	0.005
1	0.0000393	0.000157	0.000982	0.00393	0.0158	2.71	3.84	5.02	6.64	7.88
2	0.0100	0.0201	0.0506	0.103	0.211	4.61	6.00	7.38	9.21	10.6
3	0.0717	0.115	0.216	0.352	0.584	6.25	7.82	9.35	11.4	12.9
4	0.207	0.297	0.484	0.711	1.0636	7.78	9.50	11.1	13.3	14.9
5	0.412	0.554	0.831	1.15	1.61	9.24	11.1	12.8	15.1	16.8
6	0.676	0.872	1.24	1.64	2.20	10.6	12.6	14.5	16.8	18.6
7	0.990	1.24	1.69	2.17	2.83	12.0	14.1	16.0	18.5	20.3
8	1.34	1.65	2.18	2.73	3.49	13.4	15.5	17.5	20.1	22.0
9	1.73	2.09	2.70	3.33	4.17	14.7	17.0	19.0	21.7	23.6
10	2.16	2.56	3.25	3.94	4.87	16.0	18.3	20.5	23.2	25.2
11	2.60	3.05	3.82	4.58	5.58	17.2	19.7	21.9	24.7	26.8
12	3.07	3.57	4.40	5.23	6.30	18.6	21.0	23.3	26.2	28.3
13	3.57	4.11	5.01	5.90	7.04	19.8	22.4	24.7	27.7	29.8
14	4.07	4.66	5.63	6.57	7.79	21.1	23.7	26.1	29.1	31.3
15	4.60	5.23	6.26	7.26	8.55	22.3	25.0	27.5	30.6	32.8
16	5.14	5.81	6.91	7.96	9.31	23.5	26.3	28.9	32.0	34.3
17	5.70	6.41	7.56	8.67	10.1	24.8	27.6	30.2	33.4	35.7
18	6.26	7.01	8.23	9.39	10.9	26.0	28.9	31.5	34.8	37.2
19	6.84	7.63	8.91	10.1	11.7	27.2	30.1	32.9	36.2	38.6
20	7.43	8.26	9.59	10.9	12.4	28.4	31.4	34.2	37.6	40.0
21	8.03	8.90	10.3	11.6	13.2	29.6	32.7	35.5	39.0	41.4
22	8.64	9.54	11.0	12.3	14.0	30.8	33.9	36.8	40.3	42.8
23	9.26	10.2	11.0	13.1	14.9	32.0	35.2	38.1	41.6	44.2
24	9.89	10.9	12.4	13.9	15.7	33.2	36.4	39.4	43.0	45.6
25	10.5	11.5	13.1	14.6	16.5	34.4	37.7	40.7	44.3	46.9
26	11.2	12.2	13.8	15.4	17.3	35.6	38.9	41.9	45.6	48.3
27	11.8	12.9	14.6	16.2	18.1	36.7	40.1	43.2	47.0	49.7
28	12.5	13.6	15.3	16.9	18.9	37.9	41.3	44.5	48.3	51.0
29	13.1	14.3	16.1	17.7	19.8	39.1	42.6	45.7	49.6	52.3
30	13.8	15.0	16.8	18.5	20.6	40.3	43.8	47.0	50.9	53.7
40	20.7	22.2	24.4	26.5	29.1	51.8	55.8	59.3	63.7	66.8
50	28.0	29.7	32.4	34.8	37.7	63.2	67.5	71.4	76.2	79.5
60	35.5	37.5	40.5	43.2	46.5	74.4	79.1	83.3	88.4	92.0
70	43.3	45.4	48.8	51.8	55.3	85.5	90.5	95.0	100.0	104.0
80	51.2	53.5	57.2	60.4	64.3	96.6	102.0	107.0	112.0	116.0
90	59.2	61.8	65.7	69.1	73.3	108.0	113.0	118.0	124.0	128.0
100	67.3	70.1	74.2	77.9	82.4	114.0	124.0	130.0	136.0	140.0

Adapted from E. S. Pearson and H. O. Hartley, *Biometrika Tables for Statisticians*, vol. I (1962), pp. 130–131. Reprinted by permission of the Biometrika Trustees.

* For specific details on the use of this table, see page 319.

Table 8a   Critical Values of the _F_ Distribution ($\alpha = 0.05$)

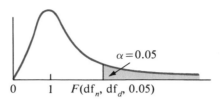

$\alpha = 0.05$

$0 \quad 1 \quad F(df_n, df_d, 0.05)$

The entries in this table are critical values of F for which the area under the curve to the right is equal to 0.05.

					Degrees of Freedom for Numerator					
	1	2	3	4	5	6	7	8	9	10
1	161	200	216	225	230	234	237	239	241	242
2	18.5	19.0	19.2	19.2	19.3	19.3	19.4	19.4	19.4	19.4
3	10.1	9.55	9.28	9.12	9.01	8.94	8.89	8.85	8.81	8.79
4	7.71	6.94	6.59	6.39	6.26	6.16	6.09	6.04	6.00	5.96
5	6.61	5.79	5.41	5.19	5.05	4.95	4.88	4.82	4.77	4.74
6	5.99	5.14	4.76	4.53	4.39	4.28	4.21	4.15	4.10	4.06
7	5.59	4.74	4.35	4.12	3.97	3.87	3.79	3.73	3.68	3.64
8	5.32	4.46	4.07	3.84	3.69	3.58	3.50	3.44	3.39	3.35
9	5.12	4.26	3.86	3.63	3.48	3.37	3.29	3.23	3.18	3.14
10	4.96	4.10	3.71	3.48	3.33	3.22	3.14	3.07	3.02	2.98
11	4.84	3.98	3.59	3.36	3.20	3.09	3.01	2.95	2.90	2.85
12	4.75	3.89	3.49	3.26	3.11	3.00	2.91	2.85	2.80	2.75
13	4.67	3.81	3.41	3.18	3.03	2.92	2.83	2.77	2.71	2.67
14	4.60	3.74	3.34	3.11	2.96	2.85	2.76	2.70	2.65	2.60
15	4.54	3.68	3.29	3.06	2.90	2.79	2.71	2.64	2.59	2.54
16	4.49	3.63	3.24	3.01	2.85	2.74	2.66	2.59	2.54	2.49
17	4.45	3.59	3.20	2.96	2.81	2.70	2.61	2.55	2.49	2.45
18	4.41	3.55	3.16	2.93	2.77	2.66	2.58	2.51	2.46	2.41
19	4.38	3.52	3.13	2.90	2.74	2.63	2.54	2.48	2.42	2.38
20	4.35	3.49	3.10	2.87	2.71	2.60	2.51	2.45	2.39	2.35
21	4.32	3.47	3.07	2.84	2.68	2.57	2.49	2.42	2.37	2.32
22	4.30	3.44	3.05	2.82	2.66	2.55	2.46	2.40	2.34	2.30
23	4.28	3.42	3.03	2.80	2.64	2.53	2.44	2.37	2.32	2.27
24	4.26	3.40	3.01	2.78	2.62	2.51	2.42	2.36	2.30	2.25
25	4.24	3.39	2.99	2.76	2.60	2.49	2.40	2.34	2.28	2.24
30	4.17	3.32	2.92	2.69	2.53	2.42	2.33	2.27	2.21	2.16
40	4.08	3.23	2.84	2.61	2.45	2.34	2.25	2.18	2.12	2.08
60	4.00	3.15	2.76	2.53	2.37	2.25	2.17	2.10	2.04	1.99
120	3.92	3.07	2.68	2.45	2.29	2.18	2.09	2.02	1.96	1.91
∞	3.84	3.00	2.60	2.37	2.21	2.10	2.01	1.94	1.88	1.83

Degrees of Freedom for Denominator

* For specific details on the use of this table, see page 345.

		12	15	20	24	30	40	60	120	∞
		Degrees of Freedom for Numerator								
Degrees of Freedom for Denominator	1	244	246	248	249	250	251	252	253	254
	2	19.4	19.4	19.4	19.5	19.5	19.5	19.5	19.5	19.5
	3	8.74	8.70	8.66	8.64	8.62	8.59	8.57	8.55	8.53
	4	5.91	5.86	5.80	5.77	5.75	5.72	5.69	5.66	5.63
	5	4.68	4.62	4.56	4.53	4.50	4.46	4.43	4.40	4.37
	6	4.00	3.94	3.87	3.84	3.81	3.77	3.74	3.70	3.67
	7	3.57	3.51	3.44	3.41	3.38	3.34	3.30	3.27	3.23
	8	3.28	3.22	3.15	3.12	3.08	3.04	3.01	2.97	2.93
	9	3.07	3.01	2.94	2.90	2.86	2.83	2.79	2.75	2.71
	10	2.91	2.85	2.77	2.74	2.70	2.66	2.62	2.58	2.54
	11	2.79	2.72	2.65	2.61	2.57	2.53	2.49	2.45	2.40
	12	2.69	2.62	2.54	2.51	2.47	2.43	2.38	2.34	2.30
	13	2.60	2.53	2.46	2.42	2.38	2.34	2.30	2.25	2.21
	14	2.53	2.46	2.39	2.35	2.31	2.27	2.22	2.18	2.13
	15	2.48	2.40	2.33	2.29	2.25	2.20	2.16	2.11	2.07
	16	2.42	2.35	2.28	2.24	2.19	2.15	2.11	2.06	2.01
	17	2.38	2.31	2.23	2.19	2.15	2.10	2.06	2.01	1.96
	18	2.34	2.27	2.19	2.15	2.11	2.06	2.02	1.97	1.92
	19	2.31	2.23	2.16	2.11	2.07	2.03	1.98	1.93	1.88
	20	2.28	2.20	2.12	2.08	2.04	1.99	1.95	1.90	1.84
	21	2.25	2.18	2.10	2.05	2.01	1.96	1.92	1.87	1.81
	22	2.23	2.15	2.07	2.03	1.98	1.94	1.89	1.84	1.78
	23	2.20	2.13	2.05	2.01	1.96	1.91	1.86	1.81	1.76
	24	2.18	2.11	2.03	1.98	1.94	1.89	1.84	1.79	1.73
	25	2.16	2.09	2.01	1.96	1.92	1.87	1.82	1.77	1.71
	30	2.09	2.01	1.93	1.89	1.84	1.79	1.74	1.68	1.62
	40	2.00	1.92	1.84	1.79	1.74	1.69	1.64	1.58	1.51
	60	1.92	1.84	1.75	1.70	1.65	1.59	1.53	1.47	1.39
	120	1.83	1.75	1.66	1.61	1.55	1.50	1.43	1.35	1.25
	∞	1.75	1.67	1.57	1.52	1.46	1.39	1.32	1.22	1.00

Table 8b   Critical Values of the *F* Distribution ($\alpha = 0.025$)

$\alpha = 0.025$

$F(df_n, df_d, 0.025)$

The entries in this table are critical values of *F* for which the area under the curve to the right is equal to 0.025.

		Degrees of Freedom for Numerator								
	1	2	3	4	5	6	7	8	9	10
1	648	800	864	900	922	937	948	957	963	969
2	38.5	39.0	39.2	39.2	39.3	39.3	39.4	39.4	39.4	39.4
3	17.4	16.0	15.4	15.1	14.9	14.7	14.6	14.5	14.5	14.4
4	12.2	10.6	9.98	9.60	9.36	9.20	9.07	8.98	8.90	8.84
5	10.0	8.43	7.76	7.39	7.15	6.98	6.85	6.76	6.68	6.62
6	8.81	7.26	6.60	6.23	5.99	5.82	5.70	5.60	5.52	5.46
7	8.07	6.54	5.89	5.52	5.29	5.12	4.99	4.90	4.82	4.76
8	7.57	6.06	5.42	5.05	4.82	4.65	4.53	4.43	4.36	4.30
9	7.21	5.71	5.08	4.72	4.48	4.32	4.20	4.10	4.03	3.96
10	6.94	5.46	4.83	4.47	4.24	4.07	3.95	3.85	3.78	3.72
11	6.72	5.26	4.63	4.28	4.04	3.88	3.76	3.66	3.59	3.53
12	6.55	5.10	4.47	4.12	3.89	3.73	3.61	3.51	3.44	3.37
13	6.41	4.97	4.35	4.00	3.77	3.60	3.48	3.39	3.31	3.25
14	6.30	4.86	4.24	3.89	3.66	3.50	3.38	3.28	3.21	3.15
15	6.20	4.77	4.15	3.80	3.58	3.41	3.29	3.20	3.12	3.06
16	6.12	4.69	4.08	3.73	3.50	3.34	3.22	3.12	3.05	2.99
17	6.04	4.62	4.01	3.66	3.44	3.28	3.16	3.06	2.98	2.92
18	5.98	4.56	3.95	3.61	3.38	3.22	3.10	3.01	2.93	2.87
19	5.92	4.51	3.90	3.56	3.33	3.17	3.05	2.96	2.88	2.82
20	5.87	4.46	3.86	3.51	3.29	3.13	3.01	2.91	2.84	2.77
21	5.83	4.42	3.82	3.48	3.25	3.09	2.97	2.87	2.80	2.73
22	5.79	4.38	3.78	3.44	3.22	3.05	2.93	2.84	2.76	2.70
23	5.75	4.35	3.75	3.41	3.18	3.02	2.90	2.81	2.73	2.67
24	5.72	4.32	3.72	3.38	3.15	2.99	2.87	2.78	2.70	2.64
25	5.69	4.29	3.69	3.35	3.13	2.97	2.85	2.75	2.68	2.61
30	5.57	4.18	3.59	3.25	3.03	2.87	2.75	2.65	2.57	2.51
40	5.42	4.05	3.46	3.13	2.90	2.74	2.62	2.53	2.45	2.39
60	5.29	3.93	3.34	3.01	2.79	2.63	2.51	2.41	2.33	2.27
120	5.15	3.80	3.23	2.89	2.67	2.52	2.39	2.30	2.22	2.16
∞	5.02	3.69	3.12	2.79	2.57	2.41	2.29	2.19	2.11	2.05

Degrees of Freedom for Denominator

* For specific details on the use of this table, see page 345.

Table 8b   (Continued)

		12	15	20	24	30	40	60	120	∞
		Degrees of Freedom for Numerator								

	12	15	20	24	30	40	60	120	∞
1	977	985	993	997	1,001	1,006	1,010	1,014	1,018
2	39.4	39.4	39.4	39.5	39.5	39.5	39.5	39.5	39.5
3	14.3	14.3	14.2	14.1	14.1	14.0	14.0	13.9	13.9
4	8.75	8.66	8.56	8.51	8.46	8.41	8.36	8.31	8.26
5	6.52	6.43	6.33	6.28	6.23	6.18	6.12	6.07	6.02
6	5.37	5.27	5.17	5.12	5.07	5.01	4.96	4.90	4.85
7	4.67	4.57	4.47	4.42	4.36	4.31	4.25	4.20	4.14
8	4.20	4.10	4.00	3.95	3.89	3.84	3.78	3.73	3.67
9	3.87	3.77	3.67	3.61	3.56	3.51	3.45	3.39	3.33
10	3.62	3.52	3.42	3.37	3.31	3.26	3.20	3.14	3.08
11	3.43	3.33	3.23	3.17	3.12	3.06	3.00	2.94	2.88
12	3.28	3.18	3.07	3.02	2.96	2.91	2.85	2.79	2.72
13	3.15	3.05	2.95	2.89	2.84	2.78	2.72	2.66	2.60
14	3.05	2.95	2.84	2.79	2.73	2.67	2.61	2.55	2.49
15	2.96	2.86	2.76	2.70	2.64	2.59	2.52	2.46	2.40
16	2.89	2.79	2.68	2.63	2.57	2.51	2.45	2.38	2.32
17	2.82	2.72	2.62	2.56	2.50	2.44	2.38	2.32	2.25
18	2.77	2.67	2.56	2.50	2.44	2.38	2.32	2.26	2.19
19	2.72	2.62	2.51	2.45	2.39	2.33	2.27	2.20	2.13
20	2.68	2.57	2.46	2.41	2.35	2.29	2.22	2.16	2.09
21	2.64	2.53	2.42	2.37	2.31	2.25	2.18	2.11	2.04
22	2.60	2.50	2.39	2.33	2.27	2.21	2.14	2.08	2.00
23	2.57	2.47	2.36	2.30	2.24	2.18	2.11	2.04	1.97
24	2.54	2.44	2.33	2.27	2.21	2.15	2.08	2.01	1.94
25	2.51	2.41	2.30	2.24	2.18	2.12	2.05	1.98	1.91
30	2.41	2.31	2.20	2.14	2.07	2.01	1.94	1.87	1.79
40	2.29	2.18	2.07	2.01	1.94	1.88	1.80	1.72	1.64
60	2.17	2.06	1.94	1.88	1.82	1.74	1.67	1.58	1.48
120	2.05	1.95	1.82	1.76	1.69	1.61	1.53	1.43	1.31
∞	1.94	1.83	1.71	1.64	1.57	1.48	1.39	1.27	1.00

Degrees of Freedom for Denominator (row labels)

From E. S. Pearson and H. O. Hartley, *Biometrika Tables for Statisticians*, vol. 1 (1958), pp. 159–163. Reprinted by permission of the Biometrika Trustees.

Table 8c   Critical Values of the *F* Distribution ($\alpha = 0.01$)

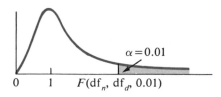

$\alpha = 0.01$

0   1 $F(df_n, df_d, 0.01)$

The entries in the table are critical values of *F* for which the area under the curve to the right is equal to 0.01.

				Degrees of Freedom for Numerator						
	1	2	3	4	5	6	7	8	9	10
1	4,052	5,000	5,403	5,625	5,764	5,859	5,928	5,982	6,023	6,056
2	98.5	99.0	99.2	99.2	99.3	99.3	99.4	99.4	99.4	99.4
3	34.1	30.8	29.5	28.7	28.2	27.9	27.7	27.5	27.3	27.2
4	21.2	18.0	16.7	16.0	15.5	15.2	15.0	14.8	14.7	14.5
5	16.3	13.3	12.1	11.4	11.0	10.7	10.5	10.3	10.2	10.1
6	13.7	10.9	9.78	9.15	8.75	8.47	8.26	8.10	7.98	7.87
7	12.2	9.55	8.45	7.85	7.46	7.19	6.99	6.84	6.72	6.62
8	11.3	8.65	7.59	7.01	6.63	6.37	6.18	6.03	5.91	5.81
9	10.6	8.02	6.99	6.42	6.06	5.80	5.61	5.47	5.35	5.26
10	10.0	7.56	6.55	5.99	5.64	5.39	5.20	5.06	4.94	4.85
11	9.65	7.21	6.22	5.67	5.32	5.07	4.89	4.74	4.63	4.54
12	9.33	6.93	5.95	5.41	5.06	4.82	4.64	4.50	4.39	4.30
13	9.07	6.70	5.74	5.21	4.86	4.62	4.44	4.30	4.19	4.10
14	8.86	6.51	5.56	5.04	4.70	4.46	4.28	4.14	4.03	3.94
15	8.68	6.36	5.42	4.89	4.56	4.32	4.14	4.00	3.89	3.80
16	8.53	6.23	5.29	4.77	4.44	4.20	4.03	3.89	3.78	3.69
17	8.40	6.11	5.19	4.67	4.34	4.10	3.93	3.79	3.68	3.59
18	8.29	6.01	5.09	4.58	4.25	4.01	3.84	3.71	3.60	3.51
19	8.19	5.93	5.01	4.50	4.17	3.94	3.77	3.63	3.52	3.43
20	8.10	5.85	4.94	4.43	4.10	3.87	3.70	3.56	3.46	3.37
21	8.02	5.78	4.87	4.37	4.04	3.81	3.64	3.51	3.40	3.31
22	7.95	5.72	4.82	4.31	3.99	3.76	3.59	3.45	3.35	3.26
23	7.88	5.66	4.76	4.26	3.94	3.71	3.54	3.41	3.30	3.21
24	7.82	5.61	4.72	4.22	3.90	3.67	3.50	3.36	3.26	3.17
25	7.77	5.57	4.68	4.18	3.86	3.63	3.46	3.32	3.22	3.13
30	7.56	5.39	4.51	4.02	3.70	3.47	3.30	3.17	3.07	2.98
40	7.31	5.18	4.31	3.83	3.51	3.29	3.12	2.99	2.89	2.80
60	7.08	4.98	4.13	3.65	3.34	3.12	2.95	2.82	2.72	2.63
120	6.85	4.79	3.95	3.48	3.17	2.96	2.79	2.66	2.56	2.47
∞	6.63	4.61	3.78	3.32	3.02	2.80	2.64	2.51	2.41	2.32

Degrees of Freedom for Denominator

Table 8c   (Continued)

		12	15	20	24	30	40	60	120	∞
					Degrees of Freedom for Numerator					
	1	6,106	6,157	6,209	6,235	6,261	6,287	6,313	6,339	6,366
	2	99.4	99.4	99.4	99.5	99.5	99.5	99.5	99.5	99.5
	3	27.1	26.9	26.7	26.6	26.5	26.4	26.3	26.2	26.1
	4	14.4	14.2	14.0	13.9	13.8	13.7	13.7	13.6	13.5
	5	9.89	9.72	9.55	9.47	9.38	9.29	9.20	9.11	9.02
	6	7.72	7.56	7.40	7.31	7.23	7.14	7.06	6.97	6.88
	7	6.47	6.31	6.16	6.07	5.99	5.91	5.82	5.74	5.65
	8	5.67	5.52	5.36	5.28	5.20	5.12	5.03	4.95	4.86
	9	5.11	4.96	4.81	4.73	4.65	4.57	4.48	4.40	4.31
	10	4.71	4.56	4.41	4.33	4.25	4.17	4.08	4.00	3.91
	11	4.40	4.25	4.10	4.02	3.94	3.86	3.78	3.69	3.60
	12	4.16	4.01	3.86	3.78	3.70	3.62	3.54	3.45	3.36
	13	3.96	3.82	3.66	3.59	3.51	3.43	3.34	3.25	3.17
	14	3.80	3.66	3.51	3.43	3.35	3.27	3.18	3.09	3.00
	15	3.67	3.52	3.37	3.29	3.21	3.13	3.05	2.96	2.87
	16	3.55	3.41	3.26	3.18	3.10	3.02	2.93	2.84	2.75
	17	3.46	3.31	3.16	3.08	3.00	2.92	2.83	2.75	2.65
	18	3.37	3.23	3.08	3.00	2.92	2.84	2.75	2.66	2.57
	19	3.30	3.15	3.00	2.92	2.84	2.76	2.67	2.58	2.49
	20	3.23	3.09	2.94	2.86	2.78	2.69	2.61	2.52	2.42
	21	3.17	3.03	2.88	2.80	2.72	2.64	2.55	2.46	2.36
	22	3.12	2.98	2.83	2.75	2.67	2.58	2.50	2.40	2.31
	23	3.07	2.93	2.78	2.70	2.62	2.54	2.45	2.35	2.26
	24	3.03	2.89	2.74	2.66	2.58	2.49	2.40	2.31	2.21
	25	2.99	2.85	2.70	2.62	2.53	2.45	2.36	2.27	2.17
	30	2.84	2.70	2.55	2.47	2.39	2.30	2.21	2.11	2.01
	40	2.66	2.52	2.37	2.29	2.20	2.11	2.02	1.92	1.80
	60	2.50	2.35	2.20	2.12	2.03	1.94	1.84	1.73	1.60
	120	2.34	2.19	2.03	1.95	1.86	1.76	1.66	1.53	1.38
	∞	2.18	2.04	1.88	1.79	1.70	1.59	1.47	1.32	1.00

Degrees of Freedom for Denominator (left vertical label)

Critical Values of the *F* Distribution ($\alpha = 0.01$)   **Table 8c**

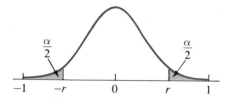

Table 9   Critical Values of *r* when $\rho = 0$

The entries in this table are the critical values of *r* for a two-tailed test at α.

    For simple correlation. df = $n - 2$, where *n* is the number of pairs of data in the sample. For a one-tailed test, the value of α is shown at the top of the table is double the value of α being used in the hypothesis test.

df α	0.10	0.05	0.02	0.01
1	0.988	0.997	1.000	1.000
2	0.900	0.950	0.980	0.990
3	0.805	0.878	0.934	0.959
4	0.729	0.811	0.882	0.917
5	0.669	0.754	0.833	0.874
6	0.662	0.707	0.789	0.834
7	0.582	0.666	0.750	0.798
8	0.549	0.632	0.716	0.765
9	0.521	0.602	0.685	0.735
10	0.497	0.576	0.658	0.708
11	0.476	0.553	0.634	0.684
12	0.458	0.532	0.612	0.661
13	0.441	0.514	0.592	0.641
14	0.426	0.497	0.574	0.623
15	0.412	0.482	0.558	0.606
16	0.400	0.468	0.542	0.590
17	0.389	0.456	0.528	0.575
18	0.378	0.444	0.516	0.561
19	0.369	0.433	0.503	0.549
20	0.360	0.423	0.492	0.537
25	0.323	0.381	0.445	0.487
30	0.296	0.349	0.409	0.449
35	0.275	0.325	0.381	0.418
40	0.257	0.304	0.358	0.393
45	0.243	0.288	0.338	0.372
50	0.231	0.273	0.322	0.354
60	0.211	0.250	0.295	0.325
70	0.195	0.232	0.274	0.302
80	0.183	0.217	0.256	0.283
90	0.173	0.205	0.242	0.267
100	0.164	0.195	0.230	0.254

From E. S. Pearson and H. O. Hartley, *Biometrika Tables for Statisticians*. vol. I (1962), p. 138. Reprinted by permission of the Biometrika Trustees.

* For specific details on the use of this table. see page 441.

Table 10   Confidence Belts for the Correlation Coefficient* $(1-\alpha=0.95)$

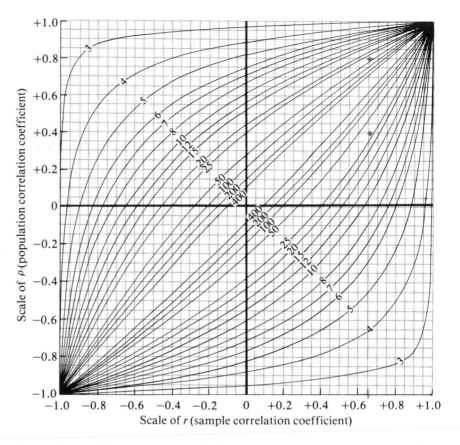

The numbers on the curves are sample sizes.

* For specific details on the use of this table, see page 442.
From E. S. Pearson and H. O. Hartley, *Biometrika Tables for Statisticians*, vol. I (1962), p. 140. Reprinted by permission of the Biometrika Trustees.

Table 11
Critical Values for the Sign Test*

The entries in this table are the critical values for the number of the least frequent sign for a two-tailed test at α for the binomial $p = 0.5$. For a one-tailed test, the value of α shown at the top of the table is double the value of α being used in the hypothesis test.

n	α 0.01	0.05	0.10	0.25	n	α 0.01	0.05	0.10	0.25
1					51	15	18	19	20
2					52	16	18	19	21
3				0	53	16	18	20	21
4				0	54	17	19	20	22
5			0	0	55	17	19	20	22
6		0	0	1	56	17	20	21	23
7		0	0	1	57	18	20	21	23
8	0	0	1	1	58	18	21	22	24
9	0	1	1	2	59	19	21	22	24
10	0	1	1	2	60	19	21	23	25
11	0	1	2	3	61	20	22	23	25
12	1	2	2	3	62	20	22	24	25
13	1	2	3	3	63	20	23	24	26
14	1	2	3	4	64	21	23	24	26
15	2	3	3	4	65	21	24	25	27
16	2	3	4	5	66	22	24	25	27
17	2	4	4	5	67	22	25	26	28
18	3	4	5	6	68	22	25	26	28
19	3	4	5	6	69	23	25	27	29
20	3	5	5	6	70	23	26	27	29
21	4	5	6	7	71	24	26	28	30
22	4	5	6	7	72	24	27	28	30
23	4	6	7	8	73	25	27	28	31
24	5	6	7	8	74	25	28	29	31
25	5	7	7	9	75	25	28	29	32
26	6	7	8	9	76	26	28	30	32
27	6	7	8	10	77	26	29	30	32
28	6	8	9	10	78	27	29	31	33
29	7	8	9	10	79	27	30	31	33
30	7	9	10	11	80	28	30	32	34
31	7	9	10	11	81	28	31	32	34
32	8	9	10	12	82	28	31	33	35
33	8	10	11	12	83	29	32	33	35
34	9	10	11	13	84	29	32	33	36
35	9	11	12	13	85	30	32	34	36
36	9	11	12	14	86	30	33	34	37
37	10	12	13	14	87	31	33	35	37
38	10	12	13	14	88	31	34	35	38
39	11	12	13	15	89	31	34	36	38
40	11	13	14	15	90	32	35	36	39
41	11	13	14	16	91	32	35	37	39
42	12	14	15	16	92	33	36	37	39
43	12	14	15	17	93	33	36	38	40
44	13	15	16	17	94	34	37	38	40
45	13	15	16	18	95	34	37	38	41
46	13	15	16	18	96	34	37	39	41
47	14	16	17	19	97	35	38	39	42
48	14	16	17	19	98	35	38	40	42
49	15	17	18	19	99	36	39	40	43
50	15	17	18	20	100	36	39	41	43

* For specific details on the use of this table, see pages 473–475.

From Wilfred J. Dixon and Frank J. Massey, Jr., *Introduction to Statistical Analysis*, 3d ed., (New York: McGraw-Hill, 1969), p. 509. Reprinted by permission.

Table 12   Critical Values of *U* in the Mann-Whitney Test*

A. The entries are the critical values of *U* for a one-tailed test at 0.025 or for a two-tailed test at 0.05.

n_2 \ n_1	1	2	3	4	5	6	7	8	9	10	11	12	13	14	15	16	17	18	19	20
1																				
2								0	0	0	0	1	1	1	1	1	2	2	2	2
3					0	1	1	2	2	3	3	4	4	5	5	6	6	7	7	8
4				0	1	2	3	4	4	5	6	7	8	9	10	11	11	12	13	13
5			0	1	2	3	5	6	7	8	9	11	12	13	14	15	17	18	19	20
6			1	2	3	5	6	8	10	11	13	14	16	17	19	21	22	24	25	27
7			1	3	5	6	8	10	12	14	16	18	20	22	24	26	28	30	32	34
8		0	2	4	6	8	10	13	15	17	19	22	24	26	29	31	34	36	38	41
9		0	2	4	7	10	12	15	17	20	23	26	28	31	34	37	39	42	45	48
10		0	3	5	8	11	14	17	20	23	26	29	33	36	39	42	45	48	52	55
11		0	3	6	9	13	16	19	23	26	30	33	37	40	44	47	51	55	58	62
12		1	4	7	11	14	18	22	26	29	33	37	41	45	49	53	57	61	65	69
13		1	4	8	12	16	20	24	28	33	37	41	45	50	54	59	63	67	72	76
14		1	5	9	13	17	22	26	31	36	40	45	50	55	59	64	67	74	78	83
15		1	5	10	14	19	24	29	34	39	44	49	54	59	64	70	75	80	85	90
16		1	6	11	15	21	26	31	37	42	47	53	59	64	70	75	81	86	92	98
17		2	6	11	17	22	28	34	39	45	51	57	63	67	75	81	87	93	99	105
18		2	7	12	18	24	30	36	42	48	55	61	67	74	80	86	93	99	106	112
19		2	7	13	19	25	32	38	45	52	58	65	72	78	85	92	99	106	113	119
20		2	8	13	20	27	34	41	48	55	62	69	76	83	90	98	105	112	119	127

B. The entries are the critical values of *U* for a one-tailed test at 0.05 or for a two-tailed test at 0.10.

n_2 \ n_1	1	2	3	4	5	6	7	8	9	10	11	12	13	14	15	16	17	18	19	20
1																			0	0
2					0	0	0	1	1	1	1	2	2	2	3	3	3	4	4	4
3			0	0	1	2	2	3	3	4	5	5	6	7	7	8	9	9	10	11
4			0	1	2	3	4	5	6	7	8	9	10	11	12	14	15	16	17	18
5		0	1	2	4	5	6	8	9	11	12	13	15	16	18	19	20	22	23	25
6		0	2	3	5	7	8	10	12	14	16	17	19	21	23	25	26	28	30	32
7		0	2	4	6	8	11	13	15	17	19	21	24	26	28	30	33	35	37	39
8		1	3	5	8	10	13	15	18	20	23	26	28	31	33	36	39	41	44	47
9		1	3	6	9	12	15	18	21	24	27	30	33	36	39	42	45	48	51	54
10		1	4	7	11	14	17	20	24	27	31	34	37	41	44	48	51	55	58	62
11		1	5	8	12	16	19	23	27	31	34	38	42	46	50	54	57	61	65	69
12		2	5	9	13	17	21	26	30	34	38	42	47	51	55	60	64	68	72	77
13		2	6	10	15	19	24	28	33	37	42	47	51	56	61	65	70	75	80	84
14		2	7	11	16	21	26	31	36	41	46	51	56	61	66	71	77	82	87	92
15		3	7	12	18	23	28	33	39	44	50	55	61	66	72	77	83	88	94	100
16		3	8	14	19	25	30	36	42	48	54	60	65	71	77	83	89	95	101	107
17		3	9	15	20	26	33	39	45	51	57	64	70	77	83	89	96	102	109	115
18		4	9	16	22	28	35	41	48	55	61	68	75	82	88	95	102	109	116	123
19	0	4	10	17	23	30	37	44	51	58	65	72	80	87	94	101	109	116	123	130
20	0	4	11	18	25	32	39	47	54	62	69	77	84	92	100	107	115	123	130	138

* For specific details on the use of this table, see page 482.

Reproduced from the *Bulletin of the Institute of Educational Research at Indiana University*, vol. 1, no 2; with the permission of the author and the publisher.

Table 13 Critical Values for Total Number of Runs (V)

The entries in this table are the critical values* for a two-tailed test using $\alpha = 0.05$. For a one-tailed test, $\alpha = 0.025$ and use only one of the critical values: the smaller critical value for a left-hand critical region, the larger for a right-hand critical region.

The larger of n_1 and n_2

The smaller of n_1 and n_2	5	6	7	8	9	10	11	12	13	14	15	16	17	18	19	20
2								2/6	2/6	2/6	2/6	2/6	2/6	2/6	2/6	2/6
3		2/8	2/8	2/8	2/8	2/8	2/8	2/8	2/8	2/8	3/8	3/8	3/8	3/8	3/8	3/8
4	2/9	2/9	2/10	3/10	3/10	3/10	3/10	3/10	3/10	3/10	3/10	4/10	4/10	4/10	4/10	4/10
5	2/10	3/10	3/11	3/11	3/12	3/12	4/12	4/12	4/12	4/12	4/12	4/12	4/12	5/12	5/12	5/12
6		3/11	3/12	3/12	4/13	4/13	4/13	4/13	5/14	5/14	5/14	5/14	5/14	5/14	6/14	6/14
7			3/13	4/13	4/14	5/14	5/14	5/14	5/15	5/15	6/15	6/16	6/16	6/16	6/16	6/16
8				4/14	5/14	5/15	5/15	6/16	6/16	6/16	6/16	6/17	7/17	7/17	7/17	7/17
9					5/15	5/16	6/16	6/16	6/17	7/17	7/18	7/18	7/18	8/18	8/18	8/18
10						6/16	6/17	7/17	7/18	7/18	7/18	8/19	8/19	8/19	8/20	9/20
11							7/17	7/18	7/19	8/19	8/19	8/20	9/20	9/20	9/21	9/21
12								7/19	8/19	8/20	8/20	9/21	9/21	9/21	10/22	10/22
13									8/20	9/20	9/21	9/21	10/22	10/22	10/23	10/23
14										9/21	9/22	10/22	10/23	10/23	11/23	11/24
15											10/22	10/23	11/23	11/24	11/24	12/25
16												11/23	11/24	11/25	12/25	12/25
17													11/25	12/25	12/26	13/26
18														12/26	13/26	13/27
19															13/27	13/27
20																14/28

* See page 487 in regard to critical values.

From C. Eisenhart and F. Swed, "Tables for testing randomness of grouping in a sequence of alternatives," *The Annals of Statistics*, vol. 14 (1943): 66–87. Reprinted by permission.

For $n_1 > 20$ or $n_2 > 20$, treat V as a normal variable with a mean and a standard deviation of

$$\mu_v = \frac{2n_1 n_2}{n_1 + n_2} + 1$$

$$\sigma_v = \sqrt{\frac{2n_1 n_2 (2n_1 n_2 - n_1 - n_2)}{(n_1 + n_2)^2 (n_1 + n_2 - 1)}}$$

Table 14   Critical Values of Spearman's Rank Correlation Coefficient

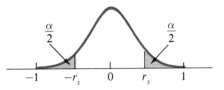

The entries in this table are the critical values of r_s for a two-tailed test at α. For a one-tailed test, the value of α shown at the top of the table is double the value of α being used in the hypothesis test.

n	$\alpha = 0.10$	$\alpha = 0.05$	$\alpha = 0.02$	$\alpha = 0.01$
5	0.900	—	—	—
6	0.829	0.886	0.943	—
7	0.714	0.786	0.893	—
8	0.643	0.738	0.833	0.881
9	0.600	0.683	0.783	0.833
10	0.564	0.648	0.745	0.794
11	0.523	0.623	0.736	0.818
12	0.497	0.591	0.703	0.780
13	0.475	0.566	0.673	0.745
14	0.457	0.545	0.646	0.716
15	0.441	0.525	0.623	0.689
16	0.425	0.507	0.601	0.666
17	0.412	0.490	0.582	0.645
18	0.399	0.476	0.564	0.625
19	0.388	0.462	0.549	0.608
20	0.377	0.450	0.534	0.591
21	0.368	0.438	0.521	0.576
22	0.359	0.428	0.508	0.562
23	0.351	0.418	0.496	0.549
24	0.343	0.409	0.485	0.537
25	0.336	0.400	0.475	0.526
26	0.329	0.392	0.465	0.515
27	0.323	0.385	0.456	0.505
28	0.317	0.377	0.448	0.496
29	0.311	0.370	0.440	0.487
30	0.305	0.364	0.432	0.478

From E. G. Olds, "Distribution of sums of squares of rank differences for small numbers of individuals," *Annals of Statistics*, vol. 9 (1938), pp. 138–148, and amended, vol. 20 (1949), pp. 117–118. Reprinted by permission.

* For specific details on the use of this table, see page 492.

Answers to Selected Exercises

Chapter 1

Section 1–3, page 11

1–2   (a) attribute   (b) continuous
        (c) discrete

1–4   (a) average cost of textbooks per student
        (b) average cost for the sample of 50 students

1–5   (c) and (d) are discrete

1–6   (c), (d), and (e) are continuous

1–7   (a), (b), and (d) are attribute

Section 1–5, page 13

1–9   (a) and (d) are statistics

Section 1–6, page 18

1–10   (b) the computer list

1–11   (a) registered voters
         (b) telephone book listings
         (d) no

1–12   A simple random sample would be very difficult to obtain if the population being sampled was extremely large and physically spread out.

Chapter Exercises, page 20

1–16   (a) American teenagers
         (b) Does your mother work? (Yes, No)
         (d) statistic

1–20   (a) data
         (b) What is the average number of times that the people in the sample go grocery shopping per week?

Chapter 2

Section 2–1, page 31

2–1

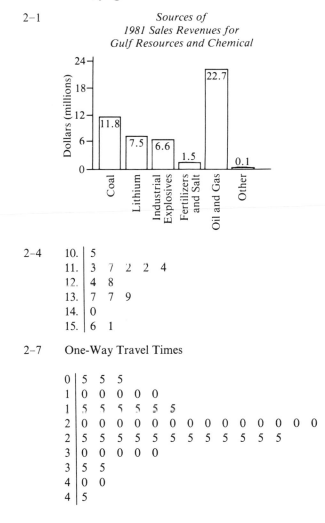

Sources of 1981 Sales Revenues for Gulf Resources and Chemical

2–4

```
10. | 5
11. | 3   7   2   2   4
12. | 4   8
13. | 7   7   9
14. | 0
15. | 6   1
```

2–7   One-Way Travel Times

```
0 | 5   5   5
1 | 0   0   0   0   0
1 | 5   5   5   5   5   5
2 | 0   0   0   0   0   0   0   0   0   0   0   0   0
2 | 5   5   5   5   5   5   5   5   5   5   5   5
3 | 0   0   0   0   0
3 | 5   5
4 | 0   0
4 | 5
```

Section 2–2, page 40

2–8　(a)

x	Tallies	Frequency
0.4	III	3
0.5	I	1
0.6	I	1
0.7	III	3
0.8	I	1
0.9		0
1.0	I	1
1.1		0
1.2	II	2
		12

(b)

Last Year's Consumer Price Index

Percent of monthly change

2–10　(a) f: 4, 14, 19, 7, 5, 3, 2, 1
　　　(b) 5　　　(c) 22, 20, 24.5

2–12　(a) f: 6, 7, 7, 8, 6, 4, 4
　　　(b) 100　　　(d) 99.5
　　　(e) cum. rel. freq.: $\frac{6}{42}, \frac{13}{42}, \frac{20}{42}, \frac{28}{42}, \frac{34}{42}, \frac{38}{42}, \frac{42}{42}$

Section 2–3, page 47

2–13　(a) 6.0　　　(b) 7　　　(c) no mode
　　　(d) 5.5

2–16　(a) 6.9　　　(b) 7　　　(c) 7　　　(d) 7.5

2–18

x	f	xf
0	1	0
1	3	3
2	8	16
3	5	15
4	3	12
$\sum$	20	46

mean: $\bar{x} = \frac{46}{20} = 2.3$

2–20　$\bar{x} = 10.8$

2–22　$\bar{x} = 2.355$

Section 2–4, page 55

2–23　(a) 7　　　　(b) $s^2 = \frac{34}{4} = 8.5$

2–26　(a) $s^2 = 3.1$　　　(c) $s = 1.8$

2–28　$\bar{x} = 12.84$ and $s = 5.29$

2–31　$s^2 = 12.2$

2–33　$s = 7.9$

2–35　$\bar{x} = 21.0$ and $s = 9.95$

Section 2–5, page 62

2–37　(a) 3.8, 5.6　　　(b) 4.7　　　(c) 3.5, 4.0, 6.9

2–39　(a) 13, 19.5　　　(b) 16.25

2–41　(a) 1.0　　　(b) 0.0
　　　(c) 2.25　　(d) −1.25

2–44　$A: z = 1.625; B: z = 1.2;$
　　　A has the higher relative position.

Section 2–6, page 67

2–46　(a) f: 7, 10, 22, 8, 7, 2, 3, 0, 1
　　　(b) $\bar{x} = 3.4$ and $s = 1.7$
　　　(d) 47; 78%
　　　(f) 56　　　(h) 98.3%

2–48　(a) 2.5%
　　　(b) 70.4 to 97.6

Chapter Exercises, page 73

2–51　(a) $22,312
　　　(b) Atlanta and Dallas-Fort Worth

2–53　(a) 55　　　(b) 100

2–55　(a) 11.8%　　　(b) 11.9%　　　(c) 1.9

2–58　(b) 1.45　　　(d) 0　　　(h) 6
　　　(j) 1.55

2–61　(a) 14.42　　　(b) 13.34

2-64 (b) median $= 3 \left[i = \dfrac{100 + 1}{2} = 50.5 \, th \right]$

mode $= 2$;

midrange $= \dfrac{1 + 11}{2} = 6$;

mean $= \dfrac{397}{100} = 3.97 = 4.0$;

midquartile $= \dfrac{2 + 5.5}{2} = 3.75$

(c) $Q_1 = 2$; $Q_3 = \dfrac{5 + 6}{2} = 5.5$

(d) $P_{15} = 1$; $P_{12} = 1$

(e) range $= 10$; $s^2 = 6.7$; $s = 2.6$

2-66

Class Limits	f	(a) Relative Frequency	(b) Cumulative Frequency
0–8	17	0.142	17
9–17	14	0.117	31
18–26	10	0.083	41
27–35	14	0.117	55
36–44	10	0.083	65
45–53	16	0.133	81
54–62	9	0.075	90
63–71	11	0.092	101
72–80	8	0.067	109
81–89	11	0.092	120
	120		

$\bar{x} = 40.75$; $s = 26.2$

2-68

(a) Classes	Frequency	(k) Cumulative Relative Frequency
10.0–10.5	15	0.15
10.6–11.1	11	0.26
11.2–11.7	7	0.33
11.8–12.3	3	0.36
12.4–12.9	6	0.42
13.0–13.5	2	0.44
13.6–14.1	14	0.58
14.2–14.7	10	0.68
14.8–15.3	27	0.95
15.4–15.9	5	1.00
	100	

$\bar{x} = 13.15$; $s = 1.91$

2-71 $\bar{x} = 46.3$ and $s = 14.7$

2-73 final

Chapter 3

Section 3–1, page 88

3-1 *Hours Studied Versus Exam Grade*

Section 3–2, page 95

3-6 (a) $r \approx \frac{2}{3}$

(b) $n = 15$; $\sum x = 81$; $\sum y = 110$;
$\sum x^2 = 487$; $\sum xy = 625$; $\sum y^2 = 842$

$$r = \dfrac{31.0}{\sqrt{49.6}\sqrt{35.33}} = 0.74$$

(c) Yes, the value $r = 0.74$ is greater than the 0.514 value from table 3–3.

3-9 (b) $r = 0.779$ (c) yes

Section 3–3, page 106

3-13 The vertical scale that is shown on figure 3–21 is not the y-axis.

3-14 (b) $\hat{y} = 3.96 + 0.625x$
(d) $7.71(7.71 \times 10)$ or 77
(e) It suggests that people who study 6 hours for the exam will average 77 on the exam.

3-16 (b) $\hat{y} = 10.1 + 0.91x$

3-18 (b) $\hat{y} = 28.9 + 14.7x$ (c) $\hat{y} = 146.5$
(e) $r = 0.843$

Chapter Exercises, page 107

3–20   0.798

3–21   (a) correlation   (c) correlation
        (e) correlation

3–22   (a) to determine whether or not there is a linear
        relationship between gallons of gas required to fill
        and number of miles travelled since last fill-up

3–24   (a) regression   (b) positive

3–29   (b) 0.788   (d) $\hat{y} = 1.73 + 0.0054x$

Chapter 4

Section 4–3, page 131

4–7   (a) $\{R, Y, G\}$ (b) no

4–8 $\{RR, RY, RG, YR, YY, YG, GR, GY, GG\}$

4–12 $\{H1, H2, H3, H4, H5, H6, T1, T2, T3, T4, T5, T6\}$

4–13   (a)

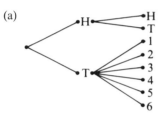

        (b) $\frac{1}{2}$

Section 4–4, page 134

4–14 $\frac{1}{4}, \frac{1}{4}, \frac{1}{2}$

4–16 $P(A) = \frac{1}{7}; P(B) = \frac{2}{7}; P(C) = \frac{4}{7}$

4–18   (b) 0.5   (c) $\frac{5}{36}$ (d) $\frac{1}{4}$
        (e) $\frac{5}{12}$

4–19   (a) 0.7   (c) 0.4

Section 4–5, page 142

4–21   (a) not mutually exclusive
        (d) mutually exclusive

4–23   (a) $P(\overline{A}) = 1 - 0.3 = 0.7$
        (c) 0.7   (d) 0.0

4–25   no

4–26   0.8

4–28   (a) 0.3   (b) 0.2

Section 4–6, page 148

4–30   (a) 0.3   (b) 0.4   (c) yes

4–32   (a), (c), and (d) are independent.

4–35   (a) $\frac{64}{343}$ (b) $\frac{4}{35}$

Section 4–7, page 154

4–38   0.7

4–40   (a) 0.15   (b) 0.65   (d) 0.5
        (e) 0.7   (f) no

4–43   (a) $\frac{20}{56}$ (b) $\frac{30}{56}$ (c) $\frac{6}{56}$

4–44   (a) $\frac{1}{2}, \frac{1}{4}, \frac{1}{8}$ (b) $\frac{9}{16}, \frac{9}{32}, \frac{9}{64}$

4–45   0.64

4–46   (a) 0.003   (c) 0.329

4–48   (b) 0.6   (d) 0.80   (e) 0.20

Chapter Exercises, page 157

4–50   (a) and (b) are true

4–52   (b) is true

4–55   (a) 0.3168   (b) 0.4659
        (c) no   (d) no

4–57   (a) 0.049   (b) 0.10175

4–59   (a) 0.20   (b) 0.33   (c) 0.50

4–62   (a) 0.3   (b) 0.7   (c) no
        (d) yes   (e) no   (f) no

4–67   (a) 0.020   (c) 0.570

4–69   0.948

4–71   (a) $\frac{8}{20}$ (c) $\frac{4}{20}$ (e) $\frac{2}{20}$
        (g) $\frac{10}{20}$ (i) $\frac{2}{4}$ (k) 0
        (m) $\frac{3}{12}$

4–75   (a) 0.0029   (c) 0.0525   (e) 0.00875

4–77   0.592

4–79   (a) 0.80   (b) 0.96   (c) 0.64

Chapter 5

Section 5–1, page 171

5-1   (a) $x = 0, 1, 2, 3,$ or 4
       (b) HHHH: 4; HHHT: 3; ...   (c) discrete

5-3 $x = n$ (bull's-eye shots);
       $x = 0, 1, 2, 3, \ldots, 7, 8$

Section 5–2, page 175

5-5   It is a probability function. (a) Each $P(x)$ is between zero and one and (b) the sum of the $P(x)$'s is exactly one.

5-7   It is a probability distribution. (a) Each $R(x)$ is between zero and one and (b) the sum is exactly one.

Section 5–3, page 180

5-9 $\mu = 2.0; \sigma = 1.0$

5-10 $\mu = 2.0; \sigma = 1.4$

5-13   (b) $\mu = 4.5; \sigma = 2.87$

Section 5–4, page 190

5-15   (a) 24   (c) 1   (e) 10
       (g) 0.0081   (i) 10   (k) 0.410

5-16   (a) See exercise 5–1(b)

(b)

x	0	1	2	3	4
$P(x)$	$\frac{1}{16}$	$\frac{4}{16}$	$\frac{6}{16}$	$\frac{4}{16}$	$\frac{1}{16}$

5-19   (a) 0.4116   (c) 0.5625   (e) 0.375

5-20   (a) 0.001   (c) 0.031   (e) 0.002

5-22   (a) yes   (b) $P(x) = \binom{5}{x} \frac{1^5}{2}; x = 0, 1, \ldots, 5$

5-23 $P(x)$ is not a binomial probability function.

5-26   0.172

5-28   0.051

Section 5–5, page 194

5-31 $\mu = 2.5$ and $\sigma = 1.1$

5-33   (a) $\mu = 7.7; \sigma = 2.7$
       (c) $\mu = 44.0; \sigma = 2.3$

Chapter Exercises, page 195

5-36 $\mu = 4.5; \sigma = 2.5$

5-39 $\mu = 5.0; \sigma = 1.58$

5-41 $P(x) = \binom{4}{x} \frac{2^x}{3} \frac{1^{4-x}}{3}; x = 0, 1, 2, 3, 4$

x	0	1	2	3	4
$P(x)$	0.012	0.099	0.296	0.395	0.198

5-43   0.132

5-45   0.833

5-48   0.196

5-50   0.889

5-53   0.972

5-56   (a) 0.733   (c) 0.6997

5-58   (a) 0.914   (b) 0.625

5-61   (a) $2^{10} = 1024$
       (b) 1, 10, 45, 120, 210, 252, 210, 120, 45, 10, 1

       (d) x $P(x) = \binom{10}{x}(0.5)^{10}; x = 0, 1, \ldots, 10$

0	0.000976
1	0.00976
2	0.0439
3	0.1172
4	0.2051
5	0.2461

       (e) $\mu = 5.0; \sigma = 1.6$

5-63 $\mu = 7.0; \sigma = 2.16$

Chapter 6

Section 6–2, page 210

6-1   a bell-shaped distribution with a mean of zero and a standard deviation of one

6-2   (a) 0.4032   (b) 0.3997
       (c) 0.4993   (d) 0.4761

6-3   (a) 0.4821   (b) 0.4949
       (c) 0.3849   (d) 0.4418

6–4 (a) $0.3849 + 0.3888 = 0.7737$
(b) $0.4599 + 0.4382 = 0.8981$

6–5 (a) 0.9132 (b) 0.2498 (c) 0.1174

6–7 (a) 0.9987 (b) 0.9394 (c) 0.2266
(d) 0.8944 (e) 0.8944

6–9 (a) 1.14 (b) 0.47 (c) 1.66
(d) 0.86 (e) 1.74 (f) 2.23

6–11 ± 1.28

6–13 (a) ± 1.65 (b) ± 2.33

Section 6–3, page 216

6–15 0.2039

6–17 (a) 0.1379 (b) 0.1190
(c) 0.6064 (d) 0.5346

6–19 (a) 0.2857 (b) 0.0002

6–21 (a) 0.1131 (c) 4.64 minutes

6–23 0.3557

6–25 (a) $\sigma = 5.32$ (c) 70.78

Section 6–4, page 222

6–27 (a) 1.645 or 1.65 (b) 2.33
(c) 1.96 (d) -1.96 (e) -2.05

Section 6–5, page 227

6–29 $P(x = 6) = 0.1822$

6–30 $P(3.5 < x < 5.5) = 0.1812$

6–33 0.0087

6–36 $P(x > 49.5) = 0.999997$

6–38 0.0287

Chapter Exercises, page 228

6–40 $z = \pm 0.84$

6–42 (a) -0.92 (b) -2.03 (c) -0.74

6–44 (a) 0.0930 (b) 0.9684

6–46 (a) 1.175 or 1.18 (b) 0.58
(c) -1.04 (d) -2.33

6–50 5.534

6–52 166.8 minutes

6–56 (a) 0.001 (b) 1.000
(c) 0.995 (d) 0.0668

6–58 (a) 0.0386 (b) 0.1762 (c) 0.9868

6–60 (a) 0.0143 (b) 619.4
(c) 107.2 (d) 755.4

Chapter 7

Section 7–1, page 239

7–1 (a) A sampling distribution of sample means is the distribution formed by the means of all possible samples of a fixed size that can be taken from a population.
(b) It is one element in the distribution of the means of all samples of size 3.
(c) There were 25 equally likely samples. Therefore, each has a probability of $\frac{1}{25}$, or 0.04.

7–3 (b)

$\tilde{x}$	$P(\tilde{x})$
0	13/125
2	31/125
4	37/125
6	31/125
8	13/125
Sum	125/125 Ⓒⓚ

(c)

$\bar{x}$	$P(\bar{x})$
0	1/125
2/3	3/125
4/3	6/125
6/3	10/125
8/3	15/125
10/3	18/125
12/3	19/125
14/3	18/125
16/3	15/125
18/3	10/125
20/3	6/125
22/3	3/125
24/3	1/125
Sum	125/125 Ⓒⓚ

Section 7–2, page 246

7–6 (a) one
     (b) $\sigma_{\bar{x}} = \sigma/\sqrt{n}$ becomes smaller as n increases

7–7 (a) approximately 50
     (b) approximately 2.5; $10/\sqrt{16}$

7–9 6.25; 5.0; 3.54; 2.5

Section 7–3, page 251

7–11 (a) 0.9974   (c) 0.9282

7–13 (a) 0.3830   (b) 0.9938
      (c) 0.3085   (d) 0.0031

7–15 (a) 0.1498   (b) 0.0089

7–17 (a) 0+   (b) 9.348

Chapter Exercises, page 253

7–19

$\bar{x}$	$P(\bar{x})$
2	1/27
8/3	3/27
10/3	6/27
4	7/27
14/3	6/27
16/3	3/27
6	1/27

7–21 (a) 0.0643   (b) 0.6103   (c) 0.9974

7–23 0.49

7–25 0.1056

7–27 0.0004

7–29 (a) 0.9759   (b) 69 to 75

7–31 0.0228

Chapter 8

Section 8–1, page 267

8–1 (a) Each answer represents a trial. The answer is either correct or incorrect, and the 14 trials are independent. Thus x is a binomial variable with $n = 14$ and $p = 0.5$.

8–3 (a) Type I error: the victim is actually alive but is thought to be dead. Type II error: the victim is dead but is thought to be alive.
     (c) Set 1; keep α as small as possible.

8–4 (a) Type A correct decision; the parachute will open and inspector okayed its use. Type II error: the parachute will not open and the inspector okayed its use.
     (c) Set 3; keep β as small as possible.

8–5 We are willing to allow the type I error to occur with a probability of (a) 0.001; (b) 0.05; (c) 0.10.

8–7 (a) reject H_0 (b) fail to reject H_0

8–9 (a) α (b) β (c) type I
     (d) type II

Section 8–2, page 278

8–11 (a) $H_0: \mu = 26$ years ($\leq$); $H_a: \mu > 26$
      (c) $H_0: \mu = 1600$ hours ($\geq$); $H_a: \mu < 1600$
      (f) $H_0: \mu = 80°$; $H_a \neq 80°$

8–13 (a) $H_a: r > a$; failure to reject the null hypothesis will result in the drug being marketed. Because of the high current mortality rate, the burden of proof should be on the old ineffective treatment.
      (b) $H_a: r < a$: failure to reject the null hypothesis will result in the new drug not being marketed. Because of the low current mortality rate, the burden of proof should be on the new drug.

8–14 (a) critical region: $z \leq -1.65$ or $z \geq +1.65$
      (c) critical region: $z \leq -1.65$
      (f) critical region: $z \geq +1.28$

8–15 $H_a: \mu > 4.9$; $z^* = 1.33$; fail to reject H_0

8–17 $H_a: \mu < 36.7$; $z^* = -2.59$; reject H_0

8–19 $H_a: \mu < 4.7$; $z^* = -1.00$; fail to reject H_0

Section 8–3, page 285

8–21 (a) 0.0694   (b) 0.1977
      (c) 0.2420   (d) 0.0174
      (e) 0.3524   (f) 0.3228

8–23 0.0548

Section 8–4, page 293

8–25 (a) 35.6   (b) 32.46 to 38.74

8–27 (a) 25.3   (b) 25.3 ± 1.01
      (c) 25.3 ± 1.33

8–29 (a) 0.77   (b) 3.63 to 5.17

8-31 $\quad n = \left[\dfrac{(2.33)(16)}{6} \right]^2 = 38.6 = 39$

8-33 $\quad$ (a) 97 $\qquad$ (b) 25

Chapter Exercises, page 295

8-35 $\quad$ (a) $\mu = 100$ $\qquad$ (b) $\mu \neq 100$ $\qquad$ (c) 0.01
$\qquad$ (d) ± 2.58 $\qquad$ (e) 100 $\qquad$ (f) 96
$\qquad$ (g) 12 $\qquad$ (h) 1.697
$\qquad$ (i) -2.357
$\qquad$ (j) fail to reject H_0

8-36 $\quad$ (a) 32.0 $\qquad$ (b) 2.4 $\qquad$ (c) 64
$\qquad$ (d) 0.90 $\qquad$ (e) 1.65 $\qquad$ (f) 0.3
$\qquad$ (g) 0.495 $\qquad$ (h) 32.495 $\qquad$ (i) 31.505

8-37 $\quad$ (a) $12 + 1.65(2.50/\sqrt{45}) = 12.61$
$\qquad$ (b) $\bar{x} = 12.49$; no, the evidence does not contradict the admissions office's claim.

8-40 $\quad$ (a) $H_a: \mu > 9$; $z* = 3.14$; reject H_0

8-43 $\quad$ (a) $H_a: \mu > 90$ (it now takes longer); $z* = 2.50$; reject H_0.

8-46 $\quad$ (a) 0.86 to 2.18 $\qquad$ (b) 1.00 to 2.04

8-49 $\quad$ (b) 2,497,000 to 3,183,000

8-51 $\quad$ 97

8-53 $\quad$ 12.64 to 13.10

8-55 $\quad$ 12.63 to 13.67

Chapter 9

Section 9-1, page 309

9-1 $\quad$ (a) 1.71 $\qquad$ (c) 2.60
$\qquad$ (e) -1.72 $\qquad$ (g) -2.47

9-2 $\quad$ (a) $t(19, 0.05)$ $\qquad$ (c) $t(18, 0.99)$
$\qquad$ (e) $\pm t(7, 0.05)$

9-3 $\quad$ 7

9-4 $\quad$ (a) -2.49 $\qquad$ (c) -0.685

9-6 $\quad$ $H_a: \mu > 8.5$; $t* = 1.66$; fail to reject H_0

9-8 $\quad$ $H_a: \mu > 58{,}950$; $t* = 2.34$; reject H_0

9-11 $\quad$ 86.0 ± 6.85

Section 9-2, page 316

9-13 $\quad$ (a) $H_a: p > 0.2$ $\qquad$ (b) 0.061
$\qquad$ (c) no $\qquad$ (d) yes
$\qquad$ (e) α decreases to 0.002

9-15 $\quad$ $H_a: p > 0.50$; $z* = 2.85$; reject H_0

9-17 $\quad$ $H_a: p < 0.60$; $z* = -2.04$; reject H_0

9-19 $\quad$ (a) 0.162, 0.438 $\qquad$ (b) 0.562, 0.838
$\qquad$ (c) 0.24, 0.76 $\qquad$ (e) 0.474, 0.526

9-21 $\quad$ 0.37 ± 0.062

9-23 $\quad$ 522

Section 9-3, page 324

9-25 $\quad$ (a) 34.8 $\qquad$ (c) 13.4
$\qquad$ (e) 12.3 $\qquad$ (g) 37.7

9-26 $\quad$ (a) $\chi^2(19, 0.05)$ $\qquad$ (c) $\chi^2(17, 0.975)$
$\qquad$ (e) $\chi^2(21, 0.95)$ and $\chi^2(21, 0.05)$
$\qquad$ (g) $\chi^2(7, 0.99)$

9-27 $\quad$ (a) 11.1 $\qquad$ (c) 9.24

9-28 $\quad$ (a) 4.58 and 19.7 $\qquad$ (c) 2.60 and 26.8

9-29 $\quad$ $H_a: \sigma^2 > (0.25)^2$; $\chi^2* = 37.24$; reject H_0

9-31 $\quad$ $H_a: \sigma < 2.1$ (lowering); $\chi^2* = 17.9$; fail to reject H_0

9-32 $\quad$ (a) $s^2 = 9.486$ $\qquad$ (b) 6.092 to 17.173

Chapter Exercises, page 327

9-35 $\quad$ (a) $\bar{x} = 3.44$; $s = 0.65$
$\qquad$ (b) $t* = -7.75$; reject H_0
$\qquad$ (c) $\chi^2* = 18.8$; fail to reject H_0

9-37 $\quad$ (a) 0.486 to 1.007
$\qquad$ (c) 0.416 to 1.337

9-39 $\quad$ \$132.40 to \$147.60

9-41 $\quad$ (a) $\bar{x} = 31.45$; $s = 8.049$
$\qquad$ (b) $H_a: \mu > 28.0$ (older); $t* = 1.92$;
$\qquad$ $t(19, 0.05) = 1.73$; reject H_0

9-43 $\quad$ 60

9-45 $\quad$ 0.0401

9-47 $\quad$ 0.660 to 0.765

9-49 $\quad$ 0.088 to 0.232

9–53   (a) H_a: $\sigma > 0.5$; $\chi^{2*} = 87.8$; reject H_0
       (b) H_a: $\mu \neq 27.5$; $t^* = 0.82$; fail to reject H_0

9–55   (a) $\bar{x} = 11.49$, $s = 0.47$
       (b) 11.23 to 11.75

Chapter 10

Section 10–1, page 337

10–1   dependent: the same set of 40 people are the sources for both sets of data

10–2   independent: a different set of trees is used as sources in the second sample

Section 10–2, page 342

10–7   (a) 0.5028   (b) 0.3830

10–9 H_a: $\mu_I - \mu_{II} \neq 0$; $z^* = -1.31$; fail to reject H_0

10–11 -0.77 to -0.43

10–13 H_a: $\mu_M - \mu_F > 2.5$; $z^* = 2.63$; reject H_0

Section 10–3, page 349

10–15 (a) 1/3.46; 3.31   (b) 1/3.70; 0.270
       (c) 2.54   (d) 1/2.45; 2.35   (e) 1.85

10–16 (a) 2.51   (c) 2.91
       (e) 2.67   (g) 1.79

10–17 (a) 1/2.54   (c) 1/3.09

10–19 H_a: $\sigma_A^2/\sigma_B^2 \neq 1$; $F^* = 2.66$; reject H_0

10–21 H_a: $\sigma_K^2/\sigma_M^2 \neq 1$; $F^* = 1.33$; fail to reject H_0

10–23 1.1 to 2.5

Section 10–4, page 356

10–25 H_a: $\mu_{II} - \mu_I \neq 0$; $t^* = 2.88$; reject H_0

10–27 H_a: $\mu_A - \mu_B > 0$; $t^* = 3.16$; reject H_0; yes, there is significant evidence to justify their hope.

10–30 $3.1 \pm 2.06\sqrt{7.704}\sqrt{0.15}$

10–32 $5.125 \pm 2.12\sqrt{10.3125}\sqrt{0.225}$

10–34 $F^* = 15.7 > F(9, 9, 0.05)$; therefore assume $\sigma_1 \neq \sigma_2$.
       H_a: $\mu_A - \mu_B > 0$; $t^* = 2.48$; reject H_0

Section 10–5, page 362

10–37 H_a: $\mu_d \neq 0$; $t^* = 1.44$; fail to reject H_0

10–40 H_a: $\mu_d > 0$ (route II is faster); $t^* = 1.916$; reject H_0

10–42 5.50 ± 4.18

Section 10–6, page 368

10–44 H_a: $p_A - p_B \neq 0$; $z^* = -1.089$
       fail to reject H_0

10–46 H_a: $p_y - p_0 \neq 0$; $z^* = 0.805$; fail to reject H_0

Chapter Exercises, page 371

10–50 H_a: $\mu_T - \mu_L \neq 0$ (there is a difference);
       $z^* = -1.80$; fail to reject H_0

10–53 H_a: $p_A - p_B \neq 0$; $z^* = -1.71$; reject H_0

10–56 (a) 0.6944   (b) 0.285 to 1.875

10–59 (a) Reject H_0; it appears that $\sigma_A \neq \sigma_B$.
       (b) Reject H_0; there is more weight loss with diet A.
       (c) 2.2 to 3.8

10–64 Reject H_0; the new material does wear longer than leather.

10–66 (a) 0.049   (b) 0.020 to 0.078

10–70 3.58 ± 12.44

10–72 H_a: $\mu_d > 0$ (improvement: $d =$ after $-$ before);
       $t^* = 3.82$; reject H_0

10–74 assumption: $\sigma_A = \sigma_B$; H_a: $\mu_B - \mu_A > 0$; $t^* = 0.82$;
       fail to reject H_0

10–76 assumption: $\sigma_A = \sigma_B$; H_a: $\mu_A - \mu_B > 0$; reject H_0

Chapter 11

Section 11–2, page 392

11–1   (a) $p_A = p_B = p_C = p_D = p_E$
       (b) χ^2
       (c) H_0: equal preference; $\chi^2 (4, 0.10) = 7.78$;
       $\chi^{2*} = 4.40$; fail to reject H_0

11–3 H_0: equally proportioned; $\chi^2(7, 0.05) = 14.1$;
       $\chi^{2*} = 11.4$; fail to reject H_0

Section 11–3, page 400

11–6 $\chi^{2*} = 3.03$; fail to reject H_0

11–9 $\chi^{2*} = 3.620$

11–12 expected values (fear) = 70.8; expected values (don't fear) = 129.2; $\chi^{2*} = 80.957$; reject H_0

Chapter Exercises, page 404

11–13 $\chi^{2*} = 8.6375$

11–16 $\chi^{2*} = 10.0$

11–18 $\chi^{2*} = 35.749$

11–20 $\chi^{2*} = 2.839$

11–23 $\chi^{2*} = 27.2369$

11–25 $\chi^{2*} = 4.7$

Chapter 12

Section 12–3, page 426

12–1 (a) The mean levels of the factor being tested are not all the same.
(c) The mean levels of the factor being tested are all the same.

12–3 (a) 0   (b) 3   (c) 16
(d) 60   (e) 1,232

12–4 (a) H_0: The means of each of the levels of the factor being tested are the same.
(c) Fail to reject H_0 would indicate that we did not find evidence that contradicted the null hypothesis.

12–5

Source	SS	df	MS
workers	17.73	2	8.865
error	25.20	12	2.1
Total	42.93	14	

$F^* = 4.22$; $F(2, 12, 0.05) = 3.89$; reject H_0: there is evidence of a difference between workers.

12–7 SS(Machines) = 20.998; SS(error) = 2.158; $F^* = 31.6$; reject H_0: there is a difference due to machine.

Chapter Exercises, page 429

12–8 SS(Typewriter) = 280.333; SS(error) = 806.333; $F^* = 3.48$; fail to reject H_0

12–10 SS(Day) = 2.2750; SS(error) = 3.5575; $F^* = 2.24$; fail to reject H_0

12–13 SS(Community) = 1618.792; SS(error) = 769.833; $F^* = 14.02$; reject H_0

12–15 SS(Brand) = 95.359; SS(error) = 126.467; $F^* = 4.776$; reject H_0

12–18 SS(Brand) = 75,047; SS(error) = 101,899; $F^* = 5.30$; reject H_0

Chapter 13

Section 13–1, page 440

13–1 Refer to the definition of the mean and the development of measures of dispersion in section 2.3.

13–2 (b) covar$(x, y) = 4.\overline{44}$
(c) $s_x = 2.981$; $s_y = 1.581$
(d) $r = 0.943$

Section 13–2, page 443

13–6 $H_a: \rho \neq 0$; critical value is 0.378; reject H_0; it appears that there is sufficient reason to reject the null hypothesis.

13–8 critical values $= \pm 0.632$; reject H_0

13–10 (a) -0.55 to 0.75   (c) 0.32 to 0.82

Section 13–3, page 449

13–12 (a) $\hat{y} = 1.0 = 0.5x$
(b) 1.5, 2.5, 3.5, 4.5, 5.5
(c) $-0.5, 0.5, -0.5, 0.5, -0.5, 0.5, -0.5, 0.5, -0.5, 0.5$
(d) $s^2 = 0.3125$

13–14 (b) $\hat{y} = 3.96 + 0.625x$
(c) (2, 5.21), (5, 7.08), (8, 8.96)
(e) 1.213

Section 13–4, page 452

13–16 (b) $\hat{y} = 2.38 + 2.66x$
(c) $H_a: \beta_1 > 0$; reject H_0

Section 13–5, page 461

13-19 (a) 7.74   (b) 7.09 to 8.39
      (c) 7.74   (d) 5.28 to 10.20

13-20 (b) 13.02 ± 2.60
      (c) 13.02 ± 8.13

Chapter Exercises, page 463

13-23 (a) Always   (c) Sometimes
      (e) Always

13-25 (b) 0.961   (d) 0.85 to 0.98

13-29 (d) $\hat{y} = 3.79 - 0.41x$ (e) 0.1236

13-30 (d) $\hat{y} = 25.4 + 0.5x$ (f) $r = 0.832$

13-32 (b) 14.29 to 20.11   (c) 24.29 to 29.61

Chapter 14

Section 14–2, page 478

14-1 (a) $M = 48$ (b) $p = P(+) = \frac{1}{2}$
      (c) $n(+) = 3$; $n = 19$; critical value is 4; therefore, reject H_0. We can conclude that the median temperature is not 48 degrees.

14-3 39 to 47

Section 14–3, page 485

14-6 H_a: average amount dispensed is not the same; $U = 14$; critical value is 23; reject H_0: they are not the same.

14-8 (a) H_0: There is no difference between the average age of men and women drivers involved in automobile accidents.
      (b) $U = 91.5$; critical value is 75; fail to reject H_0

Section 14–4, page 489

14-10 H_0: results are random; $V = 21$; critical values are 8 and 19; reject H_0; results do not appear to be random.

14-12 H_0: Increases and decreases are random.
H_a: Increases occur most frequently. $V - 8$; critical value is 3 for $\alpha = 0.025$; fail to reject H_0.

Section 14–5, page 494

14-14 H_0: independent; H_a: correlated; $r_s = 0.601$; critical values are ± 0.703; fail to reject H_0.

Chapter Exercises, page 497

14-19 sign test; $n(+) = 2$; critical value is 1; fail to reject H_0

14-21 (a) sign test; $n(+) = 4$; critical value is 5; reject H_0

14-23 $r_s = -0.358$

14-27 sign test; $n(-) = 4$; critical value is 5; reject H_0

14-29 (b) Mann-Whitney U test; $U = 47$; critical value is 27; fail to reject H_0

Appendix A

Exercises, page 510

A-1 (a) $x_1 + x_2 + x_3 + x_4$
      (b) $x_1^2 + x_2^2 + x_3^2$
      (c) $(x_1 + y_1) + (x_2 + y_2) + (x_3 + y_3)$
        $+ (x_4 + y_4) + (x_5 + y_5)$
      (d) $(x_1 + 4) + (x_2 + 4) + (x_3 + 4)$
        $+ (x_4 + 4) + (x_5 + 4)$
      (e) $x_1 y_1 + x_2 y_2 + x_3 y_3 + x_4 y_4 + x_5 y_5$
        $+ x_6 y_6 + x_7 y_7 + x_8 y_8$
      (f) $x_1^2 f_1 + x_2^2 f_2 + x_3^2 f_3 + x_4^2 f_4$

A-2 (a) $\sum_{i=1}^{6} x_i$ (b) $\sum_{i=1}^{7} x_i y_i$ (c) $\sum_{i=1}^{9} (x_i)^2$

      (d) $\sum_{i=1}^{n} (x_i - 3)$

A-4 (a) 8   (b) 68   (c) 64

A-5 (a) 8   (b) 4   (c) 12
      (d) 4   (e) 42   (f) 64
      (g) -7 (h) 32   (i) 20
      (j) 120   (k) 400

A-6 $\sum_{i=1}^{120} [0.005(12{,}000 - (i - 1)100)]$

Appendix B

Exercises, page 513

B-2 (a) Use a two-digit number to represent the results obtained. Let the first digit represent one of the coins and the second digit represent the other coin. Let an even digit indicate heads and an odd digit, tails.

Observe 10 two-digit numbers from the table. If a 16 is observed, it represents tails and heads on the two coins. One head was therefore observed. The probabilities have been preserved.

(b) A second way to simulate this experiment is to find the probabilities associated with the various possible results. The number of heads that can be seen on two coins is 0, 1, or 2. (HH, HT, TH, TT is the sample space.) P (no heads) $= \frac{1}{4}$; P(one head) $= \frac{1}{2}$; P(two heads) $= \frac{1}{4}$. Using two-digit numbers, let the numbers 00 to 24 stand for no head appeared, 25 to 74 stand for one head appeared, and 75 to 99 stand for two heads appear. The probabilities have again been preserved. Observe 10 two-digit numbers.

Appendix C

Exercises, page 515

C–1   (a) 13   (b) 9   (c) 9
      (d) 156   (e) 46   (f) 42
      (g) 103   (h) 17

C–2   (a) 8.7   (b) 42.3   (c) 49.7
      (d) 10.2   (e) 10.4   (f) 8.5
      (g) 27.4   (h) 5.6   (i) 3.0
      (j) 0.2

C–3   (a) 17.67   (b) 4.44   (c) 54.55
      (d) 102.06   (e) 93.22   (f) 18.00
      (g) 18.02   (h) 5.56   (i) 44.74
      (j) 0.67

Answers to Chapter Quizzes

Only the replacement for the word(s) in color is given. (If the statement was true, no answer is shown. If the statement was false, a replacement is given.)

Chapter 1 Quiz, page 22

1-1 Descriptive
1-6 attribute data

1-2 Inferential
1-7 discrete

1-4 sample
1-8 continuous

1-5 population
1-9 random

Chapter 2 Quiz, page 82

2-1 median
2-6 higher than

2-2 dispersion

2-3 never

2-5 zero

Chapter 3 Quiz, page 111

3-1 Regression
3-7 may have

3-2 strength of the
3-8 -1 and $+1$

3-3 $+1$ or -1
3-9 output or predicted
value

3-5 positive

Chapter 4 Quiz, page 166

4-1 any number value
between 0 and 1,
inclusive

4-4 simple
4-7 dependent
4-10 multiplication rule

4-5 seldom
4-8 complementary

4-6 sum to 1.0
4-9 mutually exclusive or
dependent

Chapter 5 Quiz, page 201

5-1 continuous
5-7 sometimes
$(p = 0.5)$

5-3 one
5-8 population

5-5 exactly two
5-9 population
parameters

5-6 binomial

Chapter 6 Quiz, page 231

6-1 its mean
6-8 some (many)

6-4 one standard
deviation

6-6 right
6-9 mutually exclusive

6-7 zero, 1
6-10 normal

Chapter 7 Quiz, page 256

7–1 is not

7–5 decreases

7–2 some (many)

7–6 approximately normal

7–3 population

7–7 sampling

7–4 divided by $\sqrt{n}$

7–8 means

Chapter 8 Quiz, page 300

8–1 Alpha

8–8 beta or 1–2

8–2 Alpha

8–9 correct decision

8–3 sample distribution of the mean

8–7 type II error

8–10 critical (rejection) region

Chapter 9 Quiz, page 332

9–2 student's t

9–7 $n - 1$

9–3 chi-square

9–9 $\sqrt{pq/n}$

9–4 to be rejected

9–10 z (normal)

9–6 t score

Chapter 10 Quiz, page 379

10–1 two independent means

10–3 F distribution

10–4 student's t distribution

10–7 is not

Chapter 11 Quiz, page 408

11–1 one less than

11–8 approximated by chi-square

11–3 expected

11–4 contingency table

11–6 test of homogeneity

Chapter 12 Quiz, page 433

12–2 mean square

12–8 mean

12–10 does not indicate

12–3 SS(factor) or MS(factor)

12–5 Reject H_0

12–9 need to

12–7 the number of factors less one

Chapter 13 Quiz, page 468

13–2 need not be

13–8 Regression

13–3 does not prove

13–9 $n - 2$

13–4 need not be

13–6 the linear correlation coefficient

Chapter 14 Quiz, page 501

14–2 t-test

14–8 power

14–3 runs test

14–10 power

14–4 assigned equal ranks

14–7 Mann-Whitney U test

Index

$\mu_{\bar{x}}$	Mean of the distribution of all possible $\bar{x}$'s
$\mu_{y\|x_0}$	Mean of all y values at the fixed value of x, x_0
μ_v	Mean number of runs for the sampling distribution of number of runs
M	Population median
n	(Sample size) number of pieces of data in one sample
$n(\ )$	Cardinal number of
$\binom{n}{r}$	Binomial coefficient or number of combinations
O	Observed value
$P(A\|B)$	Conditional probability, the probability of A given B
$P(a < x < b)$	Probability that x has a value between a and b
P_k	kth percentile
P_{75}	Price index for year 1975; also, 75th percentile
p or $P(\ )$	Theoretical probability of an event or proportion of time that a particular event occurs
p' or $P'(\ )$	Empirical (experimental) probability or a probability estimate from observed data
p^*	Pooled estimate for the proportion
Q_1	First quartile
Q_3	Third quartile
q	($q = 1 - p$) probability that an event does not occur
q'	($q' = 1 - p'$) observed proportion of the time that an event did not occur
R	Range of the data
R^2	Coefficient of determination
ρ (rho)	Population linear correlation coefficient
r	Linear correlation coefficient for the sample data or row number
r_s	Spearman's rank correlation coefficient
r_{xy}	Partial correlation
Σ (capital sigma)	Summation notation
SS ( )	Sum of squares
s^2	Sample variation
σ (lowercase sigma)	Population standard deviation
$\sigma_{\bar{x}}$	Standard deviation of the distribution of all possible $\bar{x}$'s
$\sigma_{p'}$	Standard error for proportions
$\sigma^2_{\mu r}$	Variance among the means of the r rows